Explorations with the
Texas Instruments TI-85

Explorations with the
Texas Instruments TI-85

Edited by
John G. Harvey
Department of Mathematics
University of Wisconsin
Madison, Wisconsin

John W. Kenelly
Department of Mathematical Sciences
Clemson University
Clemson, South Carolina

Ⓐ🅿

Academic Press
San Diego New York Boston
London Sydney Tokyo Toronto

ACADEMIC PRESS, INC.
A Division of Harcourt Brace & Company
525 B Street, Suite 1900
San Diego, California 92101-4495

United Kingdom Edition published by
ACADEMIC PRESS LIMITED
24-28 Oval Road, London NW1 7DX

ISBN: 0-12-329070-8

Printed in the United States of America

95 96 97   EB   9 8 7 6 5 4 3 2

Contents

Preface   *vii*

Chapter 1 Explorations in Precalculus, *John A. Dossey* **1**

  1.1 Functions on the TI-85 1
  1.2 Exponential and Logarithmic Functions 10
  1.3 Trigonometric Functions on the TI-85 15

Chapter 2 Explorations in Business Mathematics,
       Matt Hassett **23**

  2.1 Linear Systems 23
  2.2 Geometric Linear Programming 34
  2.3 The Simplex Method for Linear Programming Problems . . . 38
  2.4 Mathematics of Finance Using the SOLVER 40
  2.5 Markov Chains 48

Chapter 3 Explorations in Probability
       and Statistics, *Iris Brann Fetta* **51**

  3.1 Descriptive Statistics 52
  3.2 Probability 69
  3.3 Inferential Statistics 95
  3.4 Regression and Analysis of Variance 115
  3.5 Programming Notes 130
      References 132

Chapter 4 Explorations in Calculus I
       John Harvey and John Kenelly **135**

  4.1 Limits and Continuity 136
  4.2 Rates and Rates of Rates 143
  4.3 Curve Sketching Insights 150
  4.4 Maxima and Minima with the TI-85 155
  4.5 The Definite Integral as an Average 164
  4.6 Numerical Integration 168
  4.7 Areas and Volumes 177
      References 187

Chapter 5 Explorations in Calculus II, *Wayne Roberts* . . . **189**

  5.1 Zeroes of Functions 190
  5.2 Graphing Conic Sections 195
  5.3 Polar Graphing 202
  5.4 Parametric Equations 207
  5.5 Functions Defined by Integrals 211
  5.6 Indeterminate Forms 214
  5.7 Improper Integrals 217
  5.8 Polynomials that Approximate Functions 221

Chapter 6 Explorations in Linear Algebra, *Don LaTorre* . . **229**

  6.1 Matrices on the TI-85 230
  6.2 Systems of Linear Equations 236
  6.3 Orthogonality 251
  6.4 Eigenvalues and Eigenvectors 257

Chapter 7 Exploring Differential Equations, *T. G. Proctor* . **265**

  7.1 Euler and Improved Euler Algorithms 267
  7.2 Solution Graphs of First Order Initial Value Problems . . . 270
  7.3 First Order Input/Output Problems 277
  7.4 Second Order Problems Using Two First Order
      Differential Equations 281
  7.5 Phase Plane Solutions of Differential Equations 288
  7.6 Linear Systems of Differential Equations:
      Constant Coefficients 294
  7.7 Concluding Remarks 300
      References 305

Chapter 8 Explorations in Advanced
             Engineering Mathematics, *D. L. Kreider* **307**

  8.1 The Simple Pendulum — Elliptic Integrals 308
  8.2 Functions Defined by Integrals and Series 313
  8.3 Boundary-Value Problems 320
  8.4 Exploring Vector Ideas 325
      References 338

Appendix Introduction to the Texas Instruments TI-85   *339*

Index   *345*

Preface

In 1986, the first graphing calculator was introduced in the United States. For the first time, mathematics faculty and students could have a hand-held, inexpensive, powerful electronic tool that is "on-call" at all times. Since then, the use of graphing calculators has been incorporated into the teaching and learning of mathematics in many ways. As the use of graphing calculators grows, so do their abilities – so much so that "graphing calculator" is now a poor description of these tools and, especially, of the Texas Instruments TI-85. When compared to the graphing calculators of 1986, the TI-85 is a "supercalculator."

The TI-85 graphs functions, of course. But the TI-85 can be a useful tool in *many* different undergraduate mathematics courses because it combines function graphing with many other abilities. The purpose of this book is to explore the ways that the TI-85 can be used by both faculty and students in a variety courses. It is our hope that these explorations will encourage faculty to incorporate more fully the use of technologies into their own courses and that, in turn, they will help students to understand better the mathematics they are taught because mathematics is essential to students both in subsequent courses and in their professions.

Explorations with the Texas Instruments TI-85 has eight chapters. Each of these chapters was developed so that the reader can delve into any one of them without first working through the preceding chapters. The chapters have, however, been placed in a logical order that is much like the ordering of college mathematics courses.

The book begins with a chapter by John Dossey on precalculus mathematics and is followed immediately by Matt Hassett's chapter which explores the use of the TI-85 in business mathematics. The third chapter, Explorations in Probability and Statistics, by Iris Fetta, makes extensive use of both the statistics and list capabilities of the TI-85. Immediately following Chapter 3 are two chapters on calculus. The first calculus chapter was authored by the editors and explores ideas fundamental to the calculus: limits, the derivative and the integral, and applications of the derivative and integral. In Chapter 5, Wayne Roberts further explores the calculus, including topics from the areas of polar coordinates, vector calculus, and infinite series.

The last three chapters of the book by Don LaTorre, Gil Proctor, and Don Kreider explore more advanced topics. LaTorre's chapter explores ways in which introductory linear algebra can be restructured and enriched when using the TI-85. Proctor's chapter explores ways of using the numeric and graphing capabilities of the TI-85 in solving differential equations. And in the final chapter, Kreider explores advanced engineering applications.

This book is a mathematics book that examines ways of using an advanced graphing and numeric calculator to enhance both learning and teaching mathematics, but it covers *only* a small portion of the topics taught in the courses in which the TI-85 can be used. It does not attempt to teach the fundamental concepts of any of the topics explored; instead, it assumes that the reader has some familiarity with, or is presently studying, these topics and wants to find ways to use the TI-85 effectively and appropriately. Like most mathematics texts, we intend that a reader will not just read each section but will work through it using his or her TI-85 and then will continue to expand their knowledge of both the TI-85 and the mathematics of that section by working through the explorations that conclude the sections.

We could not have produced this book without the help of many people. First on our list are our co-authors: John Dossey, Matt Hassett, Iris Fetta, Wayne Roberts, Don LaTorre, Gil Proctor, and Don Kreider. Each of them worked hard and long on their chapters and were patient as we developed the format for the book and as the TI-85 moved into final production— even though this sometimes required them to rewrite and reformat their chapters.

Next on our list are those at Academic Press who worked so closely with us and who, too, were patient with us as we learned about desktop publishing. We would like especially to thank Chuck Glaser, Sue Purdy Pelosi, and Joe Clifford at Academic Press.

This book appears only a short while after the introduction of the Texas Instruments TI-85. Publication of this book is so timely because Texas Instruments Incorporated was willing to loan each of the authors prototype versions of the TI-85 and of the Link-85 software, and to help solve the problems encountered by the authors as they learned to use a new calculator. And so, we would like to thank Dave Stone from TI since it was he who worked most closely with us and the other authors.

Finally, we would like to thank Diane Reppert and Dee Frana of the University of Wisconsin Mathematics Department; without their hard work this book could not have been produced since they carefully TEXed all of the chapters and managed to stay sane and cheerful while working with us, the authors, and the staff at Academic Press. Our sincere thanks to the people in each of these groups.

John Harvey
University of Wisconsin
Madison, Wisconsin

John Kenelly
Clemson University
Clemson, South Carolina

September 1992

1 Explorations in Precalculus

John A. Dossey
Illinois State University

The study of precalculus is largely the study of functions. As such, this chapter will focus on exploring the abilities of the TI-85 to assist in the visualization of functions and the relationships existing between different types of functions. Central to these explorations will be the topics of graphing functions, transformations in functions related to parameter changes, and illustrative applications of the mathematics. The sections assume that the user is already acquainted with the basic TI-85 operating skills.

The material is divided into three sections. Section 1 focuses on the topics related to linear, quadratic, and other polynomial functions. Special attention is given to the interpretation of their graphs, the solution of systems from a visual perspective, transformations related to changes in their coefficients, and topics associated with rational functions. This and following sections also have explorations for the reader.

Section 2 considers topics related to exponential and logarithmic functions. In particular, the topics of geometric growth, compound interest, and growth and decay are examined from both a visual and numerical standpoint. Logarithmic functions and their properties are also examined.

Section 3 explores trigonometric functions, concentrating again on the visual aspects of problem solving where trigonometric functions are appropriate. Visual problem solving also plays a central role in this section.

1.1. Functions on the TI-85

Graphing Functions

Probably the most important contribution that graphics calculator brings to the study of precalculus is the ability to visualize the relationships inherent in the study of functions. This allows for the consideration

Explorations with the Texas Instruments TI-85
John G. Harvey and John W. Kenelly (eds.), pp. 1-22
©1993 by Academic Press, Inc.
ISBN: 0-12-329070-8

1

of functions from three different standpoints: their symbolic rules, tables of corresponding values for elements of their domain and range, and views of particular portions of their graphs. Consider for example the graph of the function $f(x) = 3x^2 + 5x + 4$ as defined for real numbers x. Using the standard window, we get the graph shown in Figure 1.1.

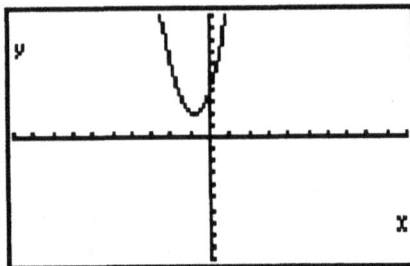

Figure 1.1. Graph of $f(x) = 3x^2 + 5x + 4$
in $[-10, 10]$ by $[-10, 10]$.

Using TRACE we can quickly verify that the $y =$ intercept is 4 and the minimum value occurs near $(-0.79, 1.92)$. As the function is quadratic, we know its graph is that of a parabola and hence the viewing window shows all of the important features of the graph. Repeated use of the ZOOM feature with box allows for a more exact description of the turning point as $(-0.833, 1.917)$. These features of the TI-85 allow for the development of strong visual analytical skills to be developed in conjunction with the use of the symbolic analytical skills that posit that the turning point for the quadratic function $g(x) = ax^2 + bx + c$ occurs at the point $(-b/2a, g(-b/2a))$, which in our case is $(-5/6, 23/12)$. The constant transformations between the symbolic and visual investigation of the function leads to strong understanding of both representations.

Linear Functions

Investigations of linear functions, $f(x) = ax + b$, often revolve around the values of the slope, a, and the y-intercept, $(0, b)$ or b. The LIST feature allows us to examine the effect of various values for the slope for the linear function $f(x) = ax + 5$.

Example 1.1

We can explore the family of linear functions, $F(x) = ax + 2$, which has a variety of slopes, but the common y-intercept of 2 as follows. First, we enter the graphing mode and use the LIST feature to help us generate some of the members of the family of linear functions with y-intercept 2. We do this by stroking the following sequence of keys. $\boxed{y(x) =}$ $\boxed{2nd}$ $\boxed{\text{LIST}}$ $\boxed{\{}$

$\boxed{1}$ $\boxed{,}$ $\boxed{2}$ $\boxed{,}$ $\boxed{3}$ $\boxed{,}$ $\boxed{4}$ $\boxed{,}$ $\boxed{5}$ $\boxed{\}}$ $\boxed{X}$ $\boxed{+}$ $\boxed{2}$ $\boxed{\text{GRAPH}}$ $\boxed{\text{GRAPH}}$. This leads to the graphing of the five functions $f(x) = 1x + 2, f(x) = 2x + 2, f(x) = 3x + 2, f(x) = 4x + 2$, and $f(x) = 5x + 2$ in succession in the graphing window. The sequential graphing shows the effect each successive addition of one to the value of the slope has on the inclination of the graph of the function.

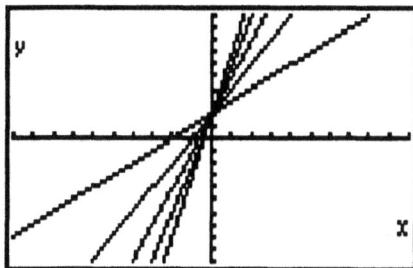

Figure 1.2. The family of functions $f(x) = ax + 2$ for $a = 1, 2, 3, 4, 5$ in $[-10, 10]$ by $[-10, 10]$.

Example 1.2

Associated with each function is its zero, r. This is the element from the domain of the function which leads to $f(r) = 0$. In the case of the linear function $f(x) = -5x + 3$, the zero of $f(x)$ is the real number r such that $f(r) = 0$. This is akin to solving $-5r + 3 = 0$. This is easily seen to be the value $r = 0.6$. Using our graphing calculator, we can locate the zero of $f(x)$ by graphing $f(x) = -5x + 3$ and locating the point where the graph of $f(x)$ crosses the x-axis. The graph of $f(x)$ combined with the use of the TRACE and the arrow keys quickly confirms the value of the root given above.

Example 1.3

Suppose the situation had called for finding the real number x such that $f(x) = 7$. This would call for examining the graphs of two functions, the original $f(x)$ and a new function $g(x) = 7$. Both of these are linear functions, so all that we need to do is to determine if they intersect, and if so, for what value of x. That is, does $f(x) = g(x)$ for some value of x. Entering $f(x) = -5x + 3$ for $y1$ and 7 for $y2$, we only need examine the value of x for which $-5x + 3 = 7$. This turns out to be the intersection of the two lines shown in Figure 1.3. The value of x is -0.8.

Quadratic Functions

The study of quadratic functions allows for the study of a number of related topics at the same time. Since the graphs of quadratic functions are parabolas, one natural topic is the study of symmetry, another is the translations that result from parameter changes in a function's symbolic rule. Symmetry is easily investigated via considering the graph of a function and the graph of the function that results when either $-x$ has been substituted for x or the $-y$ has been substituted for y in the evaluation of the function rule.

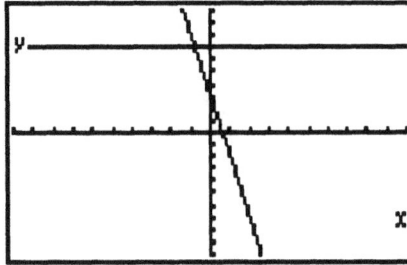

Figure 1.3. The solution to $f(x) = 7$
in $[-10, 10]$ by $[-10, 10]$.

Example 1.4

Consider the graph of the function $f(x) = 3x^2 + 1$. Consider drawing the graph of both $f(x)$ and $f(-x)$ simultaneously on your TI-85. To do this, you need to enter GRAPH and then select the FORMAT command from the graphing menu. In this, use your ▼ and ▶ keys to change SeqG to SimulG to get the functions entered to graph simultaneously rather than sequentially. Then ENTER this change and exit to enter $y1 = 3x^2 + 1$ and $y2 = 3(-x)^2 + 1$. Graphing these two shows that they apparently have the same graph. Algebraic manipulation verifies this is the fact.

A function is symmetrical with respect to the y-axis if $f(x) = f(-x)$, with respect to the x-axis if $f(x) = -f(x)$, and respect to the origin if whenever $(x, f(x))$ is on the graph, $(-x, -f(x))$ is also on the graph.

Closely related to the study of symmetry is the study of what effect the change in coefficients has on the graph of a function. Consider the graph of $f(x) = x^2$ shown in Figure 1.4. This graph could be thought of as the graph of $f(x) = ax^2 + b$ with $a = 1$ and $b = 0$. Use the LIST feature to create the graphs associated with $f(x)$ for a equal to $2, 3, 4, 5$. Make sure that you change your graphing format back to SegG. What happens? Next try with a equal to $1/2, 1/4, -1, -1/2$, and -2. Can you write a description of what happens to the graph of $f(x)$ as the absolute value of a varies, as the sign of a varies?

The experience of transforming the graph of $f(x)$ into the graphs of related functions can be continued. Consider next the effects that result from holding $a = 1$ and letting the values of b vary in $f(x) = x^2 + b$. Use the LIST feature to study the effects of letting b vary through $-4, -1, 0, 3,$ and 5. What happens?

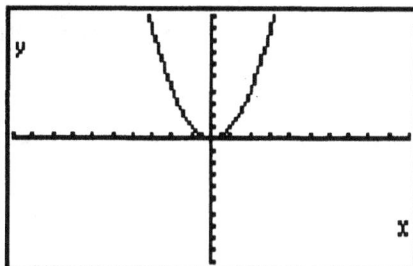

Figure 1.4. $f(x) = x^2$ in $[-10, 10]$ by $[-10, 10]$.

A third transformation of the $f(x)$ occurs when the function rule is changed to that of $f(x - h) + k$. Consider the impact that this change in parameters has on the graph of $f(x)$. Using the list feature, we can key in the following function rules: $\boxed{y1 =}$ $\boxed{(}$ $\boxed{x}$ $\boxed{-}$ $\boxed{2nd}$ $\boxed{\text{LIST}}$ $\boxed{\{}$ $\boxed{1}$ $\boxed{,}$ $\boxed{2}$ $\boxed{\}}$ $\boxed{)}$ $\boxed{x^2}$ $\boxed{+}$ $\boxed{1}$.
The resulting graphs are shown in Figure 1.5.

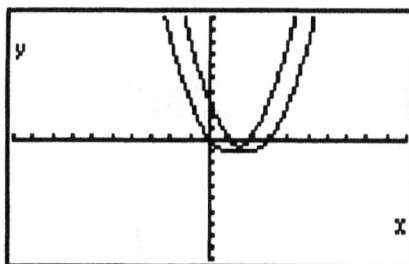

Figure 1.5. $f(x) = (x - \{1, 2\})^2 + 1$
in $[-10, 10]$ by $[-10, 10]$.

Repeat this activity again, this time allowing the value of the y-intercept to change, while holding the expression in the parentheses to $x - 1$. What happens?

Example 1.5

Consider the graph of $g(x) = -\frac{2}{3}(x - 3)^2 + 1$. Predict what will happen when you graph this function in your viewing window. What will be the features of its graph relative to the graph of $f(x) = x^2$? What will happen to its width, to its opening, to its vertex? Since $a = -\frac{2}{3}$, the graph of

$g(x)$ will open down since the sign of a is negative and it will be wider than the graph of f as the absolute value of a is less than 1. The vertex will be located at the point $(3, -1)$, as the expression $x - 3$ has the effect of moving the graph 3 units to the right. The value of b equal to 1 has the effect of sliding the graph one unit up in the vertical direction. The graph of $g(x)$ will appear as in Figure 1.6. If $g(x) = y$, this is the graph of $y - 1 = -\frac{2}{3}(x - 3)^2$.

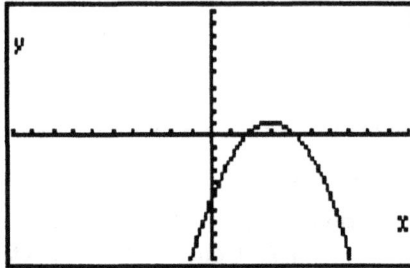

Figure 1.6. $g(x) = -\frac{2}{3}(x - 3)^2 + 1$
in $[-10, 10]$ by $[-10, 10]$.

Inverse Functions

The parametric equations feature of the TI-85 allows for the introduction of functions and their inverses in a graphical feature. If one thinks of a function establishing a correspondence between elements of one set, the domain, represented by the x-axis, and elements in a second set, the range, represented by the $f(x)$ axis. In the case of the function $f(x) = 2x + 5$, one can think of mapping the value of 2 from the domain to 9 in the range via the graph of $f(x)$ as shown in Figure 1.7. The question becomes, what mapping would reverse the actions of $f(x)$? How can a mapping that undoes the actions of $f(x)$, i.e., a $f^{-1}(x)$, be developed and illustrated? Such a function would have to map the value of 6 on the $f(x)$ axis back to the value of 0.5 on the x-axis.

This problem can be easily solved by moving from the function mode to the parametric mode on the TI-85. If we consider the function graphed in Figure 1.7, we can generate it by the graph $E(t)$, where $\boxed{xt1 =}$ $\boxed{t}$ and $\boxed{yt1 =}$ $\boxed{2}$ $\boxed{t}$ $\boxed{+}$ $\boxed{5}$. Now consider letting the y-axis serve as the domain and the x-axis as the range and graph the function $\boxed{xt2 =}$ $\boxed{2}$ $\boxed{t}$ $\boxed{+}$ $\boxed{5}$ and $\boxed{yt2 =}$ $\boxed{t}$. It is easy to see that these lines are symmetric in the line $y = x$. Thus, the graph of the second is the graph of the inverse function, as $f(f^{-1}(x)) = x$ for all real x. A little investigation shows that $f^{-1}(x) = (x - 5)/2$ and is the reflection of $f(x)$ in the line $y = x$.

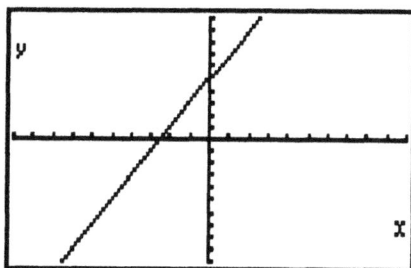

Figure 1.7. $f(x) = 2x + 5$ in $[-10, 10]$ by $[-10, 10]$.

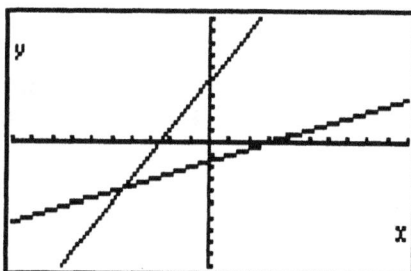

Figure 1.8. The graphs of $f(x)$ and $f^{-1}(x)$
in $[-10, 10]$ by $[-10, 10]$.

Note that in this case it is easy to see that the inverse mapping, $f^{-1}(x)$, is also a function. It is also easy to establish that this function "undoes" the mapping that $f(x)$ established. Using the same approach, but replacing the function $2x + 5$ with x^2, one can see that the "inverse mapping" is not a function. Through such investigations, the vital role played by the line $y = x$ in determining the visual image of an inverse mapping and which turn out to be inverse functions comes clear.

Rational Functions

Another topic in precalculus where graphics calculators help visualize the mathematics involved is rational functions. Functions expressed as the quotient of two polynomials often have graphic representations involving horizontal and vertical asymptotes, and thus, discontinuities.

Example 1.6

Consider the graph of the function $r(x) = \frac{6x^3 - 3x^2 - 2}{3x - 3}$ shown in Figure 1.9. This graph has a vertical asymptote at $x = 1$ and is discontinuous there. As x increases and decreases without bound, the value of the function increases without bound.

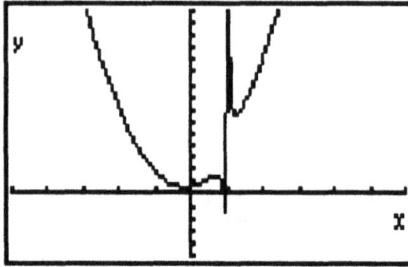

Figure 1.9. $r(x) = \frac{6x^3 - 3x^2 - 2}{3x - 3}$
in $[-5, 6]$ by $[-5, 15]$.

Completing the indicated division implied in the expression defining $r(x)$ one gets the quadratic $2x^2 + x + 1$ with a remainder term of $\frac{1}{3x-3}$. The quadratic part, $s(x)$, when graphed falls inside and above the left-hand branch of $r(x)$ and below and to the right of the right hand branch, as shown in Figure 1.10. An analysis of this relationship indicates that an

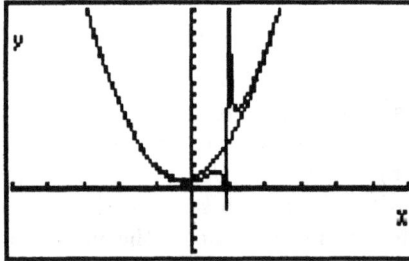

Figure 1.10. The graphs of $r(x)$ and $s(x)$
in $[-5, 6]$ by $[-5, 15]$.

interpretation of the remainder term is as the distance between the graph of $r(x)$ and $s(x)$. As x increases from the left, the value of $\frac{1}{3x-3}$ increases until 1 is reached. Here the value becomes unbounded. Beyond 1, the value of the remainder term decreases. One might ask, how far does one have to move to the right before the distance between $s(x)$ and $r(x)$ is less than .001. Setting the remainder less than this value, and solving, one finds that when $x > 332.\overline{3}$, then the vertical distance between the graphs of the two functions is less than 0.001.

Explorations with Functions

1. Create a graph which shows the family of linear functions all having slope 4 and differing y-intercepts.

2. Create a graph showing at the same time the family of linear functions through $(-2, 2)$ and the family of linear functions through $(2, 2)$. What pattern results?

3. Find the zero of the linear function $f(x) = 3x - 19$.

4. Find all of the zeros of $g(x) = 3x^2 - 11x - 704$. Give an argument that you have found all of the zeros of the function.

5. Show that the graph of $y^2 = x$ is symmetrical about the x-axis. (Hint: Since $y^2 = x$ is not a function, consider graphing it as two functions the union of whose graphs gives the graph of the relation $y^2 = x$.)

6. Is the function $f(x) = x^3$ symmetric about the origin? Argue why or why not. Support your argument with a graph.

7. Examine the expressions defining each of the following functions. Sketch a graph for the function and then use your TI-85 to verify that your sketch is correct. Do not attempt to plot points to derive your sketch. Use only what you know about the effects of the values of the coefficients have on the graph of a quadratic function.
   (a) $y1 = -3x^2$
   (b) $y1 = 2(x - 2)^2$
   (c) $y1 = .5(x + 6)^2 - 6$
   (d) $y1 = |.3(x + 2)^2| - 2$

8. Without the use of your TI-85, sketch the graphs of the following functions, then test your intuition by verifying your sketches with the TI-85. Did you agree?
   (a) $y1 = x^2 + 8x$
   (b) $y1 = x^2 - 8x$
   (c) $y1 = 2x^2 + 4x$
   (d) $y1 = -x^2 + 3x$
   (e) Write an explanation about the nature of the graphs of $ax^2 + bx$ in terms of the graph of ax^2.

9. Find the function rules for the inverses of the following functions where they exist. When the inverses do not exist, explain why.
   (a) $f(x) = 5x$
   (b) $f(x) = -x$
   (c) $f(x) = 5$
   (d) $f(x) = 3x - 2$
   (e) $f(x) = 2x^3$
   (f) $f(x) = 1 - x^2$

10. Examine the graph of the function $h(x) = \frac{x^4 + 2x^2 + x - 3}{x^2 - 3}$. Describe its asymptotes, its end behavior, and what function approximates it over the majority of its domain. What expression describes the deviation between the approximating function and the graph of $h(x)$ over the domain of h?

1.2. Exponential and Logarithmic Functions on the TI-85

The graphics calculator also assists in the investigation of situations growing from exponential growth. Consider the problem of tearing a sheet of cardboard 1 mm thick, folding it back on itself, and repeating the process over and over again. If the distance to the moon is approximately 3.91×10^{11} mm, how many repetitions of the process will it take to make a stack as high as the distance to the moon. The TI-85 allows for this process to be easily simulated. All one needs do is enter the following strokes: $\boxed{1}$ $\boxed{\times}$ $\boxed{2}$ $\boxed{\text{ENTER}}$ $\boxed{\times}$ $\boxed{2}$ and then repeatedly touch the $\boxed{\text{ENTER}}$ key until the display exceeds 3.91×10^{11}. Count the total number of times you touch $\boxed{\text{ENTER}}$ to determine the number of tears it will take. How many did you get?

Growth and Exponential Functions

The process of repeatedly tearing the cardboard above can be modeled by an exponential function, $f(x) = 2^x$, where x is the number of tears made. When you graph the function on the TI-85, you get the graph shown in Figure 1.11. This graph shows the relationship graphed over the real numbers. Clearly, it is defined for all real numbers, with a range of all positive real numbers. Thus, the range for graphs of exponential functions can be set for nonnegative reals.

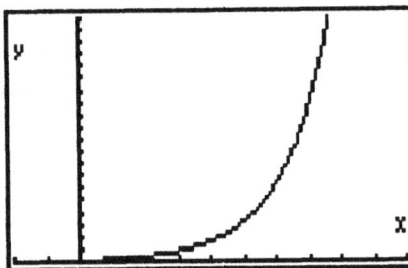

Figure 1.11. $f(x) = 2^x$ in $[-2, 10]$ by $[0, 200]$.

Investigations of the graphs of $g(x) = 5^x$, $h(x) = (.5)^x$, $m(x) = (.9)^x$, and $e(x) = (2.71828)^x$ give a picture of the commonalities and differences of exponential functions having different bases. Explore what happens when you try to evaluate $c(x) = (-2)^x$. What restrictions might we wish to place on the value we consider as a base for our exponential functions? Most generally, exponential functions are defined as functions of the form $f(x) = b^x$ where $b > 0$ and $b \neq 1$.

Example 1.7

Consider the problem of transforming exponential functions through manipulating the coefficients related to their expressions. Examine $f(x) = a(2^x)$ for the following values of a, $1, 2, 1/2$, and -1. Entering the keystrokes $\boxed{y1 =}$ $\boxed{2^{\text{nd}}}$ $\boxed{\text{LIST}}$ $\boxed{\{}$ $\boxed{1}$ $\boxed{,}$ $\boxed{2}$ $\boxed{,}$ $\boxed{.5}$ $\boxed{,}$ $\boxed{-1}$ $\boxed{\}}$ $\boxed{2}$ $\boxed{\wedge}$ $\boxed{x}$ $\boxed{\text{GRAPH}}$, we get the graphs shown in Figure 1.12. Examination of the graphs in order as they are drawn indicates that the values the coefficient takes have the same effect as they did on the graphs of polynomial functions examined in the first section of the chapter.

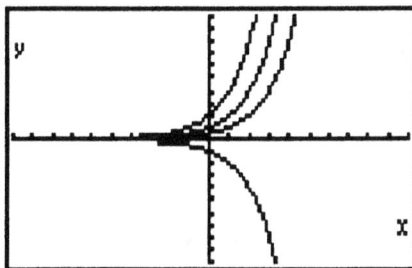

Figure 1.12. Graphs of $f(x) = \{1, 2, .5, -1\}2^x$
in $[-10, 10]$ by $[-10, 10]$.

The ability to graph exponential functions opens a world of possible explorations to discover the properties of exponential functions relative to the properties of polynomial functions. Consider the comparison of 2^x and x^2 and their relative rates of growth. For what value(s) of x are they equal, when does the one dominate the other?

Compound Interest

One of the most interesting uses of the exponential growth is in the area of compound interest. The basic formula is $A = P(1 + i)^n$, where A is the amount after n periods, P is the initial principal, i is the interest rate per period, and n is the number of interest periods for which the money is invested. For example, the calculation for $5000 invested at 5% for three years is $A = 5000(1 + .05)^3 = \$5788.13$. This can also be visualized as $A = 5000(1 + .05)(1 + .05)(1 + .05)$, with each successive multiplication by $(1 + .05)$ giving the total at the end of each successive year.

If the interest is compounded quarterly, then one-fourth of the interest is paid each of the periods. This makes the interest paid per period 1.25%, but the number of periods is quadrupled, giving $5000(1+.0125)^{12} =$ \$5803.77. Once this process is under control, any number of questions can be investigated. How fast will an amount double, what interest rate is required for an amount to double in a fixed amount of time, what is the actual yearly rate (effective rate) of interest associated with a nominal rate of 6% compounded quarterly?

What would happen if an amount P is invested at an annual interest rate of r% and compounded m times per year for t years? This would lead to the calculation of $A = P\left(1 + \frac{r}{m}\right)^{mt}$. The calculation of this for specific values of P, r, m, and t results in a variety of values. What would happen if the interest was allowed to compound instantaneously? This would take us to a limiting situation, actually $A = \lim P\left(1 + \frac{r}{m}\right)^{mt}$ as m grows without bound. Using properties of limits and setting $x = \frac{r}{m}$, this can be changed to $A = \left(\lim P(1 + x)^{\frac{1}{x}}\right)^{rt}$, where the limit is evaluated as x approaches 0. The amount in the brackets is e, the base of the natural logarithms. Hence the balance in the account approaches Pe^{rt}.

Example 1.8

How much money would \$1 invested at 100% interest grow to in one year if interest was compounded instantaneously? Using the expression above, we can evaluate the amount by stroking $\boxed{1}$ $\boxed{\times}$ $\boxed{2^{\text{nd}}}$ $\boxed{e^x}$ $\boxed{(}$ $\boxed{1}$ $\boxed{\times}$ $\boxed{1}$ $\boxed{)}$ $\boxed{\text{ENTER}}$. The result is shown to be e^1, or \$2.718.... Hence the money would not grow completely out of control, as one might guess.

Logarithmic Functions

If we reverse the roles of $f(x)$ and x in the definition of exponential functions, that is $x = b^{f(x)}$, or $f(x)$ is the power to which b must be raised in order to obtain x, we obtain a new function $f^{-1}(x)$, the inverse function to the exponential function. This new function is called the logarithm of x. The graph of the inverse function for $f(x) = e^x$ can be observed by using the parametric graphing feature of the TI-85 to graph $f(x)$ by $x1t = t$ and $y1t = e^t$. The inverse function, $f^{-1}(x)$ can be graphed by $x2t = e^t$ and $y2t = t$. The graphs for the exponential function, $f(x) = e^x$, and its inverse function, $f^{-1}(x)$, are shown in Figure 1.13.

The number $f(x)$ such that $x = b^{f(x)}$ is called the logarithm of x to the base b. $f(x) = \log_b x$ means that $x = b^{f(x)}$. Since $b^y > 0$ for all real numbers y, $\log_b x$ is defined only for $x > 0$. Most often, b takes on the values of 10 or e. In the former case we write $\log x$ and speak of common logarithms and in the latter case we write $\ln x$ and speak of natural logarithms. In almost all cases involving logarithms in business or scientific settings, natural logarithms are applied.

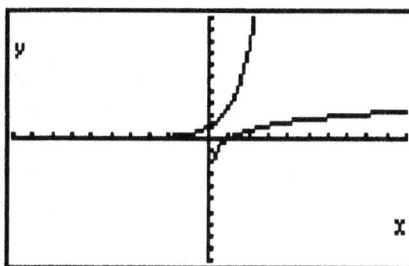

Figure 1.13. $f(x) = e^x$ and $f^{-1}(x) = \ln(x)$
in $[-10, 10]$ by $[-10, 10]$.

Applications to Growth and Decay

The most common model for dealing with growth and decay is the exponential growth model $A(t) = Ce^{kt}$, where A is the amount present at time t. C is a size of the original amount, k is the constant related to the rate of growth, and t is the time. Note that if $k > 0$ the amount grows and if $k < 0$ then the amount decays. When $t = 0$, the value reverts to the original amount, C.

Example 1.9

Suppose that the world's population, in billions, is about 5.2. Further, suppose that it is increasing at the rate of 400,000 people per day at the present. How long will it take for the world's population to double? What will be the world's population in 10 years?

These questions can be answered by first determining the rate of growth of the population. As it is increasing at the rate of 0.0004 billion per day for 365.25 days per year, the average annual rate of growth is $(0.0004)(365.25)/(5.2) = 0.028$ billions per year. Assuming this rate of growth continues, the growth model for the world's population is $A(x) = 5.2e^{(0.028)x}$. The answer to the first question, the doubling of the world's population is found by setting the amount to 10.4. Solving $10.4 = 5.2e^{(0.028)x}$ for x gives $\ln 2 = (0.028)t$ or $x = 24.76$ years. The world's population in 10 years will be the value of $A(10)$ or 6.88 billions.

Both of these questions could also be answered by analyzing the graph of $A(x)$ with the use of the TRACE and ZOOM features. Entering the graph of $A(x)$ and moving the TRACE cursor along the curve until the population reaches 10.4 billion, we find that the time required, the value of x, is about 24.77 years. Examining the second question of the world's population in 10 years, we TRACE to the graph value associated with $x = 10.0002$ to be 6.88 billion.

Example 1.10

Creepy Crawler fishing worms has discontinued advertising in *Bait Your Aqua Friends* in an attempt to control costs. They plan to reinstate advertising when sales have dropped to 75% of the present. If in one week, the sales have dropped to 95% of their initial rate, in how many days after the stopping of advertising should the company plan to resume advertising?

If we assume that the process governing the decline in advertising is a natural decay model, we can build a model that takes the form $S(x) = e^{kx}$, where k is the coefficient controlling the decay, x the number of days, and $S(x)$ the proportion of business remaining. The leads to $S(7) = .95 = e^{k7}$, or $k = -0.007$. This gives a model of $S(x) = e^{-.007x}$. Entering this function as $y1$ we can examine the graph, tracing till the proportion of original sales reaches .75. The corresponding value is 41.07 days. Carrying out the analysis symbolically, the answer obtained is 41.07 days.

Example 1.11

Find the solution to the equation $\ln x = 2x - 2^x$. As this is a mix of logarithmic, polynomial, and exponential expressions, we are unable to solve it through direct symbolic manipulation. Hence, we can change this into a function format of $f(x) = 2x - 2^x - \ln x$ and look for its zero(s). A casual visual inspection indicates the existence of at least one zero, $x = 1$. Entering the function as $y1$ and graphing on the display $[0, 10]$ by $[-10, 10]$ we get the graph shown in Figure 1.14.

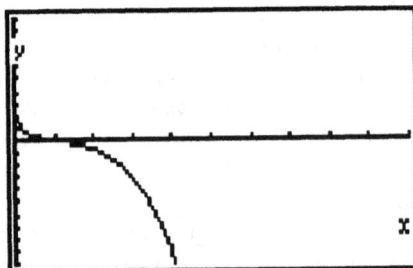

Figure 1.14. $f(x) = 2x - 2^x - \ln x$
in $[0, 10]$ by $[-10, 10]$.

An analysis of the function indicates that it will be monotone decreasing for $x > 1$, as 2^x easily exceeds the values of $2x - \ln x$ for such real numbers. Thus, $x = 1$ is the only root to the original equation.

Explorations with Exponential and Logarithmic Functions

1. Suppose a bacteria colony quadruples in 36 hours. If the initial count in the colony is 4000, how many bacteria are present at the end of 9 days?

2. If a rumor spread in a town at a rate where the number of people who know the story doubles every 30 minutes, how long will it take until everyone in a town of 75,000 knows the rumor?

3. Investigate the effect that translations of the type $f(x) = 2^{x-h}$ have on the graph of an exponential function. Write a description of the similarities and differences that exist between exponential and polynomial functions.

4. Use your ZOOM and TRACE features to solve the following equations.
   (a) $(1/2)^x = 10$
   (b) $2^x = 3^x$
   (c) $10^x = 5$

5. Find the balance in an account with an initial principal of $7500 invested at 6.5 percent compounded yearly for 6 years.

6. Suppose that $2500 is invested in an account paying interest at the nominal annual rate of 10%. Compute the balance of the account if the interest is compounded (a) annually, (b) quarterly, (c) monthly, and (d) weekly.

7. How much money, invested now, will grow to $50,000 in 10 years if the nominal interest rate is 8% and the interest is compounded monthly?

8. How much would a principal of $2000 grow to if invested for 3.5 years at 5.5%?

9. Suppose that a pain killing drug is used to calm a patient following surgery. Suppose that the pain is killed when the bloodstream contains 45 mg of the drug for each kilogram of body weight. Further, suppose the painkiller is eliminated from the body at a rate that dissipates 1/2 of the drug administered in 5 hours time. What single dose should be administered in order to numb the pain for a child weighing 50 kg for 1 hour of time?

10. What root(s) does the equation $2^x - \ln x = 2$ have?

1.3. Trigonometric Functions on the TI-85

Another area related to Precalculus where the TI-85 can assist in problem solving and visualization of the concepts deals with trigonometry. Trigonometry enters the curriculum in a variety of ways, ranging from the relationships holding between the lengths of sides and measures of angles in triangles, to the trigonometric functions seen through the unit circle model, to applications of the trigonometric functions in settings involving periodic motion.

Unit Circle and the Graphs of sin, cos, and tan

The relationship between the trigonometric functions and the unit circle are easily visualized through the use of the parametric equation graphing mode. Entering the parametric equation form for the unit circle, $\boxed{x1t =}$ $\boxed{\cos}$ $\boxed{t}$ and $\boxed{y1t}$ $\boxed{=}$ $\boxed{\sin}$ $\boxed{t}$ as one function and $\boxed{x2t =}$ $\boxed{t}$ and $\boxed{y2t =}$ $\boxed{\sin}$ $\boxed{t}$ for the second function, and setting the range parameters to tMin = 0, tMax = 2π, tStep = .1, xMin = -2, xMax = 2π, xScl = .1, yMin = -2.43, yMax = 2.43, and yScl = 1, we get a graph like that shown in Figure 1.15.

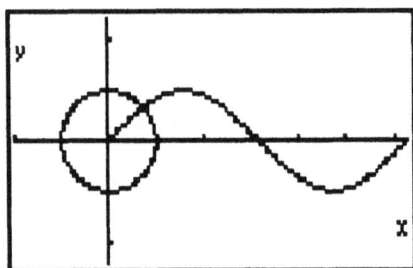

Figure 1.15. The unit circle and sin x
in $[-2, 2\pi]$ by $[-2.43, 2.43]$.

Using the TRACE feature and toggling the $\boxed{\blacktriangleright}$ and $\boxed{\blacktriangleleft}$ keys one can begin to move the cursor about the unit circle, with the display showing the values of the sine and cosine of the angle with radian measure t. To show the corresponding point on the $\sin t$ plot, one only needs to use the $\boxed{\blacktriangle}$ key. To return to the corresponding point on the unit circle, use the $\boxed{\blacktriangledown}$ key.

Translating Trigonometric Functions

In a manner like that done with polynomial and exponential-logarithmic functions, the transformations of trigonometric functions can be studied from a visual standpoint using the TI-85. In particular, students should have the opportunity to see effects of the parameter values in general equations of the form $y = A\sin(Bx + C) + D$. As we have already observed, the value of D shifts the y-intercept of the graph up and down on the vertical-axis. The value of D is called the vertical shift in trigonometric situations. The value of A expands or constricts the height of the graph of the function. The value of A is called the amplitude in trigonometric settings. B is called the period of the function. Figure 1.16 shows the effect of changes in the period of the sin function. The period of a trigonometric function is the smallest positive value of a such that $f(x+a) = f(x)$ for all x in the domain of the function. For the sin and cos functions the period is 2π radians. Changing the value of the period coefficient B in the general trigonometric equation causes the function to cycle through its period in $2\pi/|B|$ radians. Figure 1.16 shows that $\sin 2x$ goes through a full period in $\pi(2\pi/2)$ radians if $B = 2$.

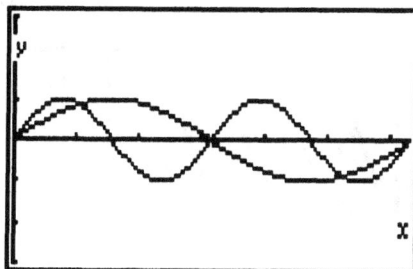

Figure 1.16. The graphs of $\sin x$ and $\sin 2x$
in $[0, 2\pi]$ by $[-3, 3]$.

The remaining parameter is C. Its value is called the phase shift for the function. It is the horizontal translation of the graph left or right from the origin. Figure 1.18 shows the shift of $\sin 2x$ right $\pi/3$ radians from the graph of $\sin 2x$ shown in Figure 1.17. In general, $\sin(x + c)$ is a shift of the graph c units to the left of its standard position. Thus, $\sin(2x - \pi/3)$ gives the graph shown in Figure 1.17 with amplitude of 2. In general the phase shift for a trigonometric function is $-C/B$.

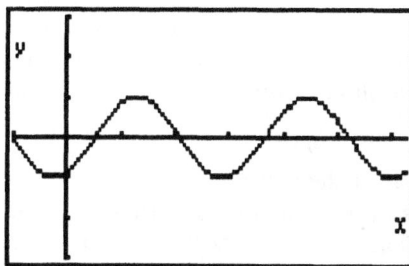

Figure 1.17. Graph of $\sin(2x - \pi/3)$
in $[-1, 2\pi]$ by $[-3, 3]$.

Example 1.12

The combination of these features leads to the ability to visualize the graph of almost any trigonometric function without having to draw a graph of it. For instance, consider the graph of $f(x) = 2\tan(3x + \pi/4) + 1$. The vertical shift of 1 indicates that the graph of the function has been translated up 1 unit from the x-axis. The amplitude of 2 indicates that the graph has been stretched 2 units in the vertical direction. As the tan function has a normal period of π radians, $f(x)$ will cycle through its period in $\pi/3$ radians. Finally the phase shift, $-C/B$ or $-\pi/12$, indicates that the graph of $2\tan(3x) + 1$ has been translated to the left $\pi/12$ units.

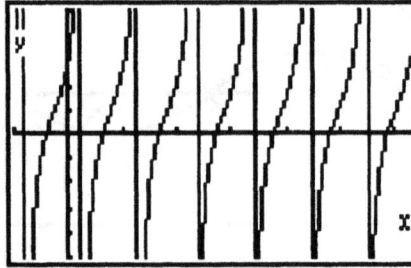

Figure 1.18. $f(x) = 2\tan(3x + \pi/4) + 1$
in $[-1, 2\pi]$ by $[-5, 5]$.

Motion Problems

The TI-85 is perhaps at its best in visualizing the action taking place in problems involving motion. Functions of the form $f(x) = A\cos Bx$ are found in situations where simple harmonic motion apply. Perhaps the most famous is that of a mass suspended on a spring attached to a fixed point. The motion of the mass, relative to its position at rest can be shown over time by considering this situation in parametric equation form. The coefficient A represents the position of the mass at the point of release and B determines the length of a complete cycle for the motion.

Example 1.13

Consider the case of the motion of an object which is pulled to a point 10 units beneath its position at rest and then released to oscillate up and down, without friction, for time $t > 0$, completing 1/2 cycle per minute. Find the equation for this motion.

Since the object was released at $t = 0$ at a point 10 units beneath its position at rest, $A = -10$. Since it completes 1/2 cycle per minute, it will complete a full cycle in 2 minutes. Thus, $B = (2\pi)/2$, or 2. The equation for the motion is $x1t = t$ and $y1t = -10\cos \pi t$. Graphically, this can be displayed as shown in Figure 1.19.

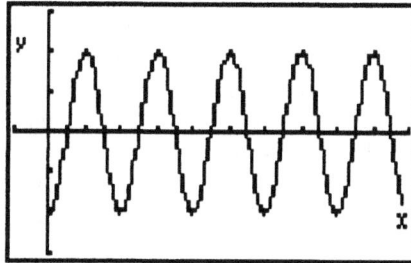

Figure 1.19. Harmonic motion of $x1t = t, y1t = -10\cos \pi t$
in $[-1, 10]$ by $[-15, 15]$.

The actual type of motion that one would get in reality would decay over time due to friction. This type of motion is called damped harmonic motion and is often modeled by equations of the form $x1t = t$ and $y1t = Ae^{-t}\cos(Bt)$. For the case in the example above, consider $x1t = t$ and $y1t = -10e^{-t}\cos \pi t$. The graph for this relation is shown in Figure 1.20.

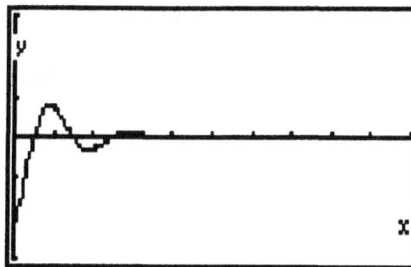

Figure 1.20. $x1t = t$ and $y1t = -10e^{-t}\cos \pi t$
in $[0, 10]$ by $[-15, 15]$.

Example 1.14

Consider the case of a baseball hit directly toward center field fence. If the fence is 12 feet high and positioned 400 feet from home plate, what initial conditions would have to obtain at the moment of contact in order for the ball to clear the fence for a home run? As several initial conditions are possible, let us consider the case where the ball is contacted with an initial velocity of 160 feet per second, is contacted at a point 4 feet above ground, is hit into a wind of 5 miles per hour, and is contacted with an initial angle of elevation of $20°$.

Over time t, the path of the ball can be seen to follow the path determined by $x1t = 160t\cos 20°$, $y1t = 160t\sin 20°$. As the ball is hit into a wind of 5 miles per hour, its motion in the horizontal direction must be corrected by the factor of -7.3 feet per second to adjust for the wind. This alters the equations to $x1t = 160t \cos 20° - 7.3t$, $y1t = 160 \sin 20°$. One other correction is needed to correct for the effects of gravity. As the ball was hit at 4 feet above the ground, its vertical acceleration due to gravity is given by $4 - 16t^2$. Adjusting the second parametric equation, controlling the vertical placement of the ball, by this, we have: $x1t = 160t \cos 20° - 7.3t$, $y1t = 160t \sin 20° + 4 - 16t^2$. The graph for this situation is given in Figure 1.21. Clearly a home run occurred in this situation, as the ball easily cleared the fence at 400 feet.

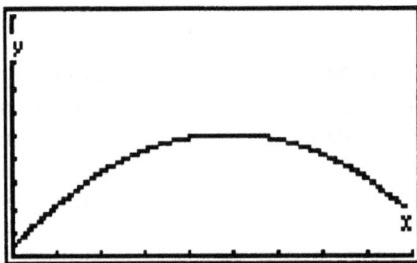

Figure 1.21. Graph of $x1t = 160t \cos 20° - 7.3t$ and $y1t = 160 \sin 20° + 4 - 16t^2$. in $[0, 450]$ by $[0, 100]$

Explorations with Trigonometric Functions

1. Develop a display for visualizing the relationship between points on the unit circle and the cos function similar to that shown in Figure 1.15.
2. Develop a display for visualizing the relationship between points on the unit circle and the tan function similar to that shown in Figure 1.15.
3. Sketch the following curves and then check your sketches by graphing them over the interval $[0, 2\pi]$.
   (a) $\sin 2x$
   (b) $-\cos x$
   (c) $3 \sin 2x$
   (d) $-3 \cos 2x$
4. Sketch the following curves and then check your sketches by graphing them over the interval $[0, 2\pi]$.
   (a) $\cos(x + .5)$
   (b) $\sin(x + \pi/2)$

    (c) $\cos(x)$

    (d) $\tan(x - \pi/4)$

5. How are the graphs of the functions in 4 (b) and 4 (c) related? Why?

6. Sketch the following curves and then check your sketches by graphing them over the interval $[0, 4\pi]$.

    (a) $5\cos(x - \pi/4) + 3$

    (b) $2\tan(x + \pi/4) - 2$

    (c) $-2\sin(2x - \pi/4) + 1$

7. In the situation with damped harmonic motion shown in Figure 1.20, for what value of t will the motion first be bounded by the lines $y = \pm.5$?

8. In Example 1.14, suppose the ball had been hit at the same height with an initial velocity of 180 feet per second, at an initial angle of 15°, and there was no wind. Would it clear the fence under these conditions?

9. Under what conditions for no wind and contacting the ball at the same height with an initial velocity of 200 feet per second, will the ball minimally clear the fence?

10. If the kicker in a collegiate football game kicks the ball from a tee at ground level with an initial velocity of 75 feet per second and at an initial angle of elevation of 60°, where will the ball come down?

2 Explorations in Business Mathematics

Matt Hassett
Arizona State University

In this chapter the expression "Business Mathematics" has a special meaning; it refers to the mathematics studied by business students in introductory mathematics courses. The typical set of required mathematics courses for college business students has two major components: Brief Calculus and Finite Mathematics. The study of Markov Chains is an application of the theory of probability. Chapter 3 of this book covers the use of the TI-85 for probability problems; students of Finite Mathematics should also read that chapter. Since Chapters 4 and 5 of this book show how to use the TI-85 in calculus, applications to Brief Calculus will not be covered here. This chapter shows how to use the TI-85 to explore Finite Mathematics. The topics covered are:

1) Linear Systems and Matrices
2) Geometric Linear Programming
3) The Simplex Method for Linear Programming Problems
4) Mathematics of Finance
5) Markov chains

Although this chapter is entitled Explorations in Business Mathematics, there are very few story problems about manufacturers of widgets. The chapter shows you how to use the TI-85 to explore and solve the mathematics problems covered in a Finite Mathematics course. The chapter begins with the study of linear systems in Section 2.1 which will do nothing more than solve systems of equations a number of different ways. The techniques used there will be applied many times in the remaining sections on linear programming and Markov chains.

2.1. Linear Systems

Particular instances of linear systems are commonly referred to as "two equations in two unknowns" or "three equations in three unknowns."

Explorations with the Texas Instruments TI-85
John G. Harvey and John W. Kenelly (eds.), pp. 23–50
©1993 by Academic Press, Inc.
ISBN: 0-12-329070-8

Below are some examples of such linear systems.

$$\begin{aligned} 2x + 4y &= 14 \\ 7x - 2y &= 1 \end{aligned} \tag{2.1}$$

$$\begin{aligned} x + y + z &= 6 \\ 2x - y + 3z &= 9 \\ x - y + z &= 4 \end{aligned} \tag{2.2}$$

Note that all of the equations in each of these systems are indeed linear. It is possible to have systems of equations that are not all linear. You may be able to solve non-linear systems using the TI-85, but the specific TI-85 applications discussed here will not work for nonlinear systems. Here is an example of a nonlinear system.

$$\begin{aligned} y &= x^2 \\ y &= x \end{aligned} \tag{2.3}$$

It is not hard to solve system 2.3; the solutions are $x = 0$, and $x = 1$. It is very important to recognize that the TI-85 operations discussed here will not solve that system directly. The operations available under SIMULT and MATRX will work only for linear systems.

Solving a system directly in SIMULT

In College Algebra students learn to solve systems like 2.1 and 2.2 using a number of different methods. Most students seem to remember best the addition method, in which multiples of the equations are added to each other to eliminate variables from equations. This takes a little time for a problem like 2.1, and can be very tedious for problems like 2.2. In Finite Mathematics, you often need to solve such systems rapidly as part of solving another problem. The SIMULT menu lets you get the solutions immediately when this is necessary. This is best illustrated by using SIMULT to solve 2.1. The system is rewritten below in a form more appropriate for the TI-85, which uses subscripted x variables instead of x and y.

$$\begin{aligned} 2x_1 + 4x_2 &= 14 \\ 7x_1 - 2x_2 &= 1 \end{aligned} \tag{2.4}$$

To solve this system, begin by pressing $\boxed{\text{SIMULT}}$. The TI-85 will prompt you for the number of equations. Press 2.

The next screen will ask for the coefficients of the first equation. The coefficient of $x1$ is 2. Enter 2 at the prompt $a1,1 =$ and press $\boxed{\text{ENTER}}$. As soon as the 2 is entered, the cursor will move down to the next prompt $a1,2 =$. Enter 4, the coefficient of $x2$ in the first equation and press $\boxed{\text{ENTER}}$.

The cursor will move down again so that you can enter 14, the constant term from the first equation.

When the first equation is entered, the screen will clear to allow you to enter the second equation. Enter the numbers 7, −2 (be sure to use the $\boxed{(-)}$ key for that instead of the $\boxed{-}$ key), and 1.

Press $\boxed{\text{SOLVE}}$ and the solution will appear immediately. The solution is

$$x_1 = 1$$
$$x_2 = 3$$

System 2.4 is easily solved with paper-and-pencil. The time saved by using SIMULT is more obvious if you solve three equations in three unknowns. System 2.2 is rewritten below in the subscripted form appropriate for the TI-85.

$$x_1 + x_2 + x_3 = 6$$
$$2x_1 - x_2 + 3x_3 = 9 \qquad (2.5)$$
$$x_1 - x_2 + x_3 = 4.$$

This system can be entered in the same way. Restart the SIMULT menu by pressing $\boxed{\text{SIMULT}}$ again, enter the number of unknowns (3), and enter the coefficients for each of the three equations. Pressing the $\boxed{\text{SOLVE}}$ key will give the solution almost immediately:

$$x_1 = 5$$
$$x_2 = 1$$
$$x_3 = 0.$$

Graphical Solutions

When linear programming is discussed in the latter part of this chapter, it will be important to look at solutions for systems. Graphical solution will be discussed in other chapters of this book, but it will be useful to review it here by solving System 2.1 graphically. To begin, you must solve each equation (by hand) for y in terms of x.

$$y = -0.5x + 3.5$$
$$y = 3.5x - 0.5 \qquad (2.6)$$

Solving the equations for y is necessary for use of the GRAPH menu on the TI-85, since the *equations* to be graphed must be entered in this form. To start work on the TI-85, press $\boxed{\text{GRAPH}}$. If you have never graphed a function on your TI-85 before, you will see a blank screen with five menu options across the bottom of the screen. Press $\boxed{\text{y(x)=}}$. You will see the prompt "y1=." Enter the right side of the first equation; i.e., $-.5x +$ 3.5, using the key $\boxed{\text{xVAR}}$ for the variable x. As soon as you finish and press $\boxed{\text{ENTER}}$, a prompt for "y2=" will appear. Enter the right side of the second equation, $3.5x - .5$. (If you have already graphed functions, the first screen will not be blank, and other functions will also be listed. Enter the two functions described above and use the $\boxed{\text{SELCT}}$ key to select them for graphing and to "unselect" all others.)

It is a good idea to set the range of the graph before displaying it. Press $\boxed{\text{RANGE}}$, and you will see the screen on which you can enter the minimum and maximum values of x and y that will appear on the graph. To include the y-intercepts of both lines and the intersection point, use

$$
\begin{aligned}
\text{xMin} &= -1 \\
\text{xMax} &= 2 \\
\text{yMin} &= -1 \\
\text{yMax} &= 4
\end{aligned}
$$

Press $\boxed{\text{GRAPH}}$, and you will see the screen below:

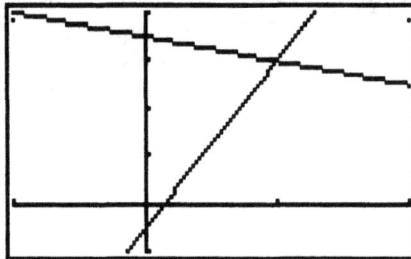

Figure 2.1. Graph of linear system (2.6).

Press $\boxed{\text{MORE}}$ key, and a new menu will appear. Press $\boxed{\text{MATH}}$ and you will see another new menu with a number of MATH options. Press $\boxed{\text{MORE}}$ again and the F5 option becomes ISECT. This option lets you find intersection points of the lines. Since the coordinates of the intersection point are the solutions for the system, ISECT will give you a way to solve the system while looking at its graph. Press $\boxed{\text{ISECT}}$, and you will see a flashing cursor that can be moved along the graph by the cursor keys. Use the right or left arrow keys to move the cursor to the intersection point, and press $\boxed{\text{ENTER}}$ twice. The coordinates of the intersection point will be displayed below the graph as $x = 1$ and $y = 3$.

Matrix Operations

When matrices are covered in Finite Mathematics, a great deal of time is spent in tedious calculations – especially for matrix multiplication. The TI-85 allows you to perform matrix operations just as if you were adding or multiplying ordinary numbers. This has advantages since matrix operations are used in many applications where it is important to get correct answers rapidly. In the next section we will review a way to solve equations using matrix operations. (If you have not already left the GRAPH menu, press $\boxed{\text{EXIT}}$ three times to leave it.)

The next example shows the ease of matrix operations on the TI-85. We will use the two matrices below:

$$C = \begin{bmatrix} 1 & 2 & 9 \\ 3 & 4 & 5 \\ 2 & 0 & 1 \end{bmatrix} \qquad D = \begin{bmatrix} 1 & -1 & 2 \\ 0 & 1 & -1 \\ 0 & 0 & 1 \end{bmatrix}.$$

The MATRX menu allows you to save these matrices in the TI-85 memory under the names C and D. To do this, press $\boxed{\text{MATRX}}$. Pressing $\boxed{\text{EDIT}}$ will start the entry process. You will be asked to type a name; type C and press $\boxed{\text{ENTER}}$. The screen will change, and the cursor will be located in the upper right-hand corner of the screen where you can type the dimension of your matrix. Press 3 $\boxed{\text{ENTER}}$ and then repeat 3 $\boxed{\text{ENTER}}$. The screen will show that you are entering a 3 × 3 matrix, and will move the cursor to a location where you can type the first entry of the matrix; i.e., the 1 at the beginning of the first row. Type 1 $\boxed{\text{ENTER}}$ and the TI-85 is ready for the next number in the first row. Type 2 $\boxed{\text{ENTER}}$ and keep going across the rows in order until the last number in the last row has been entered.

After the matrix C has been entered, press $\boxed{\text{EXIT}}$, $\boxed{\text{MATRX}}$ and $\boxed{\text{EDIT}}$ to restart the EDIT menu for the matrix D. When you have finished entering D, press $\boxed{\text{EXIT}}$ twice so that no menu is visible. You are now ready to perform matrix operations on C and D.

To add C and D, press $C+D$ $\boxed{\text{ENTER}}$. You will see the answer

$$\begin{bmatrix} [& 2 & 1 & 11 &] \\ [& 3 & 5 & 4 &] \\ [& 2 & 0 & 2 &] \end{bmatrix}.$$

To multiply C and D, type $C * D$ $\boxed{\text{ENTER}}$. The answer is

$$\begin{bmatrix} 1 & 1 & 9 \\ 3 & 1 & 7 \\ 2 & -2 & 5 \end{bmatrix}.$$

As we shall see when discussing Markov chains, some applications may require that a matrix be multiplied by itself a large number of times. Suppose you needed to multiply C by itself 4 times (i.e., $C * C * C * C = C^4$). This could be very time consuming. On the TI-85 you need only type C^4 $\boxed{\text{ENTER}}$. Try it.

In the next part of this section you will see that some equation solving applications require the use of the inverse matrix. Finding the inverse of a square matrix is easy on the TI-85. To find the inverse matrix for C, simply type

$$C^{-1}.$$

The expression may be entered easily by typing C $\boxed{x^{-1}}$. As soon as you press $\boxed{\text{ENTER}}$ the inverse matrix will appear on the screen.

Solving the system AX=B using the matrix inverse

In more advanced applied courses (like multivariate statistics), linear systems are written down in matrix form. The standard matrix form is $AX = B$, where A is the matrix of coefficients of the variables, and B is the column matrix of constant terms. For the system 2.1

$$A = \begin{bmatrix} 2 & 4 \\ 7 & -2 \end{bmatrix} \quad \text{and} \quad B = \begin{bmatrix} 14 \\ 1 \end{bmatrix}.$$

Matrix theory says that if A has an inverse you can multiply both sides of the equation $AX = B$ by the inverse of A and obtain the solution

$$X = A^{-1}B.$$

You can use the TI-85 to see this solution method in action. Enter the matrices A and B just as you did with C and D in the last section. The only real change is in the dimensions. A is 2×2 and B is 2×1. Once the two matrices have been entered, exit the menus, and type on the calculator screen

$$A^{-1}B.$$

Once this expression is typed, press $\boxed{\text{ENTER}}$ and you will see the answer in matrix form:

$$\begin{bmatrix} [& 1 &] \\ [& 3 &] \end{bmatrix}.$$

Systems which cannot be solved using SIMULT

The solution methods discussed in the first part of this section work well for systems of equations that have a unique solution for each of the variables. However, some systems have infinitely many solutions or no solution at all. The methods previously discussed will not work for such systems. The two systems below are simple examples of linear systems which do not have unique solutions.

$$x + y = 2$$
$$2x + 2y = 4 \tag{2.7}$$

$$x + y = 2$$
$$x + y = 5 \;. \tag{2.8}$$

The first system has infinitely many solutions. You can see this easily if you multiply the first equation by -2 and add it to the second one. The system becomes

$$x + y = 2$$
$$0 = 0 \;.$$

Thus the system reduces to a single equation – the equation of a straight line. Every point on the line gives a solution to that equation; so there are infinitely many solutions.

The second system clearly cannot be solved, because the values of x and y would have to add up to 2 and 5 at the same time.

To see that such systems are a problem for the TI-85, try solving them with SIMULT. In each case, you will be able to enter the coefficients of the system, but when you press $\boxed{\text{SOLVE}}$ the message ERROR 03 SINGULAR MAT will appear instead of a solution.

Other methods are necessary to handle linear systems where problems like this occur. Fortunately, the TI-85 has these methods available, as we shall see in the next section.

What to do when SIMULT won't work; elementary row operations

In discussing system 2.7 we used the familiar technique of multiplying one equation by a number and adding it to another equation. In doing this, all operations were done on the coefficients of the equations. The problem could have been solved with less writing using matrix representation of the system 2.3. The matrix of coefficients and constants of 2.7 is

$$S = \begin{bmatrix} 1 & 1 & 2 \\ 2 & 2 & 4 \end{bmatrix} \;.$$

Instead of thinking of the solution procedure as multiplying the first equation by -2 and adding it to the second equation, we multiply the first row of the matrix by -2 and add it to the second row. The result is:

$$\begin{bmatrix} 1 & 1 & 2 \\ 0 & 0 & 0 \end{bmatrix} \;.$$

This second matrix represents the final system of equations that showed us that all points on the line $x + y = 2$ were solutions to the original system.

The TI-85 has built-in commands which enable you to multiply a row of a matrix by a number, to add two rows or to add a multiple of one row to another. These commands are useful in linear programming as well as in solving systems, so we will discuss the most important of them now. We will do this first using the matrix given above for system (2.7). The technique for entering a matrix in the TI-85 and giving it a name has already been discussed. Begin by entering the matrix S above using the MATRX EDIT menu. Use the name S, and remember that the matrix is 2×3.

The built-in command for adding a multiple of a row to another row is **mRAdd**. To add -2 times the first row of S to the second row of S, type the command

$$\text{mRAdd}(-2, S, 1, 2).$$

(The command **mRAdd** is in the OPS submenu of MATRX.)

As soon as you press ENTER you will see the desired result on the screen. You should review carefully the arrangement here, so that it can be used later:

The -2 listed first is the number to be used as a row multiplier.

Next the name of the matrix is given.

The number 1 listed next shows that the first row will be multiplied by -2.

The final 2 shows that -2 times the first row will be added to row 2.

The entire command is always used in the form:

mRAdd(multiplier, matrix name, row multiplied, row added to)

The **mRAdd** command is quite useful since many problems occurring in Finite Mathematics courses require a number of operations of this type. Often students understand the basic principles but cannot do the arithmetic without occasional errors. The TI-85 removes some of the tedious calculation and frustrating mistakes.

Another command that will be useful later is **multR**. This command can be used to multiply a row by a number. The general format is:

multR(multiplier, matrix name, row)

To multiply the second row of S by .5, type multR(.5,S,2) and ENTER. Try it.

Even with these operations available, a problem that requires a large number of row operations can be time consuming. It is not easy to get a matrix to the final form which shows the solution. The TI-85 has a very useful command to help with this problem. The command is **rref**, and it reduces a matrix immediately to the final form which clearly shows the solution. To do this for the matrix S, press $\boxed{\text{MATRX}}$ $\boxed{\text{OPS}}$ $\boxed{\text{rref}}$ $\boxed{\text{EXIT}}$ $\boxed{\text{NAMES}}$ S $\boxed{\text{ENTER}}$.

You will immediately see the matrix that gives the final answer.

The command rref will solve any system, whether or not it has infinitely many solutions. A good example of a system which has a unique solution for each variable appeared in system (2.2).

$$x + y + z = 6$$
$$2x - y + 3z = 9 \qquad (2.2)$$
$$x - y + z = 4.$$

The augmented matrix (i.e., the matrix of coefficients and constants) for this system is

$$\begin{bmatrix} 1 & 1 & 1 & 6 \\ 2 & -1 & 3 & 9 \\ 1 & -1 & 1 & 4 \end{bmatrix}.$$

Go to the MATRX menu and enter this augmented matrix in the TI-85 using the name M. Then $\boxed{\text{EXIT}}$ the menu. Type rref(M) and $\boxed{\text{ENTER}}$. You will see the reduced matrix:

$$\begin{bmatrix} [& 1 & 0 & 0 & 5 &] \\ [& 0 & 1 & 0 & 1 &] \\ [& 0 & 0 & 1 & 0 &]] \end{bmatrix}.$$

This matrix shows that the final solution is $x = 5$, $y = 1$ and $z = 0$. This is the correct solution, and it is the same answer that SIMULT would give. In fact, rref will solve any system that SIMULT will solve. The advantage of rref is that it will solve systems that SIMULT cannot handle. For example, look at the system:

$$x + 2y + 3z = 11$$
$$2x + y + 3z = 13 \qquad (2.7)$$
$$x + y + 2z = 8$$

SIMULT will not solve this system. Go to the MATRX menu and enter its coefficient matrix using the name W.

$$\begin{bmatrix} 1 & 2 & 3 & 11 \\ 2 & 1 & 3 & 13 \\ 1 & 1 & 2 & 8 \end{bmatrix}.$$

Now EXIT the menu. Type rref(W) and ENTER . You will see:

$$\begin{bmatrix} [& 1 & 0 & 1 & 5 &] \\ [& 0 & 1 & 1 & 3 &] \\ [& 0 & 0 & 0 & 0 &]] \end{bmatrix}$$

If this final matrix is converted back to a system of equations, the result is

$$x + \quad +z = 5$$
$$y + z = 3 .$$
$$0 = 0$$

This system can be solved for x and y:

$$x = 5 - z$$
$$y = 3 - z$$

It is easily seen that there are infinitely many solutions. If you substitute any value for z, you will get a corresponding solution for x and y. Using $z = 0$, you get the solution $z = 0$, $x = 5$ and $y = 3$. Using $z = 1$, the solution becomes $z = 1$, $x = 4$ and $y = 2$.

Summary of linear system solving on the TI-85.

The TI-85 has built-in menus and functions to solve most of the problems on linear systems that are encountered in a Finite Mathematics course. The most important are:

1. The SIMULT menu, which will solve a system that has a unique solution and also has the same number of unknowns and equations (e.g., three equations in three unknowns or four equations in four unknowns).

2. For two equations in two unknowns, the student can graph the two lines using GRAPH and then find the solution graphically as the intersection point of the lines.

3. Using the TI-85 capability to find the inverse of a matrix, the student can solve the system $AX = B$ by finding

$$X = A^{-1}B .$$

As with SIMULT, this will work only if the solution is unique and only if the number of equations equals the number of unknowns.

4. The student can use operations of row multiplication and addition to reduce the coefficient matrix of the system step-by-step to a reduced matrix that shows the final answer. If the step-by-step analysis is not

required, the matrix can be reduced in one step using the command **rref**. This will always work.

The capabilities to solve systems quickly, to graph them, to find inverses and to use row operations are important because they can be used for much more than simply solving linear systems. In the following sections, we will use these capabilities in the solution of linear programming problems and the analysis of Markov chains.

Explorations with Linear Systems

1. Three simultaneous systems are given for solution below:

$$2x + 4y = 6$$
$$4x + 8y = 12 \tag{2.8}$$

$$2x + 4y = 6$$
$$4x + 8y = 15 \tag{2.9}$$

$$2x + 4y = 6$$
$$4x + 6y = 12 \tag{2.10}$$

   a. Try to solve each system using SIMULT. (Only one will be solvable this way.)

   b. Graph each system and explain why only one can be solved by SIMULT.

   c. One of the systems has infinitely many solutions. Use your graphical results from b) to explain which system this is, and then solve it using **rref**.

2. Two simultaneous systems are given for solution below:

$$-2.0x + y = 1$$
$$-2.2x + y = 1 \tag{2.11}$$

$$x + y = 1$$
$$-x + y = 1 \tag{2.12}$$

   a. Verify that each system has $(0, 1)$ as a solution.

   b. Change the constant term 1 to 2 in the first equation of each system, and solve the new systems.

$$-2.0x + y = 2$$
$$-2.2x + y = 1 \tag{2.13}$$

$$x + y = 2$$
$$-x + y = 1 \quad \quad (2.14)$$

c. The solution in one system did not change very much when the constant changed. The other changed a great deal. Explore each system by graphing it and then graphing the changed system in which the 1 is changed to 2. This should enable you to explain the difference in the amount of change.

NOTE: The study of the effect of changes in the coefficients on the final solution is called *sensitivity analysis*.

2.2. Geometric Linear Programming

Business mathematics texts usually discuss linear programming. The solution of linear programming problems requires a great deal of paper-and-pencil work for either graphing or extended calculations. In this section and the following one, we will show by example how much of this work can be done on the TI-85. The example we will use is the solution of the following problem:

Maximize
$$Z = 2x + 3y.$$

Subject to
$$4x + y \leq 6$$
$$x + 2y \leq 5$$
$$x, y \geq 0.$$

We will not attempt to teach the procedure for graphical solution here. We will assume that you are familiar with the basic procedure, but wish to learn how to use the TI-85 to execute it. The steps in the procedure are reviewed below:

1) Graph the system of inequalities.

2) Find the corner points of the graph from step (1).

3) Evaluate Z at each corner point. The largest value of Z found at a corner point is the maximum.

Graphing the system of inequalities

The graph of the solution is bounded by the lines

$$4x + y = 6$$
$$x + 2y = 5$$
$$x = 0$$
$$y = 0.$$

These lines are obtained by turning each inequality in the system into an equality. The last two lines are very easy to graph; they are the coordinate axes. Before using the TI-85, solve each of the first two equations for y in terms of x using algebra.

$$y = -4.0x + 6.0$$
$$y = -0.5x + 2.5.$$

These expressions for y are in the proper form for the GRAPH menu. Press $\boxed{\text{GRAPH}}$ and then $\boxed{y(x) =}$. Then enter the two expressions:

$$y_1 = -4.0x + 6.0$$
$$y_2 = -0.5x + 2.5.$$

To set the range conveniently, press $\boxed{\text{EXIT}}$ and $\boxed{\text{RANGE}}$. Then choose

$$
\begin{aligned}
\text{xMin} &= -1 \\
\text{xMax} &= 2 \\
\text{xScl} &= 1 \\
\text{yMin} &= -1 \\
\text{yMax} &= 4 \\
\text{yScl} &= 1
\end{aligned}
$$

Press $\boxed{\text{GRAPH}}$ and you will see this display:

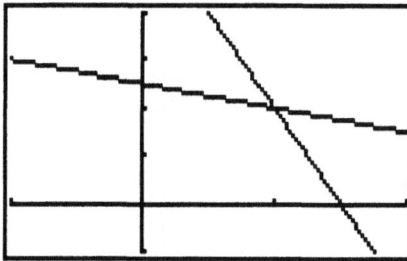

Figure 2.2. Graph of region for linear programming problem.

The region you are seeking is bounded by the four lines in the display. Press $\boxed{\blacktriangleright}$ and you will see a $+$ in the middle of the region. Below the display you will see the coordinates of the point indicated by the $+$. (It is wise to check that the coordinates of that point actually satisfy the inequalities in the system to assure that you have picked the correct part of the graph as the solution set.)

Finding the corner points

You can find the coordinates of the corner points without ever leaving the GRAPH menu. The first corner point requires nothing more than looking at the graph; it is $(0,0)$. To find the other corner points, we will use two different menu options:

a. Press TRACE . A flashing cursor will appear on the uppermost line segment. Press ▶ until the cursor has moved all the way down the line to the x-axis. The screen coordinates will show $x = 1.5$ and $y = 0$. This is the x-intercept of the first line. The second corner point found is $(1.5, 0)$.

b. TRACE can be used again to find the intersection of the second line and the y-axis. Press ▲ and the tracing cursor will jump to the second line. The ◀ key will move you along that line to the y-axis. The third corner point is the y-intercept $(0, 2.5)$.

c. We used TRACE to look for the intercepts because those are easy to check (or simply to find) by hand. Checking is important, because tracing may bring you near a point but not exactly onto it. A more powerful option is ISECT, which solves for the intersection point of two graphed functions. Press EXIT to return to the menu, and then press MORE and MATH . Press MORE again, and ISECT . You will see tracing cursor appear on the graph. Use the right arrow key to move down the graph until you appear to be at the intersection point. Press ENTER twice, and the TI-85 will solve for the coordinates of the intersection point. That corner point is $(1, 2)$.

This problem has been chosen so that the numerical answers are simple and easy to check mentally. Other problems may have more complicated numerical answers, but the corner points can be found in the same way – use TRACE and hand checking or ISECT. The corner points for this problem are

Table 2.1. Corner points of the region.

x	y
0.0	0.0
0.0	2.5
1.5	0.0
1.0	2.0

Finding the maximum value

The maximum for Z occurs at one of the four corner points listed above. The problem can be finished by adding another column to the table for Z values.

Table 2.2. Values of Z at the cornerpoints.

x	y	$Z = 2x + 3y$
0.0	0.0	0.0
0.0	2.5	7.5
1.5	0.0	3.0
1.0	2.0	8.0

The maximum value is $Z = 8$; it occurs at the corner point $x = 1$, $y = 2$.

The numbers here are so easy to work with that most students will not need the TI-85 to evaluate Z. In more complicated problems it is advisable to $\boxed{\text{EXIT}}$ from the GRAPH menu and use the calculator to do the Z calculations.

Sensitivity analysis

The speed of the TI-85 makes it very easy to redo the calculations if one of the numbers input needs to be changed. When mathematicians solve real applied problems, they often look at the effect of changes in one or more of the coefficients. Suppose, for example, that you were not sure about the coefficients in the problem just solved, and thought that the second inequality in the system just solved might have to be changed to give the new system:

Maximize
$$Z = 2x + 3y.$$

Subject to
$$4x + y \leq 6$$
$$2x + 2y \leq 5$$
$$x, y \geq 0.$$

The new problem can be solved very quickly on the TI-85. Enter the GRAPH menu, and press $\boxed{y(x) =}$. Cursor down to the $y2 =$ equation and change the expression there to the new solution for y

$$y2 = -x + 2.5.$$

Now graph the solution. Only one corner point has changed. The point at $(1, 2)$ has changed to a new corner point that can be found using ISECT. It is $(1.1667, 1.3333)$. The change in corner points leads to a change in only one row of the table.

Table 2.3. Computed z-values for new problem.

x	y	$Z = 2x + 3y$
0.0	0.0	0.0
0.0	2.5	7.5
1.5	0.0	3.0
1.667	1.3333	6.333

However, this one change changes the final answer to the maximum problem. This reworking of the problem shows the importance of checking the coefficients. This type of study of the effects of changes in the problem is called a *sensitivity analysis*. It is not hard to do with a powerful calculator but can take a great deal of time when problems must be reworked using pencil-and-paper.

Explorations with Linear Programming

1. Solve graphically:
   Maximize
   $$Z = x + y.$$
   Subject to
   $$2x + y \leq 4$$
   $$x + y \leq 3$$
   $$x, y \geq 0.$$

Note that the solution for Z is just the sum of the x- and y-coordinates of the corner point where this sum is maximized.

2. Change the constant 4 in the first constraint to 5. How does this change the maximum value of Z?

3. Change the 3 in the second constraint to 4. How does this change the maximum value of Z?

4. Would it make any sense to change the 3 in the second constraint to −1? (Graph the region and look carefully at which points are inside it.)

2.3. The Simplex Method for Linear Programming Problems

The geometric method used above works only for problems that can be graphed in two dimensions. These problems can only have two variables. There is a more general solution procedure called the Simplex Method which does not rely on graphing. It will work for any number of variables and inequalities. The simplex method requires some time and thought, and we will not try to derive the procedure here or to explain it in detail. However, the matrix operations available on the TI-85 can be used to save time and increase accuracy in calculations.

Pivoting on the TI-85

The simplex method consists of a series of special matrix operations called *pivots*. A linear programming problem can be written in matrix form. The student who has begun to study linear programming will recognize that the problem solved in Section 2.2 is represented by the matrix

$$\begin{bmatrix} 4 & 1 & 1 & 0 & 0 & 6 \\ 1 & 2 & 0 & 1 & 0 & 5 \\ -2 & -3 & 0 & 0 & 1 & 0 \end{bmatrix}$$

Even if you have not studied linear programming you can understand what a pivot operation is. First you pick an entry in the matrix to be the *pivot element*. We will pick the entry 2 which is in the second row and second column. (The second row is now referred to as the *pivot row*, and the second column is called the *pivot column*.) The pivot operation has two steps:

a) Multiply the pivot row by one over the value of the pivot element, so the pivot element becomes 1. In this case we would multiply the second row by .5.

b) Add multiples of the pivot row to the remaining rows until all other entries in the pivot column are zero.

This can be done easily using the elementary row operations in the MATRX menu on the TI-85. (The elementary row operations were discussed in Section 2.2.) Begin by entering the MATRX menu and choosing $\boxed{\text{EDIT}}$. Name the matrix to be entered LP; its dimensions are 3 × 6.

When the matrix has been entered, press $\boxed{\text{EXIT}}$ twice to leave the MATRX menu. To pivot, use the following steps.

a) Type the command

$$\text{multR}(.5, LP, 2).$$

This will make the pivot element 1. The matrix below will appear on the display.

```
[[   4.0    1.0  1.0  0.0  0.0  6.0  ]
 [   0.5    1.0  0.0  0.5  0.0  2.5  ]
 [  -2.0   -3.0  0.0  0.0  1.0  0.0  ]]
```

You will need to do more work on this matrix. If you see that it is correct, type

$$\text{Ans} \longrightarrow L$$

using the $\boxed{\text{STO>}}$ key to type the arrow. This will save your work in a new matrix L.

b) Type the command

$$\text{mRadd}(-1, L, 2, 1) \longrightarrow L.$$

This will add -1 times row 2 to row 1 and zero out the element in the first row and second column. Then type

$$\text{mRadd}(3, L, 2, 3) \longrightarrow L.$$

This will add 3 times row 2 to row 3 and zero out the element in the third row and second column. The final matrix is

$$\begin{bmatrix} 3.5 & 0.0 & 1.0 & -0.5 & 0.0 & 3.5 \\ 0.5 & 1.0 & 0.0 & 0.5 & 0.0 & 2.5 \\ -0.5 & 0.0 & 0.0 & 1.5 & 1.0 & 7.5 \end{bmatrix}$$

There is much more to linear programming than performing pivot operations, but these basic operations are very time-consuming when done by hand. The TI-85 will remove the monotony of tedious calculations and allow you to concentrate on more important issues.

2.4. Mathematics of Finance Using the SOLVER

Finite Mathematics courses typically contain a unit on Mathematics of Finance. This unit covers interest calculations which most people recognize clearly as business mathematics. The calculations done here are also commonly referred to as the mathematics of the "Time Value of Money."

The Basic Equation $PV = FV/(1 + I)^\wedge N$

The fundamental idea in time value of money is that money grows over time because it earns interest. If you can earn 10% interest per year and invest $100 today, you will have $110 at the end of a year – the original $100 plus $10 interest. If you leave the money invested for a second year, you will earn $11 interest and end up with $121 in two years. The amount you start with in the present is called the Present Value, abbreviated PV. The amount you end up with one or two years in the future is called a Future Value, abbreviated FV. If the interest rate is represented by I and the number of years is represented by N, the basic equation relating present value and future value is

$$FV = PV(1 + I)^N .$$

Future value calculations can be done directly on the calculator without using any menu features. For example, to see what $100 would grow to in 2 years at 10% interest, you can simply type in 100 $\boxed{X}$ 1.1 $\boxed{\wedge}$ 2 $\boxed{\text{ENTER}}$. The calculator will show the answer 121.

The time value equation can be solved for PV in terms of FV. The solution is

$$PV = \frac{FV}{(1 + I)^N} .$$

Present value calculations tell you how much money to invest now if you need a certain amount in the future. It is easy to see from the example above that if you need $121 in 2 years and you can earn 10% interest, you need to invest $100 now. This could have been done as a present value calculation on the calculator by directly dividing 121 by 1.1^2.

Even though simple calculations can be done directly in this manner, much more powerful analysis is possible if the advanced features of the TI-85 are used.

Equations; time value of money in the SOLVER

One very nice feature of the TI-85 is the capability to put useful equations in memory. Let's do this with the basic time value equation. Type

$$PV = FV/(1 + I)\boxed{\wedge}N\boxed{\text{ENTER}} .$$

This will store the expression $FV/(1 + I)^\wedge N$ under the name PV, and calculate the value of that expression whenever you type PV. To see this work, you must first tell the TI-85 the vaues of FV, I and N that you want to use. This can be done using the $\boxed{\text{STO}\blacktriangleright}$ key. To repeat the calculation of the present value of 121 in 2 years at 10%, type

$$121 \quad \boxed{\text{STO}\blacktriangleright} \quad FV \quad \boxed{\text{ENTER}}$$
$$2 \quad \boxed{\text{STO}\blacktriangleright} \quad N \quad \boxed{\text{ENTER}}$$
$$.10 \quad \boxed{\text{STO}\blacktriangleright} \quad I \quad \boxed{\text{ENTER}}$$

Now if you type PV $\boxed{\text{ENTER}}$, you will see the correct present value answer of 100. This illustrates how the TI-85 works with formulas. If you type the name of a variable which represents a formula (like PV), the TI-85 evaluates the formula.

The ability to work directly with equations is useful, but it is even better to work in the SOLVER on the TI-85. This is true for two reasons. First, the SOLVER makes the process of assigning values to the variables easier. Second, and more important, the SOLVER lets you solve the equation for any variable, not just PV. To begin work in the SOLVER, press SOLVER . You will then see a new screen which is mostly blank. The cursor will be at the top, in a space reserved for the name of the equation you are working with. If there is already an equation name in this space, press the CLEAR key and it will become blank. At the bottom of the screen are menu key indicators which show the names of equations which are in memory. If you see PV here, press the F key underneath it and the name PV will appear at the top of the screen. (If you see five other names and not PV, press MORE to look through memory further and find PV.) Once PV is at the top of the screen after the prompt "eqn:," press ENTER.

As soon as you do this, you will see on the screen

$$\text{exp}=\text{PV}$$
$$\text{exp}=$$
$$\text{FV}=121$$
$$\text{I}=.1$$
$$\text{N}=2$$
$$\text{bound}=\{-1e99, 1e99\}$$

GRAPH   RANGE   ZOOM   TRACE   SOLVE

The first line is a reminder that the expression you are working with is PV. The next lines show the values most recently assigned to the variables in the expression for PV. To see how the solver works, we will look at a new problem. Suppose you want to begin saving now to have $5000 for tuition for an MBA in two years. A bank will guarantee you 7% interest. How much should you put in the bank now? In terms of our variables, the problem is:

GIVEN: $FV = 5000, I = .07, N = 2$
TO FIND: PV

To enter the new values of the variables, use the ▼ key to move the cursor down to FV. Then type 5000. Move the cursor down again to I and type .07. Finish by entering 2 for N.

To solve for a variable, you must move the cursor back to the line on which that variable appears. Move the cursor to the line which displays exp=. This is the line containing the old PV number, and you want to solve on it for the new PV number.

The SOLVE menu option is above the F5 key. Press $\boxed{\text{SOLVE}}$. A small moving line will appear in the upper right hand corner of the screen. This is the SOLVER's busy signal; it shows that a calculation is in progress. In a few seconds you will see the correct answer, 4367.1936413661. You should invest \$4367.19 or 4367.20, depending on what the bank will do when it rounds off your interest earned.

To check this answer, re-enter the rounded amount of 4367.19 on the top line, and move the cursor down one line to FV. Press $\boxed{\text{SOLVE}}$ again. This will cause the SOLVER to solve for FV. The answer is 4999.995831, which does round up to 5000.

Notice what the SOLVER does that the ordinary calculator does not do. Even though the equation you are using is written down in a form which is solved for PV, the TI-85 will solve for FV without requiring you to put in a new equation. The TI-85 will also solve for I, the interest rate or N, the number of periods. Finite Mathematics textbooks give separate algebraic formulas for each of FV, N and I. Although these are of interest mathematically, it is much more convenient to enter just one formula and solve for any variable. You are now able to solve each of the following practical problems:

I. **Effective interest rate.** Your financial planner offers you an investment in which you invest \$1000 and get back \$2000 in 12 years. He says this must be a good investment because you can double your money. What rate of interest will you earn if you make this investment?

   Solution: Use the PV equation on the SOLVER. The values to enter are PV=1000 (enter this next to exp=), FV=2000 and N=12. Move the cursor to I and press $\boxed{\text{SOLVE}}$. The interest rate is approximately 5.95%. Does that impress you?

II. **Required term.** You put \$1000 in an account earning 12% interest. How long will it take for your account to reach \$3000?

   Solution: Enter 3000 as FV and .12 as I. The value of PV should be 1000 from the last example; change it to 1000 if this is not the case. Solve for N. The answer is 9.6940354129408. Your money will triple in less than 10 years at 12% interest.

Annuities on the TI-85

We have just discussed the present value and future value of a single payment. Many payments in life are made on a repeated basis; e.g., payments on a mortgage or a car loan are made every month. An *annuity* is a series of repeated equal payments. The simplest type of annuity is a series of payments of 1. Suppose you were to receive \$1 at the end of each year for 3 years. This is an ordinary annuity for three years. If you were planning to save each dollar in a bank account earning 10%, it would make sense to talk about the future value of the annuity. The calculation of future value could be done separately for each year and the results added. The future value would be:

$$1 \cdot (1.1)^2 + 1 \cdot (1.1) + 1 = 3.31 \, .$$

This is relatively easy to calculate, since there are only three terms, but an annuity for twenty years would be a problem. In finite mathematics courses, geometric series methods are used to produce simple formulas for future and present values of annuities. The general formula given there for the future value of an ordinary annuity of payments of 1 is

$$s_{n\rceil_i} = \frac{(1+i)^n - 1}{i} \, .$$

This equation can be entered in the TI-85. Type

$$sn = ((1 + I)^\wedge N - 1)/I \, .$$

To check the correctness of the equation entered, redo the last problem. Enter the SOLVER and use sn as the expression to be solved. Enter $I = .1$ and $N = 3$. Then solve for the value of exp. The answer is 3.31.

The last calculation used the TI-85 to show that

$$s_{3\rceil_{.1}} = 3.31 \, .$$

A much more appropriate problem for the calculator would be to find the future value of \$1 deposits made for 30 years into an account earning 10% interest. Enter $N = 30$ in the solver, leaving $I = .10$. Then solve for exp to find the future value. The answer is approximately \$164.49.

It is natural to ask why all of these calculations would be done for investments of \$1 per year – most mortgages and car loan payments are a bit higher. The reason for doing the mathematics this way is really historical. Before the advent of sophisticated calculators, people who did future value calculations for annuities used tables of values for sn. You did not calculate a value; you looked it up in a table of \$1 annuities. If you had an annuity of \$100 payments, you used the fact that its future value was 100 times the future value of an annuity of \$1 payments. The general equation in use was really

$$FVANN = PMTANN * sn$$

where PMTANN stands for the annuity payment made and FVANN represents the future value of the annuity using that payment. Using the result of the last calculation, the future value of an annuity of $100 per year at the end of each of the next 30 years would have a future value of

$$100 s_{\overline{30}|.10} = 16,449.40\,.$$

The TI-85 allows you to use previously defined variables like sn in new equations. This means that you can enter the equation for FVANN into the TI-85 as it appears above and use it to solve any annuity problem. To see this, EXIT the SOLVER and type the above equation (FVANN = PMTANN ∗ sn) in the TI-85. Then go to the SOLVER menu and use the equation for FVANN. Enter a payment value of PMTANN = 100 with $I = .1$ and $N = 30$. Solve for the value of exp and you will find the answer just given – approximately $16,449.40.

The present value of an annuity can be calculated in the same way. The present value of an annuity of payments of 1 for n periods at a rate of i is

$$a_{\overline{n}|i} = \frac{1}{i}\left(1 - \frac{1}{(1+i)^n}\right)\,.$$

Enter this equation in the TI-85 by pressing $\boxed{\text{EXIT}}$ and typing

$$an = (1/I) \ast (1 - 1/(1 + I)^\wedge N)\,.$$

Check the entry by using the solver to calculate the equation for an with $I = .1$ and $N = 30$. The answer should be approximately 9.4269.

The present value of any other payment PMTANN can be found using the same multiplication principle that was applied to future values. The present value of a 30 year annuity of $100 payments at 10% should be 100(9.4269) or $942.69.

This can also be done directly by the TI-85. $\boxed{\text{EXIT}}$ the solver and type the general equation

$$\text{PVANN} = \text{PMTANN} \ast an\,.$$

Then return to the solver menu and select this equation. Solve it for the value of exp using PMTANN = 100, $N = 30$ and $I = .10$. The answer will be approximately 942.69, as expected.

It might be useful to summarize what has been done with annuities. The formulas given here enable you to find:

      Present value for annuities with payments of 1;

      Present value for an annuity with payment PMTANN;

      Future value for annuities with payments of 1;

      Future value for an annuity with payment PMTANN.

Since the calculator can find present or future values for any annuity, there is no intrinsic need to calculate values for annuities of 1 separately. However, the expressions *sn* and *an* have been used for so long that they are often stated separately in textbooks, and it is useful to be able to calculate them separately.

The TI-85 will solve the annuity equations for any one of the variables. Thus you could solve an annuity for its effective interest rate. This will not usually be requested in a Finite Mathematics Course, since the mathematics behind the calculation is too advanced for that course. However, the solutions obtained here are of practical interest in finance. The next two topics covered are beyond the basic business mathematics course but would enable you to turn your TI-85 into a true financial calculator.

General time value of money on the TI-85

If you buy a financial calculator, you will usually find keys marked i, n, PV, PMT and FV. These keys can be used to solve any of the problems discussed in this section. The basic equation which relates all these quantities is

$$PV = PMT \cdot a_{n1_i} + \frac{FV}{(1+i)^n}.$$

If you wish to enter this equation and make your TI-85 a financial calculator, enter the equation below into your calculator. *If you do so, you will change the definition of PV used earlier, but you will have a more powerful financial calculator.* The equation is

$$PV = PMT * an + FV/(1+I)^\wedge N.$$

Enter this equation and then go to the SOLVER menu. Select the PV equation which you have just entered. You can now solve any of the previous types of problem.

a) **Basic PV and FV problems.** One of the first problems in this chapter was to find the present value of 121 two years from now at 10% interest. Enter 0 for PMT (no regular payment is made), .1 for I, 2 for N and 121 for FV. Return the cursor to exp = and press $\boxed{\text{SOLVE}}$. The answer is 100.

Note that the key step in elementary PV-FV problems is to set PMT=0 since no regular payment is made.

b) **Annuity problems.** Another problem solved previously was the future value at 10% of an annuity of three payments of 1 at year end. Change the value of PV to 0 by entering 0 at exp =. Then enter

PMT $= 1$ and $N = 3$. Move the cursor to FV and press F5 for SOLVE. The answer is -3.31.

The answer in Section 5.3 was 3.31, so the magnitude of the answer agrees with the prior results. Why is the number negative? If you think about the equation above, you will see that the numbers you entered require a solution of the equation

$$0 = 1 \cdot a_{\overline{3}|.10} + \frac{FV}{(1+.10)^3}$$

for FV. Solving for FV will clearly give a negative number as solution. This may seem too complicated, but your financial calculator will do exactly the same thing. The TI-BA35 financial calculator gives the same answer of -3.31 for this problem.

c) **Loan calculations.** The new PV equation is ideal for loan analysis. Suppose you need to borrow \$100,000 to buy a house. Mortgage loan rates are now 8.5%, and you want to get a thirty year loan. You would like to calculate the monthly payment on the loan. The entries you need to make are:

> 100000 for PV (the amount of the loan),
>
> .085/12 for I (the monthly rate),
>
> 12×30 for N (the total number of months),
>
> 0 for FV (the loan pays off after 30 years).

Return the cursor to PMT and press $\boxed{\text{SOLVE}}$. The required monthly payment is 768.91.

Although the loan analysis above will not be discussed in most mathematics courses for college business students, it is the kind of practical problem for which those courses prepare you. The equation solving power of the TI-85 allows you to solve real loan problems without having to buy a financial calculator.

Explorations with Time Value of Money

All problems deal with the loan in (c) above – a 360 month loan for \$100,000 at a monthly rate of .085/12.

1. What happens to the monthly payment if the stated rate increases from 8.5% to 9.0%? How about 9.5%, 10%, 11% and 16%? Rates were as high as 16% in the late 1970s. What effect do you think that this had on home purchases?

2. Return to the original interest rate of .085/12 and the original payment of 768.91. The expression FV represents the balance due on the loan at any month N. Solve for the loan balance in months 12, 60, 180, 240 and 300. During which periods is the loan being paid off most rapidly?

2.5. Markov Chains

The theory of Markov chains uses matrices to solve problems in probability. A simple example of such a problem arises in the study of how political party affiliations change over time. Suppose a country has two political parties D and R. In every election year, 70% of the members of D remain in their party and 30% move to R. In the party R, 20% of the members move to D and 80% stay in R. The probabilites involved here can be written in a matrix:

$$T = \begin{bmatrix} .7 & .3 \\ .2 & .8 \end{bmatrix}$$

The matrix T is called a one period transition matrix. The first row shows the probabilities for a D either remaining a D (.7) or becoming an R (.3). The second row shows the probabilities of an R either becoming a D (.2) or remaining an R (.8).

You would expect that over time there will be more members of R than members of D because the members of R are less likely to switch. The theory of Markov chains tells you that in the long run the country's population will reach a steady state in which it will have the same percent of D and R in every election. There are two ways to solve for those long run percentages. Both are quite simple to implement on the TI-85.

Method 1. This is the easiest method to use. It relies on the fact that if you multiply T by itself a large number of times, the resulting product has the desired percentages in its rows. This can be done easily on the TI-85. For example, to multiply T by itself 50 times, enter the matrix T and type

$$T\char`^50.$$

You will see the result

$$T = \begin{bmatrix} .4 & .6 \\ .4 & .6 \end{bmatrix}.$$

Note that the two rows are identical. In the long run the population will consist of 40% D and 60% R.

This is easy to do for any problem. Type in the matrix and take a large power of it. If the rows are not the same, try a larger power. This technique should work whenever the probability matrix has no zero entries.

It is not the method most commonly presented in textbooks for obvious reasons. If a student does not have a TI-85, it is unrealistic to think of finding very large powers of a matrix by hand.

Method 2. This method takes more work, but most textbooks use it. You must solve the system of equations given by:

$$[x \ y] \begin{bmatrix} .7 & .3 \\ .2 & .8 \end{bmatrix} = [x \ y].$$

The value of x gives the long term percentage for D, and the value of y gives the long term percentage for R. If you multiply the matrices and collect terms, this system of equations can be simplified to

$$.7x + .2y = x$$
$$.3x + .8y = y.$$

Another simplification gives

$$-.3x + .2y = 0$$
$$.3x - .2y = 0.$$

This system reduces to one equation

$$.3x - .2y = 0.$$

The system has infinitely many solutions, but we know that the solutions must add up to 1. Thus we can solve simultaneously the new system

$$x + y = 1$$
$$.3x - .2y = 0.$$

The SIMULT menu (which was discussed in Section 2.1) will solve this system. The answer found will be the same as the matrix multiplication answer; i.e., $x = .4$ and $y = .6$.

This method is easier to implement for a person who does not have a calculator. The steps are:

1) Set up the system

$$[x \ y] \ T = [x \ y].$$

2) Multiply out the system by hand, collect terms and simplify to obtain a new system of equations.

3) The new system of equations will have infinitely many solutions. The SIMULT menu will not solve it, but the **rref** command will reduce it to show the solution equations.

4) The variables must sum to 1. Add this equation to those found in (3). The resulting system can be solved in SIMULT.

Both Method 2 and Method 1 will work for larger problems involving three or more probabilities. Which one is best? Method 1 is easier to implement on the TI-85, but you may have some concern about the effects of roundoff errror in the process. Method 2 takes more time, but is also more likely to be required in class. Given the possibility of roundoff error, you might wish to solve problems both ways and compare answers.

It is important to make one final observation, and to present a challenge. A few books set up the general 2×2 problem in which only the variables x and y are present. They then derive a simple algebraic formula which always gives the solution for x and y. The challenge to the reader is to derive the general solution formula. This will only work for 2×2 problems; the general methods outlined above are still necessary for larger problems. However, it is true that for the original $x - y$ problem, some thinking could give a better solution than the calculator methods. This is important to remember. Even though the TI-85 is a very useful mathematical tool, there are still times when some analysis will give an even simpler method. It is important that the calculator not become a substitute for thinking.

3 Explorations in Probability and Statistics

Iris Brann Fetta
Clemson University

Statistics deals with collecting, organizing, analyzing, displaying and interpreting information and with forming conclusions concerning the source of that information. Read a newspaper, watch television, listen to any sports report or look at any magazine. The chances are that you will see a graph or table that presents you with data or with a conclusion that has been made as a result of the collection of data. Important decisions are made each day using the results of statistical surveys. Any decision that is made must have an associated measure of the reliability of that decision. An understanding of the theory of *probability* is therefore necessary. The chances are that whatever career you choose, you will need to know how to apply the basic concepts of these topics.

This chapter explores the use of the TI-85's built-in statistics applications and provides programs designed to enhance the capabilities of the calculator in exploring many of the topics in an introductory course in probability and statistics. It is not the aim of this chapter to teach you these topics, and you should consult a statistics textbook for more complete explanations when you need them. The programs in this chapter are not intended to give quick solutions, but are designed to help you think about the mathematics in a way that will give you greater insight into the underlying concepts. The chapter is divided into four major sections: descriptive statistics, probability, inferential statistics, and regression.

Section 3.1, Descriptive Statistics, gives methods for entering and editing statistical data and discusses procedures for obtaining numerical descriptive measures. Instructions for drawing histograms, control charts and scatter plots are included.

Section 3.2, Probability, begins with a discussion of the calculator's random number generator and simulation techniques. The remainder of

Explorations with the Texas Instruments TI-85
John G. Harvey and John W. Kenelly (eds.), pp. 51-134
(c)1993 by Academic Press, Inc.
ISBN: 0-12-329070-8

this section explores the binomial, Poisson, normal and Student t probability distributions with programs that will allow you to obtain probabilities for and line graphs of these discrete and continuous probability distributions.

Inferential Statistics, Section 3.3, begins with a program that generates random samples from exponential, uniform and normal populations to illustrate the sampling distribution of the sample mean. Then, applications for the calculator to the inferential statistics topics of confidence interval estimation and hypothesis testing are given. A program is provided that constructs a normal quantile plot so as to assess whether data can be considered as normally distributed.

Section 3.4, Regression and Analysis of Variance, begins with a program for exploring the least squares method in linear regression. Residual analysis, a topic usually explored only with the aid of computers, is discussed. Other regression models available in the statistics menu of the TI-85 are discussed, and the section concludes with an application to comparing the means of several populations using one-way analysis of variance techniques.

The TI-85 has a port to let you communicate by a cable with another TI-85 or with a PC or Macintosh computer. Many statistical data sets are available on computer disks and can be transmitted to your TI-85. Consult Chapter 19 of the TI-85 *Graphics Calculator Guidebook* for the proper procedure.

3.1. Descriptive Statistics

Press $\boxed{\text{STAT}}$ to access the statistical features of the TI-85 and notice the five menus: CALC, EDIT, FCST, DRAW, and VARS. The CALC menu is used to provide numerical descriptive measures and to calculate regression coefficients. The EDIT menu allows you to enter, edit, sort, and clear numerical information, while the DRAW menu provides access to the four built-in statistical graphs and allows you to store and recall graphic objects to the screen. Predicted x or y values for regression analysis are obtained using the FCST menu. The VARS menu gives access to summary statistics and regression variables. Before any of these commands can be used, data must be entered into the calculator in the form of a list.

Entering and Editing Data

There are three methods that can be used to enter, store and manipulate statistical data in the TI-85. To use a list of entered data, you must assign that list a name. List names can be at most eight characters in length, and the list names are case sensitive.

Let's use the points scored in Super Bowls I – XXVI (1967 – 1992) to explore these three data entry methods (Table I).

Table I. Super Bowls I - XXVI.

Game	Winner	Points	Loser	Points	Total Points
I	Green Bay	35	Kansas City	10	45
II	Green Bay	33	Oakland	14	47
III	New York	16	Baltimore	7	23
IV	Kansas City	23	Minnesota	7	30
V	Baltimore	16	Dallas	13	29
VI	Dallas	24	Miami	3	27
VII	Miami	14	Washington	7	21
VIII	Miami	24	Minnesota	7	31
IX	Pittsburgh	16	Minnesota	6	22
X	Pittsburgh	21	Dallas	17	38
XI	Oakland	32	Minnesota	14	46
XII	Dallas	27	Denver	10	37
XIII	Pittsburgh	35	Dallas	31	66
XIV	Pittsburgh	31	Los Angeles	19	50
XV	Oakland	27	Philadelphia	10	37
XVI	San Francisco	26	Cincinnati	21	47
XVII	Washington	27	Miami	17	44
XVIII	LA Raiders	38	Washington	9	47
XIX	San Francisco	38	Miami	16	54
XX	Chicago	46	New England	10	56
XXI	Giants	39	Denver	20	59
XXII	Washington	42	Denver	10	52
XXIII	San Francisco	20	Cincinnati	16	36
XXIV	San Francisco	55	Denver	10	65
XXV	Giants	20	Buffalo	19	39
XXVI	Washington	37	Buffalo	24	61

Method 1: Data is entered in the form of a list and stored or used from a command line on the home screen. Access the LIST menu and begin the list with $[\{]$. Enter the winner points scored, beginning with 35. Separate each value from the next with $[\ ,\ ]$ and end the list with $[\}]$. In order to use the data in statistical calculations, you must store the list to a name by pressing $\boxed{\text{STO>}}$ followed by the name (say SBWIN) and $\boxed{\text{ENTER}}$. (If you press $\boxed{\text{ENTER}}$ before you store the list, you can recall it to the command line with $\boxed{\text{ENTRY}}$.) Activate the menu key $\boxed{\text{NAMES}}$ and notice that SBWIN now appears in the LIST menu. Press $\boxed{\text{SBWIN}}$ $\boxed{\text{ENTER}}$ and see {35 33 16 23 16 24 1...}. You can scroll through the data by pressing and holding $\boxed{\blacktriangleright}$. To verify that you have entered all the winning

scores, press `OPS` `dimL` `EXIT` `NAMES` `SBWIN` `ENTER` to obtain 26, the number of values in the list.

Method 2: Data is entered, edited, and stored in the LIST editor element by element. Let's input the points scored by the loser using this method. Access the LIST EDIT menu and name the list SBLOSE. Press `ENTER` and input 10 in the e1 position. Enter the data values one by one, using `ENTER` or `▼` to move to the next element. Press `LIST` `NAMES` and verify that SBLOSE now appears on the menu. `EXIT` `EXIT` returns you to the home screen.

Method 3: Data is entered or edited as a pair of lists in the STAT editor as data points. Access the STAT editor with `STAT` `EDIT` and the list selection screen will appear. If you have not used any existing lists in statistical calculations, the names xStat and yStat will appear as the xlist Name and ylist Name. Otherwise, the names of the most recently used lists will appear in the list name positions. The menu appearing at the bottom of the display screen will contain xStat, yStat, and the lists, in alphabetical order, that you have assigned names. You may enter *one-variable data* (*x*-values) or *two-variable data* (*x* and *y*-values) in the statistical memory location. The *frequency* of a value is the number of times that value occurs in a set of data. When entering one-variable data, the *y*'s represent the frequencies of the *x* data values. When entering two-variable data, the *y*-values represent the second coordinates of the data points (x, y).

Enter the xlist Name by selecting an existing list from the menu (use `MORE` to access additional list names), use the existing name, or type in the name of a new or existing list. Let's use this method to enter the total points scored in the Super Bowl, so type in SBTPTS as the new xlist Name. For purposes of this example, press `ENTER` to select the name yStat for the ylist Name. (If there is unwanted data in the *x*, *y* positions of the statistics editor, it can be deleted with the instruction `CLRxy`.) Since this is one-variable data, enter the first value, 45, in the x1 position and press `ENTER` `ENTER` to move to the next data position. (This accepts the default value of 1 in the *y*-value position.) Continue in this manner to enter the remaining 25 values. Press `QUIT` to leave the STAT EDIT menu.

Recall that when entering one-variable data, the *y*-values represent the frequency of occurrence of *x*-data values. Notice that the value 37 occurs twice and 47 occurs three times in this data set. You could have entered a frequency of 2 for the value 37 and a frequency of 3 for the value 47 instead of entering these values in different *x* positions with each having a frequency of 1. However, frequencies other than 1 are *not* reflected in the xlist. Therefore, if you enter a list with frequencies greater than 1, all calculations *must* be performed in the STAT mode with both the xlist and the associated ylist. This method of entry is not advised if you will be using the individual lists.

Lists of the same length may be added, subtracted, multiplied, divided, and compared term-by-term using math functions. If you had not been exploring the methods of data entry in this example, it would have been much easier to obtain the list SBTPTS with the operation $\boxed{\text{SBWIN}}$ $\boxed{+}$ $\boxed{\text{SBLOSE}}$ $\boxed{\text{STO}>}$ SBTPTS. You can also use $\boxed{\text{SUM}}$ in the LIST OPS menu to obtain the sum of the elements in any list. Verify that the sum of the total points scored in all the Super Bowl games is 1109 by entering $\boxed{\text{SUM}}$ $\boxed{\text{SBTPT}}$. Check your entry of the other lists by verifying that $\boxed{\text{SUM}}$ $\boxed{\text{SBWIN}}$ = 762 and $\boxed{\text{SUM}}$ $\boxed{\text{SBLOSE}}$ = 347.

If you notice an incorrect value while viewing the data, you can edit the value using either $\boxed{\text{LIST}}$ $\boxed{\text{EDIT}}$ or $\boxed{\text{STAT}}$ $\boxed{\text{EDIT}}$. Follow one of the procedures used to enter data, position the blinking cursor over the *value* to be changed and type in the correct number. If you want to delete a data *value* while in the LIST EDIT menu, position the cursor on the x-value you wish to eliminate and press $\boxed{\text{DELi}}$. The remaining data values automatically shift up one x-position. $\boxed{\text{INSi}}$ inserts a data position at the location of the cursor. To delete a data *point* while in the STAT EDIT menu, position the cursor on either the x or y-value of the point you wish to eliminate and press $\boxed{\text{DELi}}$. The remaining data points automatically shift up one position. $\boxed{\text{INSi}}$ inserts a new data position at the location of the cursor.

Numerical Descriptive Measures

Now that these data are entered, let's see what information is revealed about the points scored in the Super Bowl. The versatile TI-85 offers you several methods of accessing one-variable statistics. From the home screen, access the list of built-in instructions with $\boxed{\text{CATALOG}}$, press O to bring you quickly to the beginning of the names beginning with this letter, and position the cursor next to OneVar. Press $\boxed{\text{ENTER}}$ to copy the name to the home screen and use the $\boxed{\text{LIST}}$ $\boxed{\text{NAMES}}$ menu to place SBWIN after the OneVar instruction. Pressing $\boxed{\text{ENTER}}$ displays the message DONE telling you that the one-variable statistics have been calculated and stored in the results variables $\bar{x}$, σ_x, s_x, Σx, Σx^2 and n. The calculated values of these variables can be assessed with $\boxed{\text{STAT}}$ $\boxed{\text{VARS}}$. Verify that the mean number of points scored by the winning team in the Super Bowl is 29.308. If you consider this data as the population of Super Bowls, the standard deviation is $\sigma_x = 10.045$. If you consider it as a sample of all Super Bowls past, present and future, the standard deviation is $s_x = 10.244$. To calculate the *variance* for a set of data, square the standard deviation. For instance, pressing $\boxed{s_x}$ $\boxed{x^2}$ $\boxed{\text{ENTER}}$ gives the sample variance 104.942.

How about the points scored by the losers? Let's explore a second method of obtaining the summary statistics with the data SBLOSE. You can access the values in the results variables from the variables menu. You must first, however, calculate these values by choosing OneVar from the CATALOG and entering the name SBLOSE following the instruction. After this calculation, press $\boxed{\text{VARS}}$ $\boxed{\text{MORE}}$ $\boxed{\text{MORE}}$ $\boxed{\text{STAT}}$ and use $\boxed{\blacktriangledown}$ to locate $\bar{x}$. (Notice there is no symbol for μ, the population mean, since the calculation formula is the same as that for the sample mean.) Copy the name to the home screen with $\boxed{\text{ENTER}}$ and use $\boxed{\text{ENTER}}$ once more to display the value 13.346. Proceed in the same manner to find $\sigma_x = 6.330$ or $s_x = 6.456$.

A third way to obtain one-variable statistics is to calculate and display them from within the STAT menu. Press $\boxed{\text{STAT}}$ $\boxed{\text{CALC}}$, load the lists SBTPTS and yStat, and press $\boxed{\text{1-Var}}$ to see all the summary statistics displayed at once. Notice that the mean number of total points scored, 42.653, is the sum of the mean number of points scored by the winners and losers. Also note that *neither* the sample standard deviation (13.151) *nor* population standard deviation (12.896) is the sum of the individual standard deviations.

The lists xStat and yStat always contain the data used in the *previous* statistical calculation. Calculated results variables are always for the lists in xStat and yStat. If you change either of these lists, the results variables are *not* recalculated until you activate the 1-Var or OneVar instruction.

Suppose you want to sort this data. The TI-85 offers two types of sort instructions: sortA or sortD in the LIST menu and SORTX or SORTY in the statistics menu. The difference in these is that sortA reorders the x values in *ascending* order from smallest to largest without automatically storing the sorted list in memory while SORTX sorts in ascending order the elements of both the xlist and ylist based on the x values, and the lists in memory are reordered correspondingly. The LIST menu instruction sortD sorts only the x values in *descending* numerical order from largest to smallest while the STAT menu command SORTY sorts both the x and y lists in *ascending* order based on the y values. Because you want to keep the original order of these three sets of data for further explorations, rename one of your lists, say SBWIN, to TEMP with the keystrokes SBWIN $\boxed{\text{STO>}}$ TEMP, and explore the different sorting commands with the list TEMP.

Press $\boxed{\text{LIST}}$ $\boxed{\{}$ 1 $\boxed{,}$ 2 $\boxed{,}$ 3 $\boxed{\}}$ $\boxed{\text{STO>}}$ L $\boxed{\text{ENTER}}$. Press $\boxed{\text{STAT}}$ $\boxed{\text{CALC}}$ and enter L as the xlist Name. Press $\boxed{\text{yStat}}$ $\boxed{\text{ENTER}}$ to insert yStat as the ylist Name. Since you entered list L from the command line and not from within the STAT editor, the current yStat list probably will not have the correct length or frequencies for your x values. For instance, press $\boxed{\text{1-VAR}}$, and see the message ERROR 12 DIM MISMATCH message. Press $\boxed{\text{QUIT}}$ to return to the home screen. Now press $\boxed{\text{STAT}}$ $\boxed{\text{EDIT}}$ $\boxed{\text{ENTER}}$

$\boxed{\text{ENTER}}$ (accepting **L** and yStat as the xlist and ylist Names), and see the warning message list length mismatch telling you that the lists are of unequal length. If you press $\boxed{\text{CONT}}$, additional x positions are filled with 0's or y positions are filled with 1's in the list of shorter length. *This, of course, will change your data.* Press $\boxed{\text{CONT}}$, and scroll through the data to see that twenty-three 0's have been added to list **L**. You could at this point use $\boxed{\text{DELi}}$ to delete the unwanted data points. It is usually more efficient, when you receive the warning message, to choose $\boxed{\text{EXIT}}$ instead of $\boxed{\text{CONT}}$ and create a yStat list that contains the correct frequencies.

The list of frequencies you create could be given any name, but let's continue calling it yStat for the purpose of this illustration. Create the proper dimension for the list by entering the number of data values, 3, followed by $\boxed{\text{STO>}}$ $\boxed{\text{dimL}}$ $\boxed{\text{yStat}}$. Enter the frequency 1 in each of the positions of yStat with the instructions $\boxed{\text{Fill}}$ 1 $\boxed{\text{,}}$ $\boxed{\text{yStat}}$. (The instruction to dimension the length of a list, dimL, and the instruction Fill are found in the CATALOG or the LIST OPS menu.) You could also create the yStat list using either Method 1 or Method 2 for entering data. Note that if you originally entered the one-variable data in the STAT EDIT mode using Method 3 and you have not changed yStat, the yStat list will automatically consist of the proper frequencies for your x values. Practice these procedures by entering the list TRY = $\{1, 2, 3, 4, 5\}$, create a yStat list of length 5, and find the one-variable statistics for TRY from the STAT CALC menu.

To delete a list you no longer wish to use, press $\boxed{\text{MEM}}$ $\boxed{\text{Delet}}$ $\boxed{\text{LIST}}$ to access the lists you have named. Position the arrow by the list you wish to eliminate, and press $\boxed{\text{ENTER}}$. The list is deleted from memory. Eliminate lists L, TEMP, and TRY, and exit the delete menu with $\boxed{\text{QUIT}}$.

Another numerical measure of a set of data is the *median*. Unlike the mean, which can be associated with the center of mass of a data distribution, the median measures the geometric center. One-half of the data values lie to the left of the median and one-half lie to the right of the median. Therefore, the median divides the data into two equal portions, the bottom 50% and the top 50%. The median is also known as the 50th percentile. In describing the position of a particular data value in relation to the other measurements in a data set, you are using a numerical descriptive measure called a *percentile*. Percentiles divide the data into 100 equal parts. When your data set is arranged in order from smallest to largest, the pth percentile is a number (that may or may not be one of the data values) that divides the bottom p% from the top (1−p)% of the data. [7] The following program, PCNTILE, will sort one-variable data and calculate percentiles. *Before entering PCNTILE and other programs in this chapter, refer to Section 3.5 Programming Notes.*

```
 PROGRAM: PCNTILE
:Input "Data list?",D
:sortA D
:Ans→D
:OneVar D
:Lbl R
:Input "Percentile?",P
:.01*P*n→W
:iPart W→I:fPart W→F
:If F>0
:Goto A
:I+1→J
:D(I)+D(J)→G
:G/2→P
:Goto B
:Lbl A
:D(I+1)→P
:Lbl B
:Disp P
:Goto R
```

Program 3.1.

Suppose you wish to know by what score 90% of the Super Bowls were won. The 90th percentile is a value such that 90% of the winning scores fall at or below that number. Note that 0.9(26) = 23.4. Only 23 of the values fall at or below the 23rd sorted score, so choose the 24th ordered value for the 90th percentile. Enter sortA SBWIN and use ▶ to see that the ordered score of the 24th game is 42. Execute program PCNTILE and enter SBWIN at the Data list? prompt. At the Percentile? prompt, enter 90, and the value of the 90th percentile, 42, is displayed. Enter 50 at the Percentile? prompt to find that the median of the winning scores is 27. To exit the program, press [QUIT]. If you wish to access the list of sorted data used in program PCNTILE, it is stored in list D. What are the 90th percentiles and medians for the losing and total points scores?

Histograms

A *histogram* is a statistical graph used to depict one-variable statistical data that has been grouped into classes. The classes appear on the horizontal axis and the frequencies of the classes appear on the vertical axis. Let's construct a histogram for the total points scored in the Super Bowl. You must begin by telling the calculator where you wish to view the graph. The smallest value in the list SBTPTS is found with the LIST OPS menu instruction min. This value can be entered directly in the xMin position of the range with the keystrokes $\boxed{\text{GRAPH}}$ $\boxed{\text{RANGE}}$ $\boxed{\text{LIST}}$ $\boxed{\text{OPS}}$ $\boxed{\text{min}}$ $\boxed{\text{LIST}}$ $\boxed{\text{NAMES}}$ $\boxed{\text{SBTPT}}$ $\boxed{\text{ENTER}}$. When you move to the next position in the range, min SBTPTS is evaluated to 21. Using the instruction max SBTPTS, enter 66 as xMax.

The range parameter xScl determines the width of each of the classes of the histogram. As a general rule, the number of classes in any histogram should be between 5 and 20, and the width of each class should be the same to avoid visual distortion. The rule used to determine the equal class width is

$$\text{class width} = \frac{\text{largest data value} - \text{smallest data value}}{\text{number of classes}}.$$

The value obtained from this formula should be rounded up so that all data values will be included in the histogram. The TI-85 will give an error message if the class width is greater than 63.

Let's arbitrarily choose 6 classes for the SBTPTS data. Position the cursor at the xScl position in the range and enter (66 - 21)/6. Round this value to 8. (If you constructed the histogram with xScl = 7.5, the largest data value would fall on the right-hand boundary of the last interval and would not be included in the visible histogram. The calculator will place any data value falling on the right boundary of a class in the next class.) Since the frequencies, the number of data values in each of the classes, is always nonnegative, yMin should be 0. However, to avoid the "bottom" of the histogram being covered by the menu, set yMin to −2. Until the histogram is drawn, you are not sure how many data values will be in the class with the largest frequency. Let's enter a guess and set yMax to 10. If you wish to use the tic marks on the vertical axis as an aid when counting the frequencies, you should set yScl to an appropriate value. Exit the range menu with $\boxed{\text{QUIT}}$.

The instruction to draw the histogram, Hist, can be accessed from the CATALOG, your CUSTOM menu, or with the keystrokes $\boxed{\text{STAT}}$ $\boxed{\text{DRAW}}$ $\boxed{\text{HIST}}$. If you use the STAT menu instruction HIST, the histogram will be drawn using the data in the current xlist. If you use the CATALOG instruction Hist, you must enter the name of the data list following the instruction before pressing ENTER. Be sure that all functions have been turned off with FnOff and enter HIST SBTPT to draw the histogram. Notice that you cannot see all of the last rectangle (class). Why? With a class width of 8, you are grouping the total points scored in the Super Bowl into the following *frequency distribution*:

Class Intervals	Frequency
21 to 29	4
29 to 37	4
37 to 45	5
45 to 53	7
53 to 61	3
61 to 69	3

Since the last class ends at 69, this should be the value chosen for xMax. In general, xMax should equal xMin $+ k*$ xScl when the histogram is composed of k classes. It is acceptable to increase xMax or lower xMin, but doing so will leave blank spaces on the left and right of the viewing screen. If you cannot see the top of the highest rectangle or if there is too much blank space at the top of the viewing window, reset yMax. Exit the range menu with QUIT, recall the last command, Hist SBTPTS, to the screen with ENTRY, and redraw the histogram with ENTER.

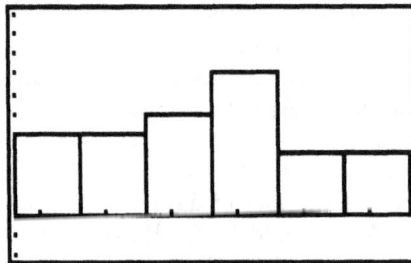

Figure 3.1. Histogram of SBTPTS.

How do you determine the frequencies? Since the trace command does not work with statistical graphs drawn by the TI-85, use the cross-hair cursor to count the frequencies (heights) of the bars (classes). The cross-hair cursor appears on the graphics screen when you press ►. Move it to the top of each of the rectangles with the cursor keys. Round the displayed y-value to the nearest whole number to obtain the frequency of the class. You should find that the frequencies are, from left to right, 4, 4, 5, 7, 3, and 3. The sum of the frequencies is 26, the number of data values.

A histogram is called *symmetric* if you could fold it over an imaginary line drawn through the mean of the data set and have one half of the graph fall *exactly* on the other half. When there is perfect symmetry in the graph, the mean and median will coincide. Since the mean of the SBTPTS data is 42.654 and the median is 44.5, the histogram is not symmetric. Look at the shape of this histogram. Even though it is not symmetric, it appears to have a larger proportion of data values near its center than in the "tails" of the graph. Store this graph with the keystrokes $\boxed{\text{STAT}}$ $\boxed{\text{DRAW}}$ $\boxed{\text{MORE}}$ $\boxed{\text{STPIC}}$ TPTS $\boxed{\text{ENTER}}$.

Now let's look at a histogram of six classes for the winning points scored in the Super Bowl. Reset the *x*-range to xMin = 14, xMax = 56, xScl = 7, press $\boxed{\text{QUIT}}$, and draw the histogram with $\boxed{\text{HIST}}$ $\boxed{\text{SBWIN}}$. This histogram appears to have a larger proportion of data values in the left tail and therefore has a long tail extending to the right (indicating a small proportion of relatively large values). When a distribution is not symmetric, it is said to be *skewed*. Because of the long right tail, the histogram for the winning points scored in the Super Bowl is called *rightward* or *positively skewed*. A histogram is called *leftward* or *negatively skewed* if it has a larger proportion of data values in the right tail and consequently, a long tail extending to the left.

Clear the graphics screen with $\boxed{\text{ClDrw}}$ $\boxed{\text{ENTER}}$, and recall the graph of the total points scored in the Super Bowl with $\boxed{\text{STAT}}$ $\boxed{\text{DRAW}}$ $\boxed{\text{MORE}}$ $\boxed{\text{RCPIC}}$ $\boxed{\text{TPTS}}$ $\boxed{\text{ENTER}}$. In cases such as the SBTPTS data where a visual determination of the longer tail is difficult, skewness may be determined by the relationship of the mean of a data set to its median.

Shape of Histogram	Relationship of Mean to Median
symmetric	mean=median
positively or rightward skewed	mean>median
negatively or leftward skewed	mean<median

Recall that the mean of the SBTPTS data is 42.654 and the median is 44.5. Thus, the total points scored in the Super Bowl data is slightly negatively (or leftward) skewed. Explore the concept of skewness by redrawing the histogram with 4 and then 8 classes. (You should reset xMin to 21 since the range parameter settings for the picture TPTS are *not* stored with the graph.) Notice that changing the number of intervals into which the data is grouped will not change the skewness of the set of data, though the histogram may look different.

Applications of the Standard Deviation

It is often useful to compute the percentage of data values falling within one, two and three standard deviations of the mean of a set of data. Reset the *x*-range parameters to xMin = 21, xMax = 69, xScl = 8, and recall the histogram TPTS. Press $\boxed{\blacktriangleright}$, use the cross-hair cursor to approximately locate the interval $\overline{x} \pm s_x = (29.50, 55.81)$, and roughly estimate the number of data values falling in this interval. Reset xMin to 3, xMax to 83, and redraw the histogram. Estimate the number of data values falling in the intervals $\overline{x} \pm 2s_x = (16.35, 68.96)$ and $\overline{x} \pm 3s_x = (3.20, 82.11)$. Instead of using the graph to obtain estimates of the number of data values falling within one, two and three standard deviations of the mean, you could look at the data and tally the *exact* number of values falling in each of these intervals. Why not have the TI-85 do this for you?

Program DATADISP constructs a histogram of six intervals showing the dispersion of one-variable statistical data within one, two, and three standard deviations of the mean of the data. The program sets xMin = $\overline{x} - 3s_x$, xMax = $\overline{x} + 3s_x$, xScl = s_x and draws a histogram whose classes give the intervals $\overline{x} \pm s_x$, $\overline{x} \pm 2s_x$, and $\overline{x} \pm 3s_x$. The histogram will always consist of six intervals since it counts three standard deviations on either side of the mean. The height of each bar can be found using the cross-hair cursor and represents the number (frequency) of data values falling in each of the indicated intervals. (For some data sets, you may find it necessary to reset yMax to a larger number if you cannot see the top of the highest rectangle.)

```
PROGRAM: DATADISP
:Input "Data list?",D
:OneVar D
:x̄-3Sx→xMin
:x̄+3Sx→xMax
:Sx→xScl
: -2→yMin
:.7n→yMax
:1→yScl
:Hist D
```

Program 3.2.

Execute program DATADISP for the SBTPTS list to obtain the histogram in Figure 3.2. Notice that there are no data values outside the interval $(\overline{x} - 2s_x, \overline{x} + 2s_x)$. To see the cross-hair coordinates more easily when you are finding the frequencies, press $\boxed{\text{CLEAR}}$ to hide the menu at the bottom of the graphics screen. Verify that the frequencies of the classes are

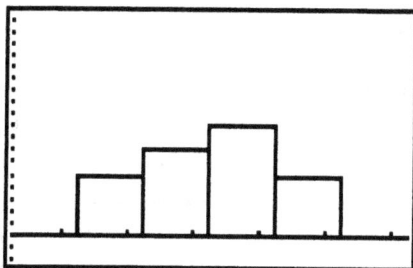

Figure 3.2. Dispersion of SBTPTS.

$0, 5, 7, 9, 5$ and 0. Thus, there are $7 + 9 = 16$ data values in the interval $\bar{x} \pm s_x$, $5 + 7 + 9 + 5 = 26$ values in the interval $\bar{x} \pm 2s_x$ and $0 + 5 + 7 + 9 + 5 + 0 = 26$ values in the interval $\bar{x} \pm 3s_x$.

The *Empirical Rule* tells you that for mound (bell)-shaped data, approximately 68% of the measurements fall within one standard deviation of the mean, approximately 95% of the data fall within two standard deviations of the mean, and approximately 99.7% of the data fall within three standard deviations of the mean. The percentages of data in these respective intervals for SBTPTS are 61.5%, 100% and 100%. We therefore see, as indicated in the histogram, that this data is not mound-shaped but skewed. Are either of the data sets SBWIN or SBLOSE approximately mound-shaped?

Scatter Plots

One of the best graphical displays of two-variable data is obtained with a *scatter plot* or *scatter diagram* showing the data points (x, y). Information about the data set is obtained by looking at the scatter plot for a pattern and obvious deviations from that pattern. Measurements of a variable taken at regular intervals of time are *time series*. While the study of time series is beyond the scope of this chapter, let's look at a simple example that plots the Super Bowl points as a function of the number of the Super Bowl in which they were scored.

The easiest way to create the list of Super Bowl numbers is to use the sequence function found in the LIST OPS menu. From the home screen, key in [LIST] [OPS] [MORE] [seq] [ALPHA] A [,] [ALPHA] A [,] 1 [,] 26 [,] 1 [ )] [STO>] T [ENTER] to generate the list {1 2 3 ... 26} and store it as T. Let's construct a scatter plot of the points scored by the winner of the Super Bowl as a function of the number of the Super Bowl. To draw a good scatter plot, you must first set the viewing window so that you can clearly see *all* the data points. Avoid having points on the "edge" of the viewing window. Now, the horizontal axis variable, time, is between 1 and 26. Press $\boxed{\text{GRAPH}}$ $\boxed{\text{RANGE}}$ and set xMin to 0 and xMax to 28. Since the SBWIN data ranges from 14 through 55, set y-range values such as yMin = 5 and yMax = 60. Also set xScl and yScl each to 0 so that the tic marks do not interfere with your view of the data points. To draw the scatter plot, make sure all equations are turned off with $\boxed{\text{FnOff}}$ and enter $\boxed{\text{Scatter}}$ $\boxed{\text{(}}$ $\boxed{\text{ALPHA}}$ T $\boxed{\text{,}}$ $\boxed{\text{SBWIN}}$.

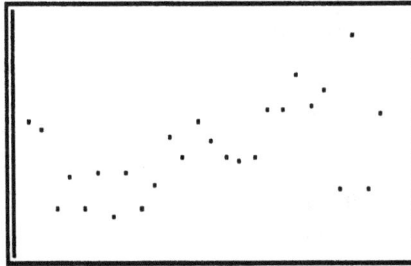

Figure 3.3. Scatter plot of year vs SBWIN.

There seem to be three clusters of points that follow an overall trend of increase for Super Bowls III through XXII. Super Bowls XXIII and XXV are definitely out of line with this trend, however. What do you feel could account for the significantly fewer points scored by the winners in these two games?

You can obtain another graph of this two-variable data, one that is a more popular form of time series representation, with the xyline instruction. This graph connects two-variable data points, in the order in which they

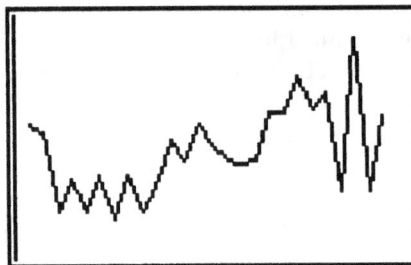

Figure 3.4. Time series for year vs SBWIN.

are entered, with line segments. Use $\boxed{\text{CATALOG}}$ $\boxed{\text{xyline}}$ to copy the instruction to the command line. To tell the TI-85 the data you are using, enter $\boxed{\text{(}}$ $\boxed{\text{ALPHA}}$ T $\boxed{\text{,}}$ $\boxed{\text{SBWIN}}$ following the instruction xyline, and press $\boxed{\text{ENTER}}$.

Let's now compare the points scored by the winners and losers of the Super Bowls. Enter $\boxed{\texttt{min}}$ $\boxed{\texttt{SBLOSE}}$ (or use the temporary sortA instruction) to see that the minimum points scored by the loser of the Super Bowls is 3. Since the maximum number of points scored by the winner is 55, set y-range parameters such as yMin = 0 and yMax = 60. Graph the time series for SBLOSE with $\boxed{\texttt{xyline}}$ $\boxed{\texttt{(}}$ $\boxed{\texttt{ALPHA}}$ T $\boxed{\texttt{,}}$ $\boxed{\texttt{SBLOSE}}$ $\boxed{\texttt{ENTER}}$. Return to the home screen, remember that $\boxed{\texttt{ENTRY}}$ returns the last-entered instruction to the screen for editing, and overlay the graph for the points scored by the winning team with the instruction xyline (T, SBWIN). Use the cross-hair cursor to determine the "closest" Super Bowl in terms of points scored. Refer to Table I to check your guess. Which seems to vary more — points scored by the winner or points scored by the loser?

Figure 3.5. Time series for SBWIN and SBLOSE.

Another way to compare the points scored by the winners and losers of the Super Bowls is to consider the differences in the points scored by the winning and losing teams. Compute the difference data and store it for use with the keystrokes $\boxed{\texttt{SBWIN}}$ $-$ $\boxed{\texttt{SBLOSE}}$ $\boxed{\texttt{STO>}}$ SBDIFF. Clear the graphics screen with $\boxed{\texttt{ClDrw}}$ $\boxed{\texttt{ENTER}}$, and construct a scatter plot for the differences with the instruction Scatter (T, SBDIFF) or an xyline plot with the instruction xyline (T, SBDIFF). How many of the 26 Super Bowls resulted in "close" games?

Certain calculations can help you examine the time plot more closely. If the distribution of difference points is not extremely skewed, there should be approximately 95% of the data within the interval $\bar{x} \pm 2s_x$. Use program DATADISP with the list SBDIFF and find that the percentages of data values falling within one, two and three standard deviations of the mean are respectively 65.4%, 96.2% and 100%. Since these percentages are close to the Empirical Rule values 68%, 95% and 99.7%, you can place the *control limits* $\bar{x} \pm 2s_x$ on the scatter diagram and note that no more than about 5% of the data should fall outside the control limits if the distribution remains the same over time. A variable that continues to have the same distribution when observed over time is said to be *in control*.

Are the point differences in the Super Bowl in control? Let's construct a control chart for the SBDIFF data. Calculate the one-variable statistics for the data of interest with $\boxed{\text{OneVar}}$ $\boxed{\text{SBDIFF}}$. Press $\boxed{\text{GRAPH}}$ $\boxed{\text{y(x)=}}$, and use the STAT VARS menu to enter $\bar{x} - 2s_x$ in $y1$, $\bar{x} + 2s_x$ in $y2$ and $\bar{x}$ in $y3$. Since time (number of the game) is being plotted on the horizontal axis, set xMin = 0 and xMax = 28. Use y-range values so that you can see all the data points and the three horizontal lines. One possibility is yMin = -10 and yMax = 50. Graph the control chart with the instruction Scatter (T, SBDIFF).

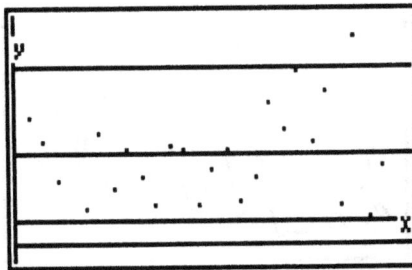

Figure 3.6. Control chart for SBDIFF.

Notice that the point difference seems to be in control since only one point, Super Bowl XXIV, is outside the control limits. Data values called *outliers* may represent faulty measurements in recording observations or may be valid measurements that, for one reason or another, differ markedly from the others in the set. Super Bowl XXIV represents an outlier for this data in the sense that it was a one-sided game when compared to the rest of the Super Bowl games.

The control chart instructions given above use the CATALOG instruction OneVar. When entered from the home screen, OneVar stores the values of the results variables as x values. That is why you use the control limits $\bar{x} \pm 2s_x$ even though SBDIFF is being plotted on the y-axis. You should *not* use the instruction 1-Var in the STAT DRAW menu because one-variable statistics are calculated and stored with x being the independent variable in the STAT menu. A control chart has x equal to time with the control limits calculated for the variable being plotted on the vertical axis. Also, when two-variable data is entered in the STAT menu, the instruction 1-Var does not give valid results. If you want to construct the control chart from the STAT DRAW menu, press $\boxed{\text{STAT}}$ $\boxed{\text{CALC}}$, enter **T** as the xlist Name, SBDIFF as the ylist Name and press $\boxed{\text{LinR}}$. This action causes the one-variable statistics to be calculated for *both* the xlist (T) and ylist (SBDIFF). Press $\boxed{\text{GRAPH}}$ $\boxed{\text{y(x)=}}$, and use the STAT VARS menu to enter $\bar{y} - 2s_y$ in $y1$, $\bar{y} + 2s_y$ in $y2$ and $\bar{y}$ in $y3$. Graph the control chart with $\boxed{\text{STAT}}$ $\boxed{\text{DRAW}}$ $\boxed{\text{SCAT}}$.

Realize that this type of control chart is based on a variable that is assumed to have a (roughly) mound-shaped distribution. This is not true for all data sets. Scatter plots and the instruction LinR will be discussed again in Section 3.4.

Explorations with Descriptive Statistics

1. As listed in *The 1992 Information Please Almanac*, the lengths of term (in years) for the 14 previous Chief Justices of the Supreme Court of the United States are 5, 4, 34, 28, 8, 14, 21, 10, 8, 11, 4, 7, 15, 17.

   a) Construct a histogram of 5 classes of equal integer width for this data.

   b) Determine the mean and median and identify the skewness for this data.

   c) Find the exact percentage of data falling within one, two and three standard deviations of the mean for the lengths of term of the Chief Justices.

   d) The five-number summary for the spread of a data set is the minimum value, the 25th percentile, the median, the 75th percentile, and the maximum value. Compute the five-number summary for the Chief Justice data.

2. Choose your preferred method of entering statistical data and input the 19 guide prices of near-mint condition Mickey Mantle baseball cards as listed in the March, 1992 issue of the *Beckett Baseball Card Monthly*. (See Table II.)

   a) After you finish searching your parents' attic for Mickey Mantle baseball cards, store this list to the name MANTLE. What is the mean guide price for a Mickey Mantle baseball card? What do you feel accounts for the large standard deviation? Find the median for this data set. Find and interpret the 25th percentile.

   b) Which card is an outlier for this set of data? Edit the data by eliminating this value and recompute the mean, median, standard deviation and 25th percentile. Which of these numerical descriptive measures are most affected by the presence of the outlier?

Table II. Prices of Mickey Mantle baseball cards.

Year	Card Series	Price
1952	Topps	$25000
1953	Bowman Color	1750
1953	Topps	2500
1954	Bowman	850
1955	Bowman	450
1956	Topps	900
1957	Topps	850
1958	Topps	525
1959	Topps	410
1960	Topps	340
1961	Topps	350
1962	Topps	425
1963	Topps	360
1964	Topps	195
1965	Topps	410
1966	Topps	185
1967	Topps	210
1968	Topps	185
1969	Topps	200

3. According to the Bureau of Labor Statistics [6], the unemployment rate has varied over time as

Year	1929	1932	1945	1950	1970	1975	1980	1985	1989	1990	1991
Rate	3.2	23.6	1.9	5.3	4.9	8.5	7.1	7.2	5.3	5.5	6.6

a) Construct a control chart for this data. Do you see any patterns in the scatter plot? Does the unemployment rate seem to be in or out of control?

b) What assumption have you made in using control limits that are two standard deviations on either side of the mean of the data? Use program DATADISP to (roughly) check if this is a valid assumption.

4. Table III gives the earnings distribution of year-round, full-time workers by sex for persons 15 years old and over as of March 1990. [6]

a) Use the LIST OPS instruction seq to construct the list of group numbers and store it as GRP. Store the earnings list for women in FEM and the earnings list for men in MALE.

b) Set range parameters that allow the points for *both* sets of data to be clearly seen on one graph.

c) Use the xyline instruction to construct a plot of GRP (x) versus FEM (y).

d) With the data for the women still on the screen, use the xyline instruction to overlay a plot of GRP (x) versus MALE (y).

e) Discuss what information is revealed by these graphs with a class-mate of the opposite sex.

Table III. Earnings distribution of year-round, full-time workers by sex.

GroupNumber	EarningsGroup	Percent Distribution Women	Men
1	$2499 or less	1.4	1.1
2	$2500 – $499	1.4	0.7
3	$5000 – $7499	4.1	1.8
4	$7500 – $9999	6.3	2.9
5	$10,000 – $14,999	20.1	10.2
6	$15,000 – $19,999	20.9	12.9
7	$20,000 – $24,999	16.5	13.4
8	$25,000 – $49,999	26.3	42.1
9	$50,000 and over	3.0	14.9

5. The number of occurrences of each letter of the English alphabet was counted on one randomly chosen page of a statistics text. [3]

a) What is the most-used letter? the least-used letter?

b) What percentage of the letters are vowels?

c) Why do you think the letters T, R, S, L, N and E are the ones most often chosen on the *Wheel of Fortune* game show?

d) Find and interpret the five number summary for this data.

e) Are most of the letters used rarely or frequently?

f) Identify any outliers.

Table IV. Percentage occurrence of letters.

A	7.9	H	5.6	O	6.5	V	0.8
B	1.8	I	6.6	P	2.1	W	1.6
C	2.6	J	0.1	Q	0.09	X	0.15
D	3.3	K	0.4	R	6.5	Y	2.4
E	13.7	L	3.7	S	7.3	Z	0.06
F	2.7	M	2.6	T	10.0		
G	2.0	N	8.0	U	1.5		

3.2. Probability

Probability may be regarded as a numerical measure of the chance that a certain outcome (event) of an experiment will occur. The probability of an event E is denoted by $P(E)$, a real number between 0 and 1, and indicates the likelihood that E will happen. The closer the probability is to 1, the more likely the event is to happen, and the closer the probability is to 0, the less likely the event is to occur.

Simulation Techniques

The *frequency* of an outcome is the number of times it occurs in repetitions of an experiment. When the frequency is divided by the number of repetitions, a fraction, called the *relative frequency* of the outcome, results. Probability gives the *relative frequency* with which an event is *expected* to occur. The *Law of Large Numbers* states that if an experiment is repeated again and again under identical conditions, the relative frequency of an event will approach the theoretical probability of that event. However, many replications of the experiment under identical conditions may be difficult or impossible to perform.

Simulation is the process of representing an experiment with a model. The simulation technique has the advantage over the actual experiment in that many identical repetitions can be performed quite efficiently with the aid of a computer or in this case, your calculator. Once simulations are performed, you can compare the outcomes of a large number of trials to the "theoretical" results. Simulation techniques usually involve *random numbers*; that is, numbers that are chosen in such a way that each one is equally likely to be the one selected. Most statistics texts include a table of random numbers. The TI-85 has its own built-in program for generating pseudo-random numbers on the interval $[0, 1)$. A "true" random number generator on the interval $[0, 1)$ would select each real number in that interval with equal probability. Since it is impossible to simulate with the precision of real numbers, random number generators in calculators and computers generate pseudo-random numbers; that is, random numbers that have a fixed number of decimal places. These pseudo-random outcomes look and behave for the most part like theoretical random numbers.

Twelve-digit pseudo-random numbers on the interval $[0, 1)$ may be obtained with the TI-85's random number generating function rand. Access this function in the CATALOG, your CUSTOM menu, or with the keystrokes MATH PROB rand . The letters rand should appear on the display screen. Press ENTER several times and you will see some calculator-generated random numbers between 0.000000000000 and 0.999999999999. The first random number that is generated by the TI-85 is dependent on the value stored in the calculator memory. You may "seed" the random number generator (simulate randomly choosing a position in a table of random numbers) by storing any whole number value to rand.

To simulate the experiment of tossing one fair coin, let the outcome "tails" be represented by 0 and the outcome "heads" be denoted by the number 1. Since rand yields a random number x such that $0 \leq x < 1$, $0 \leq 2x < 2$. Also notice that if $0 \leq x < 0.5$, $0 \leq 2x < 1$, and when $0.5 \leq x < 1$, it is true that $1 \leq 2x < 2$. Two commands that will be useful in this section are found in the MATH NUM menu. These are iPart, which returns the integer portion of a number, and int, which returns the greatest integer contained in a number. These two commands may be used interchangeably for the discussions in this section since you are dealing only with nonnegative whole numbers in probability simulations.

Key in either $\boxed{\text{iPart}}$ 2 $\boxed{\text{rand}}$ or $\boxed{\text{int}}$ 2 $\boxed{\text{rand}}$ and press $\boxed{\text{ENTER}}$ several times. Notice that the only values displayed are 0 and 1. Press $\boxed{\text{ENTER}}$ twenty more times, recording the number of heads (1's) that appear. Divide by 20 to obtain the relative frequency of heads. Press $\boxed{\text{ENTER}}$ thirty more times, again recording the number of heads that appear. Divide the total number of heads in the fifty trials by 50 to obtain the relative frequency of heads. What value do you find the relative frequency of heads approaching?

Let's now look at a program that will allow the calculator to toss the coin for you. There are several programs in this section that all have the same first steps. The following program will be used as a subroutine of those programs to avoid having to enter the same information at the beginning of each one. To avoid clutter in the menu, programs that are used only as subroutines will start with Z. This will cause them to appear at the end of the alphabetically listed program menu.

Enter program ZFSTEP, a subroutine that clears the graphics screen, turns off entered functions, and prompts for N, the number of times an experiment is repeated.

```
PROGRAM: ZFSTEP
:FnOff
:ClDrw
:Input  "N=",N
:Return
```

Program 3.3.

Program COINTOSS simulates the experiment of tossing a fair coin N times and draws a histogram of the results as the coin is being tossed.

```
PROGRAM: COINTOSS
:ZFSTEP
:.8N→yMax
:int (yMax/5)→yScl
:For(K,1,N,1)
:K→dimL COIN
:iPart 2rand→COIN(K)
:Hist COIN
:End
```

Program 3.4.

For the coin toss simulation, the outcome 0 represents tossing a tail, and 1 represents tossing a head. If you wish to see the actual outcomes (results of the N coin tosses) that are generated by this program, they are stored in list COIN. Before use of program COINTOSS, set the range parameters xMin = 0, xMax = 2, xScl = 1 and yMin = −1. Execute this program for $N = 10$, 30 and 50, and observe the histograms. The height of the vertical bar on the left represents the total number of tails that are obtained, and the height of the vertical bar on the right represents the total number of heads that are obtained in the N tosses. Since you will get either a tail or a head on each toss, these two heights must sum to N. Each time you run the program, use either the cross-hair cursor, the y-axis tic marks, or the instruction $\boxed{\text{sum}}$ $\boxed{\text{COIN}}$ to determine the height of the bar representing the number of 1's. Divide the height by N to find the relative frequencies of heads for the 10, 30, and 50 tosses of the "calculator coin." You will find it helpful to remember that once you initially load the program and there are no intervening command line keystrokes, pressing $\boxed{\text{ENTER}}$ will cause the program to execute again.

Compare your results with those you obtained earlier using the random number generator. Would you expect them to be exactly the same? What is your prediction for the probability of obtaining a head when the calculator coin is tossed? Another activity possible with program COINTOSS is to execute the program for several different values of N, each time recording whether you get more tails (0's) or heads (1's). Each time tails "wins," score 1 for T. Each time heads "wins," score 1 for H. If it is a tie, each receives 1/2 of a point. What would you expect as the final result if you play this game many times? Do you think your conjecture depends on N, the number of times the coin is tossed?

The number of times you can toss the coin is dependent upon the available memory in your calculator. If you receive the message ERROR 15 MEMORY, you have run out of available memory, and you should delete the list COIN. Stored pictures occupy a good bit of memory. Press $\boxed{\text{MEM}}$ $\boxed{\text{DELET}}$ $\boxed{\text{MORE}}$ $\boxed{\text{MORE}}$ $\boxed{\text{PIC}}$, and delete any stored graphs you no longer want to keep.

We have seen that iPart 2rand produces the values 0 and 1. Enter on the home screen the simulation step $\boxed{\text{iPart}}$ $\boxed{(}$ 6 $\boxed{\text{rand}}$ $\boxed{+}$ 1 $\boxed{)}$. Press $\boxed{\text{ENTER}}$ several times and note that this simulation produces the values 1, 2, 3, 4, 5, or 6. What are the possible outcomes when a die (singular of dice) is rolled one time and you look at the number of dots on the upturned face of the die? If the die is *fair* (i.e., each face has an equal chance of being the upturned face), you should have an equal chance of getting one, two, three, four, five or six dots. The random number generator iPart (6rand + 1) simulates a single toss of a fair die.

How would you simulate the toss of a pair of dice so that you could find the sum of the number of dots on the upturned faces? The possible sums are 2, 3, 4, 5, 6, 7, 8, 9, 10, 11, 12. Since iPart 6rand + 1 represents the outcomes 1, 2, 3, 4, 5, 6 for the toss of one die, iPart 6rand + 1 can also be used to represent the outcomes for the toss of the other die. The outcomes for the sum of the number of dots on the upturned faces of both dice are obtained with the simulation (iPart 6rand + 1) + (iPart 6rand + 1). Enter this simulation step on the home screen and press $\boxed{\text{ENTER}}$ many times. Notice that the numbers range between 2 and 12. If the dice are fair, $P(1) = P(2) = P(3) = P(4) = P(5) = P(6) = 1/6$ for each die. However, are the sums 2, 3, 4, 5, 6, 7, 8, 9, 10, 11, 12 equally likely? Let's explore and see.

Recall that one approach to determining probabilities experimentally is to perform many repetitions of the experiment (under identical conditions) and determine the *relative frequency* of the event; that is, the proportion of times the event occurs. Program 3.5, RELFREQ, simulates N identical repetitions of an experiment, computes the relative frequency of a particular outcome, EVENT, and graphs the relative frequency of that event as a function of the number of times the experiment is performed. A running count of the relative frequency of the event under consideration is kept in list R. You should enter your guess for the theoretical probability of the outcome when prompted with $P(E)=$. Your guess is stored in $y1$ and graphs as a horizontal line. If your guess for the theoretical probability is correct, you should see the relative frequency graph approaching the theoretical probability line as the number of repetitions increases.

The program steps :Disp S(K), :0→ Z, :Lbl D, :IS>(Z, 10), and :Goto D display the outcomes for the sum of the number of dots appearing on the

```
 PROGRAM: RELFREQ
:ZFSTEP
:N→dimL S:N→dimL R
:N→dimL T
:Fill(0,R)
:seq(A,A,1,N,1)→T
:0→xMin:N→xMax
:N/5→xScl:.1→yScl
:-.05→yMin:.5→yMax
:Input "EVENT E=",E
:Input "P(E)=",P
:y1=P
:0→K:0→F
:Lbl L
:K+1→K
:iPart 6rand+1+iPart 6rand+1→S(K)
:Disp S(K)
:0→Z
:Lbl D
:IS>(Z,10)
:Goto D
:If S(K)==E
:F+1→F
:F/K→R(K)
:If K<N
:Goto L
:xyline (T,R)
:Pause
:Disp "Relative frequency="
:Disp R(N)
```

Program 3.5.

two upturned faces of the dice and slow the display of those values for easier viewing. If you wish more delay between appearance of values, change the 10 in : IS>(Z, 10) to a larger number. If you do not wish to observe the values, these steps may be omitted. The simulation results are stored by the program in list *S*.

Now, back to the question: "Are the sums 2, 3, 4, 5, 6, 7, 8, 9, 10, 11, 12 equally likely?" Execute program RELFREQ to simulate the experiment of tossing two fair dice N times and observe the number of times the event "the sum of the number of dots appearing on the two upturned faces is seven" occurs. The program will first ask for N. Roll the dice 25 times by keying in 25 ENTER. You are next prompted with EVENT E=. Since you are observing the number of times the sum is seven, enter 7. Next the program will display P(E)=, asking you for your guess of $P(7)$. If the sums are equally likely, $P(7)$ should be 1/11 since there are 11 possible sums, so enter 1/11. The program will now simulate tossing the dice and count the number of times the outcome "seven" occurs as the sum. Each time a sum of seven is obtained, the frequency counter F increases by one. Before the relative frequency graph is drawn, the outcomes of the experiment are displayed. Watch carefully and count the number of times you see a 7. After you finish viewing the graph, press ENTER to have the program display the relative frequency of the event. Repeat the experiment for fifty tosses of the dice. One possible graph with relative frequency 0.2 is given in Figure 3.7. Your graph may not look exactly like this one because you are using random numbers. Is this value of the relative frequency or the one you obtained close to 1/11? Maybe, maybe not. Remember that the relative frequency approaches the theoretical probability as the number of trials gets larger and larger. Execute program RELFREQ three more times for $N = 60$, 75, and 100. (If your available memory permits, you could use larger values than these.) Each time you execute the program, input $E = 7$ and $P(E) = 1/11$. It looks like $P(7) \neq 1/11$, doesn't it? Based on the relative frequencies you obtained in each execution of the program, what is your guess for the probability of obtaining a sum of 7 when two fair dice are rolled?

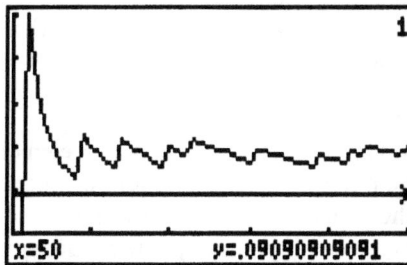

Figure 3.7. Relative frequency graph for 50 tosses of two dice and $P(7) = 1/11$.

Actually, there are 36 different equally-likely outcomes for the results of the toss of two dice. Each outcome is represented as an ordered pair (x, y) where x is the number of dots appearing on the upturned face of one of the dice and y is the number of dots appearing on the upturned face of the other die. To obtain a sum of 7, you would consider the outcomes $(1, 6)$, $(6, 1)$, $(2, 5)$, $(5, 2)$, $(3, 4)$ and $(4, 3)$. The theoretical probability is therefore $6/36 = 1/6$. Rerun program RELFREQ for $N = 50$ and $N = 75$ with an input of $1/6$ when requested for the probability of the event 7. One possible graph for $N = 75$ with relative frequency 0.16 appears in Figure 3.8. Are you convinced?

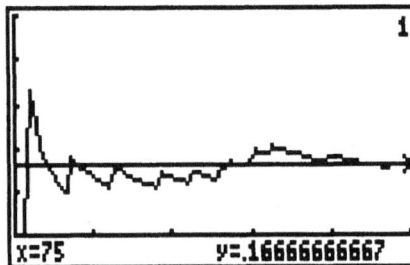

Figure 3.8. Relative frequency graph for 75 tosses of two dice and $P(7) = 1/6$.

The sums 2, 3, 4, 5, 6, 7, 8, 9, 10, 11, 12 are not equally-likely. In fact, the sum of 2 only has a 1/36 chance of happening, and 7 is the most likely sum with $P(7) = 6/36 = 1/6$.

Program RELFREQ may be modified quite easily to simulate other experiments. Appropriate changes in yMax and the simulation step involving rand, the random number generator, should be made. You will be asked to make these changes and perform three other simulations in the explorations at the end of this section.

Permutations and Combinations

The method of sampling, that is, how the elements of a sample are chosen, influences the number of outcomes for an experiment. If you sample with replacement, the element chosen is returned to the population before the next element is selected. If you sample *without replacement*, the chosen element is not returned to the population before the next one is chosen. When the *order* in which the elements are chosen is important, the order in which the elements are selected must be considered. For instance, when tossing a coin two times, HT is a different basic outcome than TH since a different face appears as a result of each toss. If the order in which the elements are listed is not important, the order in which they are selected should not be considered. For example, if three people are to be chosen from a group of 8 people to form a committee, the committee is the same regardless of how the people chosen are arranged.

A *permutation* is an *ordered* arrangement *without repetition of elements* of a set of distinct objects. The formula for the number of permutations of n different objects chosen r at a time is $nPr = n!/(n-r)!$ For the set $\{a, b, c\}$, the permutations are ab, ba, ac, ca, bc, and cb with $3P2 = 6$. The permutation formula is in your TI-85 in the $\boxed{\text{MATH}}$ $\boxed{\text{PROB}}$ menu. To use your calculator to find the number of permutations of n objects chosen r at a time, enter n followed by $\boxed{\text{nPr}}$ and r, and press $\boxed{\text{ENTER}}$. Verify that you obtain $3P2 = 6$.

A *combination* is a selection of the distinct objects of a set without regard to order. As with permutations, repetitions of elements are not allowed. The essential difference between permutations and combinations is that combinations ignore the order in which the objects are chosen. The formula for the number of combinations of n different objects chosen r at a time is $nCr = \frac{n!}{r!(n-r)!}$. For the set $\{a, b, c\}$, the combinations are ab, ac, and bc with $3C2 = 3$. The combination formula is in the $\boxed{\text{MATH}}$ $\boxed{\text{PROB}}$ menu. To use your calculator to find the number of combinations of n objects chosen r at a time, enter n followed by $\boxed{\text{nCr}}$ and r, and press $\boxed{\text{ENTER}}$. Verify that you obtain $3C2 = 3$. In your TI-85, the functions nPr and nCr are defined only for whole number values of n and r where $r \leq n$.

Binomial Probabilities

Many problems in probability involve situations in which an experiment with two outcomes is repeated many times. The two outcomes are usually called *success* and *failure*, and each repetition of the experiment is called a *trial*. In such an experiment, you are usually interested in the probability that an event will occur a certain number of times in a specified number of trials.

A *binomial experiment* is one that is characterized by the following:

a) the experiment consists of a fixed number N of identical trials,

b) there are only two possible outcomes of each trial, success or failure,

c) the probability of success on each trial, P, remains constant from trial to trial,

d) the trials are independent. That is, the outcome of any one trial does not influence or affect the outcome of any other trial.

One example of a binomial experiment is tossing a coin a fixed number of times and counting the total number of heads obtained. Success is obtaining a head, failure a tail, and the probability of a head remains the same from toss to toss. The coin can't remember the results of previous flips, so the trials (tosses) are independent. Let's use the calculator to simulate the outcomes of the experiment of tossing a fair coin 4 times and counting the number of heads observed.

Recall that the simulation step iPart 2rand produces the values 0 (tails) and 1 (heads) for the experiment of tossing one coin. The instruction iPart 2rand + iPart 2rand + iPart 2rand + iPart 2rand produces the values 0, 1, 2, 3, and 4, simulating the number of heads obtained in the toss of a fair coin four times. Press PRGM EDIT RELFR ENTER, and use ▼ to go the the eighth line of the program. Edit the line so that 1 is stored in yMax. Next, go to line 15, and replace the die toss simulation step with :iPart 2rand + iPart 2rand + iPart 2rand + iPart 2rand→ S(K). Execute program RELFREQ with input of $N = 50$ to estimate the probability of obtaining two heads when four fair coins are tossed fifty times. The event of interest is obtaining two heads, so enter 2 at the EVENT E= prompt. What should you enter for $P(E)$? One way of obtaining two heads in four tosses is the outcome HHTT. Another is HTTH. How many ways can the two heads be placed in the four positions? The answer is 4C2 = 6. Since the tosses are independent for a fair coin, P(HHTT) = P(HTHT) = $\cdots$ = P(TTHH) = $(.5)^4$. Thus, the probability of two heads should be $6(.5)^4$ = .375. Enter .375 for $P(E)$ and observe the results. Repeat these steps for $N = 75$ and $N = 1000$. Are your relative frequencies approaching the theoretical probability? They should, but remember that randomness is not predictable and 50, 75, and 100 tosses are still a long way from repeating the experiment infinitely many times!

Simulation steps involving iPart 2rand will simulate binomial probabilities when the probability of success on each trial is 1/2. Program REL-FREQ could be modified to simulate probabilities for experiments where the probability of success on each trial is other than 1/2.

```
PROGRAM: ZBINOFML
:(N nCr x)(P^x)(1-P)^(N-x)→B
:Return
```

Program 3.6.

Now that you have a good understanding of the binomial experiment, let's use a formula to calculate the (theoretical) binomial probability of x successes in N trials where P is the probability of success on any one trial.

This formula will be used as a subroutine in Program 3.7, but you can easily modify it to output specific binomial probabilities.

A *random variable* is a numerical outcome whose value depends on chance. The number of successes obtained in N trials of a binomial experiment is a random variable whose possible values are $0, 1, 2, \ldots, N$. The *probability distribution* of a random variable is a table, formula or graph that shows the population of values of the random variable and the associated probabilities of those values. Program 3.7, BINOGRAF, computes

---

```
 PROGRAM: BINOGRAF
:FnOff
:ClDrw
:Prompt N,P
:N+1→dimL BINP
:0→xMin:N+1→xMax
:N/5+.5→xScl:0→yMin
:int (N*P)→x
:If N*P>.5N
:round(N*P,0)→x
:ZBINOFML
:B+.05→yMax
:0→x
:Lbl L
:ZBINOFML
:B→BINP(x+1)
:Line(x+.5,0,x+.5,BINP(x+1))
:x+1→x
:If x≤N
:Goto L
:Pause
:N*P→μ
:√(μ(1-P))→σ
:Disp "μ=",μ
:Disp "σ=",σ
:round(BINP,4)
:Ans→BINP
:Disp "BINP=",BINP
```

Program 3.7.

probabilities for the number of successes in a binomial experiment in addition to drawing a *line graph* of the probability distribution. The program also outputs the mean or *expected number of successes*, $E(x)$, and the variance of the random variable. Probability distributions give the probabilities associated with all possible values of the random variable. For that reason, the symbol μ is also used to represent the expected number of successes for the experiment, and the symbol σ^2 is used to represent the variance. For a binomial experiment consisting of N trials with probability P of success on any one trial, $\mu = E(x) = N*P$, and the variance is $\sigma^2 = N*P*(1-P)$. The expected number of successes is also called the expected value of the variable.

Line 16 of program BINOGRAF draws a vertical line between the horizontal axis and each value of y, the probability for that value of x. Unfortunately, the line for $P(x = 0)$ is hidden by the x-axis if the values $x = 0, 1, \ldots, N$ are used in drawing the graph. Therefore, the entire graph is shifted to the right 0.5 units by the addition of .5 in line 16. Realize that the cross-hair cursor will give incorrect x-values if you activate it with line graphs drawn by this program.

Suppose a baseball player has a batting average of .300. What is the probability distribution for x, the number of hits his next 5 times at bat? Check the conditions for a binomial experiment: the experiment consists of 5 trials, either the player gets a hit or he doesn't, and .300 is given as the constant probability of a hit. You will have to make the assumptions that the five times at bat are under nearly the same conditions and the times at bat are independent. Otherwise, the results of the binomial formula may not be valid. Execute program BINOGRAF with $N = 5$ and $P = .3$ to obtain

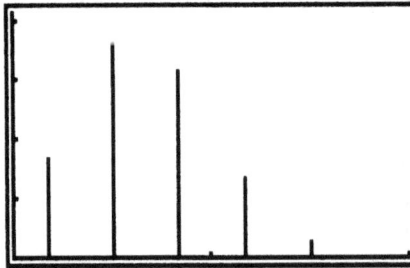

Figure 3.9. Line graph of binomial distribution
with $N = 5$ and $P = 0.3$.

What is the probability the player gets exactly 3 hits his next five times at bat? After you finish viewing the graph of the probability distribution, press ⟮ENTER⟯ and the list of probabilities is displayed. Since there is no 0th

position in a list, $P(x = 0)$ is in the first position of the list, $P(x = 1)$ is in the second position of the list, etc. The value in the fourth position of the list of probabilities BINP is $P(x = 3) = .1323$. You can access this value by entering BINP from the LIST NAMES menu and scrolling with ▶ to the fourth element or with the keystrokes BINP (4). Still another way to obtain $P(x = 3)$ is to enter 3 STO> x, and execute program BINOFML. After the Done message is displayed to indicate the probability has been calculated by the binomial formula and stored in variable B, enter ALPHA B to see .1323. (Note that you do not have to store 5 in N and 0.3 in P since these values have not been changed since the execution of BINOGRAF.)

It is cumbersome to try to remember that the probability that $x = a$ is stored in position $a + 1$ of the probability list BINP. Program 3.8, CUMPROB, gives an individual probability or will accumulate probabilities between a first value (F) and a last value (L) according to the value of x, not the position of $P(x)$ in the list BINP.

```
PROGRAM: CUMPROB
:Input "list of probabilities",P
:Prompt F,L
:0→T
:For(K,F+1,L+1,1)
:P(K)+T→T
:End
:Disp T
```

Program 3.8.

Execute program CUMPROB and input BINP as the list of probabilities. To find the probability of 3 hits, input $F = 3$ and $L = 3$ to obtain $P(x = 3) = .1323$. Is the value .1323 reasonable? How many hits would you expect the player to get if .300 is his batting average? For the player's next five times at bat, you would expect him to get a hit $\mu = 1.5$ times if the experiment were repeated many times under identical conditions. Since you would expect 1.5 hits, the probability of 3 hits should not be very large. Of course, 1.5 times is not a possible number of hits. Recall that μ is interpreted as a long term average or mean number of successes for the population of possible x-values.

Let's consider another example. Suppose you are taking a true-false test consisting of 20 questions. What is the probability distribution for the number of questions you answer correctly if you have not studied and only guess the answer to each question? Execute program BINOGRAF for $N = 20$ and $P = .5$.

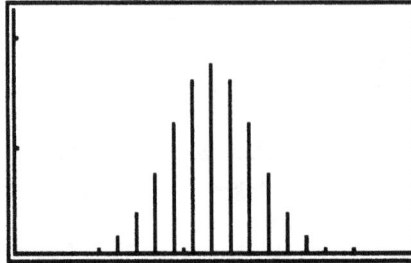

Figure 3.10. Line graph of binomial distribution
with $N = 20$ and $P = 0.5$.

Notice that the mean number of correct answers is $\mu = 10$. (That's a score of 50% on the test.) Also note that the mean occurs at the x value corresponding to the "tallest line," and this distribution is symmetric about that value. (Press $\boxed{\text{GRAPH}}$ if you wish to recall the line graph to the screen.)

What is the probability you score 80% on the quiz? Enter $F = 16$ and $L = 16$ in program CUMPROB and obtain 0.0046. What is the probability that you make 80% *or better* on the quiz? Execute program CUMPROB with $F = 16$ and $L = 20$ to obtain a probability of 0.0059. Your chances aren't too good, are they?

Now suppose that you have studied for the test and have a probability of 0.8 of obtaining the correct answer to any question. How does studying affect the distribution of possible scores on the test? Execute program BINOGRAF for $N = 20$ and $P = 0.8$.

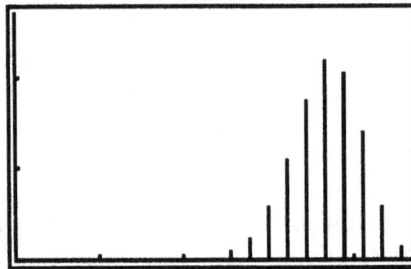

Figure 3.11. Line graph of binomial distribution
with $N = 20$ and $P = 0.8$.

The graph has moved to the right, indicating that larger values of x are now more probable. The mean is $\mu = 16$. What is the probability that you obtain 16 correct answers? Run program CUMPROB with $F = 16$ and $L = 16$ to obtain $P(x = 16) = .2182$. You studied and have only about a 22% chance of making an 80% on the test? This doesn't seem right! Think for a minute and realize that this probability represents the chance of obtaining *exactly* 20 correct answers. You probably are interested in the probability of making at *least* 80% on the test; that is, obtaining 16 *or more* correct answers. Press [ENTER] to execute program CUMPROB, and input $F = 16$ and $L = 20$ to obtain a probability of .6296. That's a better than even chance!

Poisson Probabilities

A random variable that counts the number of successes that occur in a specific period of time, area, or volume is the Poisson random variable. Other characteristics of a Poisson experiment are:

a) the probability that an event occurs in any specified unit of time, area, or volume is the same for all units,

b) the occurrences of the events are independent across the units,

c) the mean number of occurrences per unit is given by λ.

The probability distribution for a Poisson random variable is given by the formula

$$p(x) = \frac{\lambda^x e^{-\lambda}}{x!} \quad \text{for} \quad x = 0, 1, 2, 3, \ldots .$$

Program 3.9, POISGRAF, will construct a line graph of the Poisson probability distribution for $0 \leq x \leq 5\lambda$, output the mean and variance of the Poisson random variable, and store the Poisson probabilities in the list POSP. (For the same reason given in the discussion following program BINOGRAF, the line graph drawn by this program is shifted to the right 0.5 units.) For easier entry of program POISGRAF, refer to programming note 9 in Section 3.5 for instructions on editing program BINOGRAF.

Suppose the number of absences on Friday in your statistics class has a Poisson distribution with a mean of 8. Use program POISGRAF to construct a graph of the probability distribution of the number of Friday absences, and find the probability that no more than 6 students are absent next Friday. Enter 8 when prompted for λ, and observe the line graph given in Figure 3.12. Press [ENTER] and observe μ, σ, and POSP, the list of Poisson probabilities. To find the probability that no more than six students are absent next Friday, use program CUMPROB with $F = 0$ and $L = 6$ to obtain 0.3133. What is the probability that more than 8 students are absent a week from Friday?

```
 PROGRAM: POISGRAF
:FnOff
:ClDrw
:Input  "λ=",L
:0→xMin:5L+1→xMax
:L+.5→xScl:0→yMin
:int L→R
:(L^L)(e^-L/R!)→B
:B+.05→yMax
:0→x
:Lbl H
:int 5L→U
:U+1→dimL POSP
:(L^x)(e^-L/x!)→B
:B→POSP(x+1)
:Line(x+.5,0,x+.5,POSP(x+1))
:x+1→x
:If x≤U
:Goto H
:Pause
:L→µ
:√L→σ
:Disp  "µ=",µ
:Disp  "σ=",σ
:round(POSP,4)
:Ans→POSP
:Disp  "POSP=",POSP
```

Program 3.9.

This probability equals $P(x = 9) + P(x = 10) + P(x = 11) + \ldots$. Using the fact that the probability of an event plus the probability of its complement sum to 1,

$$P(x > 8) = 1 - [P(x = 0) + P(x = 1) + \cdots + P(x = 8)]$$
$$= 1 - .5925 = .4075\,.$$

Construct graphs for the Poisson distribution for $\lambda = .5, 1.35, 3.5, 10,$ and 25. What is the change in the skewness of the graph as λ increases?

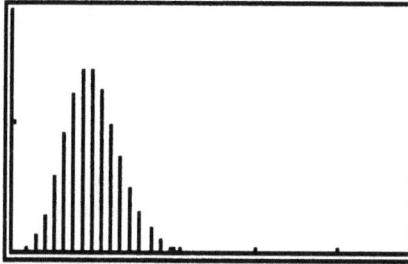

Figure 3.12. Line graph of Poisson probability distribution, $\lambda = 8$.

The Poisson distribution with $\lambda = N * P$ is often used to approximate the binomial distribution when N is large and P is small. Construct a graph of the binomial distribution for $N = 50$ and $P = 0.1$ with program BINOGRAF. (It will take a moment to determine the 51 probabilities, so be patient.) Notice that $\mu = 5$ and $\sigma = 2.1213$. Press $\boxed{\text{GRAPH}}$ $\boxed{\text{STAT}}$ $\boxed{\text{DRAW}}$ $\boxed{\text{MORE}}$ $\boxed{\text{STPIC}}$, and name the picture $P1$. Let's look at the graph of the Poisson distribution with $\lambda = 5$ on the same graphics screen to see how "close" these two distributions are. However, if you execute program POISGRAF, the range parameters will be set to different values, and you will not get a good visual comparison. Let's edit program POISGRAF so that the viewing window will be the same for both graphs. Press $\boxed{\text{RANGE}}$ and notice that in the current settings, x ranges from 0 to 51 and y from 0 to .2349246. Recall the contents of the program POISGRAF to a new program, say COMP, and edit COMP so that xMax = 51 and yMax = .2349246. Execute program COMP for $\lambda = 5$ and note that $\mu = 5$ and $\sigma = 2.2361$. Recall the Poisson graph to the screen with $\boxed{\text{GRAPH}}$, and press $\boxed{\text{GRAPH}}$ $\boxed{\text{STAT}}$ $\boxed{\text{DRAW}}$ $\boxed{\text{MORE}}$ $\boxed{\text{RCPIC}}$ $\boxed{\text{P1}}$ to overlay the binomial graph.

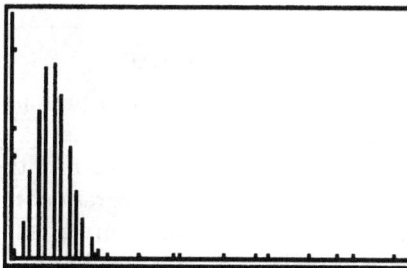

Figure 3.13. Binomial distribution, $N = 50$, $P = 0.1$ and
Poisson distribution, $\lambda = 5$.

Yes, your calculator is operating properly. You can see no visible difference in the two graphs in this viewing window! To further emphasize that this is a good approximation, use program CUMPROB to find $P(1 \leq x \leq 8)$ for the lists BINP and POSP. The Poisson approximation becomes even better as N increases.

Normal Probabilities

The binomial and Poisson random variables are *discrete* because they assume a countable number of values. A random variable whose outcomes are associated with an interval of points on the real number line is called a *continuous* random variable. One of the most important continuous distributions is the *normal* probability distribution. The defining equation of the normal probability distribution with mean μ and standard deviation σ is

$$f(x) = \frac{1}{\sigma\sqrt{2\pi}}e^{-(x-\mu)^2/2\sigma^2} \quad \text{for} \quad -\infty < x < \infty.$$

The graph of the normal distribution is symmetric about μ and has the shape of a bell. Let's explore the graph of this widely-used probability distribution with program NMGRAF.

```
PROGRAM: NMGRAF
:FnOff
:Prompt μ,σ
:μ-3σ→xMin:μ+3σ→xMax
:σ→xScl
:-.05→yMin:(σ√(2π))⁻¹+.05→yMax
:y1=(σ√(2π))⁻¹e^-.5((x-μ)/σ)²
:DispG
```

Program 3.10.

Execute program NMGRAF, and input, when prompted by the program, $\mu = 5$ and $\sigma = 1$. After viewing the graph, press RANGE and note that xMin = 2, xMax = 8, and yMax = .4489. Return to the home screen with EXIT, and press ENTER to activate the program once more. Input $\mu = 9$ and $\sigma = 2$. After viewing this normal graph, press RANGE and notice that the range parameters are xMin = 3, xMax = 15, and yMax = .2495. To compare these two graphs in a common viewing window, set xMin = 2 and yMax = .4489. Press the *menu key* $\boxed{\text{GRAPH}}$ to draw the normal graph for $\mu = 9$ and $\sigma = 2$. Use $\boxed{\text{STPIC}}$ to store the picture as P2, press $\boxed{\text{GRAPH}}$ $\boxed{\text{y(x)=}}$, and replace σ with 1 and μ with 5 in y, the equation of the normal density function. Press $\boxed{\text{GRAPH}}$ and then the menu key $\boxed{\text{GRAPH}}$. Recall P2 with $\boxed{\text{RCPIC}}$, and compare the graphs. What happens to the graph as μ changes? What is the change in the graph as σ varies? Graph the probability distribution for $\mu = 0$ and $\sigma = 1$. This normal random variable with this distribution is called the *standard normal random variable* and is commonly denoted by the letter z. Explore the normal distribution with program NMGRAF using different means and standard deviations.

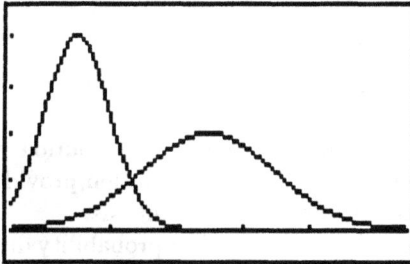

Figure 3.14. Normal distribution graphs for $\mu = 5$, $\sigma = 1$ and $\mu = 9$, $\sigma = 2$.

Another reason that the normal distribution is important is that it provides, in many cases, approximations to other distributions, including the binomial and Poisson distributions. You can overlay the graph of the normal distribution on the graphs generated by program BINOGRAF and POISGRAF with Program 3.11. This program uses the values of μ and σ that are determined by the discrete probability line graph programs and therefore should be used only after the execution of programs BINOGRAF or POISGRAF. (The normal graph is shifted 0.5 units to the right because the discrete probability graphs are shifted by that amount.)

```
PROGRAM: NMOVLY
:DrawF (σ√(2π))⁻¹e^-.5((x-μ-.5)/σ)²
:DispG
```

Program 3.11.

Execute program BINOGRAF with $N = 20$ and $P = .5$, and then execute program NMOVLY.

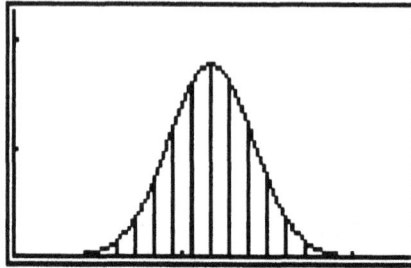

Figure 3.15. Binomial distribution, $N = 20$, $P = .5$ and normal distribution $\mu = 10$, $\sigma = \sqrt{5}$.

Repeat this experiment for the binomial distribution with $N = 30$ and $P = .25$, the binomial distribution with $N = 20$ and $P = .9$, the Poisson distribution with $\lambda = 1$ and the Poisson distribution with $\lambda = 10$. Form a conjecture as to when the normal distribution provides a useful approximation of the binomial or Poisson distributions.

Because the normal distribution is a probability distribution for a continuous random variable, the area bounded by the x-axis and the graph of the normal curve is 1. The probability that x is in the interval from a to b is defined as $\int_a^b f(x)dx$. Refer to Chapter 2 of this manual and use integration techniques to find $P(2.5 \leq x \leq 3.46)$ for the normal random variable x with mean $\mu = 0$ and standard deviation $\sigma = 1$.

Even though you can find probabilities associated with the normal distribution using the integral function in your TI-85, you will find it more efficient to use Program 3.13. Program 3.12, ZAREA, is a subroutine of Program 3.13 and uses an approximating form [1] of the integral of the normal probability function to calculate $P(x > a)$ if $a \geq \mu$. This probability is the area under the normal curve to the right of the value a and is also known as the *tail area*. By the symmetry of the normal distribution about its mean, the tail area equals $P(x < a)$ if $a < \mu$. You will obtain the probabilities given in tables of normal probabilities for inputs of $\mu = 0$ and $\sigma = 1$. With program NPRB, there is no need for the normal table in your textbook!

Execute program NPRB with $\mu = 0$, $\sigma = 1$, and $a = 1$ to find the probability that the standard normal random variable z is greater than 1. Did you obtain a tail area of .1587? What is the probability that z is between 0 and 1? Execute program NPRB with $\mu = 0$, $\sigma = 1$ and $a = 0$ to find $P(z > 0) = .5$. Wouldn't you expect this area since the mean of the standard normal random variable z is zero? Because $P(0 \leq z < \infty) = .5$ and $P(z > 1) = .1587$, $P(0 \leq z \leq 1) = .5 - .1587 = .3413$.

```
 PROGRAM: ZAREA
:abs z→x
:If x>5.5
:Goto L
:.5e^-((83x^3+351x²+562x)/(703+165x))→AREA
:Goto E
:Lbl L
:e^(-x²/2-.94/x²)/(x√(2π))→AREA
:Lbl E
:Return
```

Program 3.12.

```
 PROGRAM: NPRB
:Prompt μ,σ
:Input "a=",A
:(A-μ)/σ→z
:ZAREA
:Disp "tail area is",AREA
```

Program 3.13.

Consider another example: Find $P(32 \leq x \leq 50)$ for the normal random variable x with mean $\mu = 25$ and variance $\sigma^2 = 134$. How large should this probability be? The total area to the right of the mean is 0.5, so $P(32 \leq x \leq 50) < 0.5$. Use program NMGRAF to graph the normal distribution for $\mu = 25$, $\sigma = \sqrt{134}$, and use TRACE to approximately locate $x = 32$ and $x = 50$. Visually estimate what proportion of the total area is between these two values. To find the exact answer, execute program NPRB twice, once with $a = 32$ and again with $a = 50$. When prompted for σ, enter $\sqrt{134}$. The difference in tail areas, $.2727 - .0154$, equals $.2573 = P(32 \leq x \leq 50)$.

What percentage of the total area under the standard normal distribution is within three standard deviations of its mean? Since $\mu = 0$ and $\sigma = 1$, you are to find $P(-3 \leq z \leq 3)$. Execute program NPRB with input $a = 3$, and observe $P(z > 3) = .0013$. Since $P(z < -3) = .0013$ by the symmetry of the normal distribution,

$$P(-3 \le z \le 3) = 1 - (.0013 + .0013) = 1 - .0026 = .9974 \,.$$

Now you know how the Empirical Rule value 99.7% is obtained!

Suppose you wish to work in reverse and find the value of a when given $P(x \ge a)$ for a normally distributed random variable x. Program 3.15, NVAL, uses an approximation formula [1] to output the value of the standard normal variable z such that the area under the normal probability function to the right of z is the known tail area P. The approximating formula is in subroutine ZVAL, Program 3.14. If you input in program NVAL a probability $P > .5$, the program halts and displays the message Done reminding you that any tail area must be less than or equal to half of the total area under the normal curve.

```
PROGRAM: ZVAL
:-ln (2P)→y
:√((4y^4+100y^3+205y²)/
: (2y^3+56y^2+192y+131))→z
:Return
```

Program 3.14.

```
PROGRAM: NVAL
:Input "tail area=",P
:If P>.5
:Stop
:ZVAL
:Disp "a=",z
```

Program 3.15.

Use program NVAL to find the value, a, of the standard normal variable z such that $P(z > a) = 0.025$. Since the area to the right of a is the tail area, input .025 in NVAL and observe that $a = 1.9601$.

Consider another example: The scores on a national aptitude test are normally distributed with mean 700 and standard deviation 150. What score is the 90th percentile for the test? The 90th percentile will have 90% of the scores falling to the left of it and 10% of the scores to the right of it. Therefore, the tail area is 0.10. Enter this value in program NVAL to find $a = 1.2816$. This value, however, is a value of the standard normal variable z. The formula to convert between a normally distributed random variable with mean μ and standard deviation σ and the standard normal random

variable is the *z-score* formula $z = (x - \mu)/\sigma$. Thus, the 90th percentile is
$x = z\sigma + \mu = 1.2816(150) + 700 = 892.24$.

You, not the calculator program, must determine when the value a given by program NVAL is negative. Suppose you wish to know the value of z, say a, such that $P(z > a) = .83162$. Since $.83162 > .5$, input the tail area $1 - .83162 = .16838$ in program NVAL to obtain the output $a = .9607$. Think for a moment and realize that $P(z > .9607)$ must be less than .5 since the area to the right of the mean 0 is .5. Thus, the value of a must be negative. The answer is $a = -.9607$.

T Probabilities

You have seen that the shape and location of the normal distribution graph depends on the parameters μ and σ. The standard normal variable z has $\mu = 0$ and $\sigma = 1$, and when a variable x has a normal distribution with mean μ and standard deviation σ, the z-score, $z = (x - \mu)/s$, has the standard normal distribution. As we will see in the next section, it is often the case that a statistical problem involves making a decision about the mean of a population. When census results giving the exact value of a population mean μ are not available, the best estimate of μ is the sample mean $\bar{x}$. When a random sample of size n is chosen from a population that has a normal distribution with mean μ and unknown variance σ, the distribution of the statistic $\frac{\bar{x}-\mu}{s/\sqrt{n}}$ is the t or *Student's t distribution*. Like the standard normal distribution, the shape of the t distribution is that of a mound or bell symmetric about the mean 0. Unlike z, there is a different t distribution for each sample size. The t distribution is also more variable than the standard normal distribution since the random quantity s appears in the formula rather than the fixed parameter σ.

A particular t distribution is specified by giving its *degrees of freedom*. The number of degrees of freedom is a function of the random quantities in the statistic that has the t distribution. In particular, the number of degrees of freedom, *df*, for the statistic $\frac{\bar{x}-\mu}{s/\sqrt{n}}$ is $n-1$. For small samples, the standard deviation s can vary quite a bit. As the sample size increases, more information is gained about the population. In fact, for most applications, a sample size of $n = 30$ or more will yield values for the t statistic that are very close to the values of the standard normal variable z.

Program 3.17, TVAL, calculates values of a variable having the t distribution. For user-input values of the degrees of freedom and the tail area, the program will output the value a such that the tail area equals the probability that t is greater than or equal to a. Any tail area is assumed to be between 0 and 0.5. Program 3.16, ZTNUM, contains a calculation formula [8] that will be used as a subroutine of several other programs in this chapter.

```
PROGRAM: ZTNUM
:z(8F+3)/(8F+1)→A
:√(F(e^(A²/F)-1))→z
```

Program 3.16.

```
PROGRAM: TVAL
:Input "df=",F
:Input "tail area=",P
:If P>.5
:Stop
:ZVAL
:ZTNUM
:Disp "a=",z
```

Program 3.17.

Use program TVAL to verify that for inputs of a tail area = .05 and 10 degrees of freedom, the output is $a = 1.813$. What should happen to the value of a if the number of degrees of freedom increases to 20? Since the graph of the t distribution for 20 degrees of freedom is not as variable as the graph of the t distribution for 10 degrees of freedom, a should be smaller. Execute TVAL for 20 degrees of freedom and a tail area of .05 to find $a = 1.725$.

Execute program NVAL to find the value of a such that $P(z > a) = .05$. Now execute program TVAL for increasingly larger degrees of freedom. Do you find the values of a you obtain from program TVAL approaching the value obtained from program NVAL?

Explorations with Probability

1. The eighteenth-century French naturalist Buffon tossed a coin 4040 times and obtained 2048 heads (relative frequency .5069). Around 1900, the English statistician Pearson heroically tossed a coin 24,000 times and recorded 12,012 heads (relative frequency .5005). While imprisoned by the Germans during World War II, the English mathematician John Kerrich tossed a coin 10,000 times and obtained 5067 heads (relative frequency = .5067). [4]

a) Write a program, say RECORD, that will allow you to toss the calculator coin many times. Begin by recalling the contents of program COINTOSS to your new program.

b) Edit the program steps to the following:

: Input " N = " , N
: N → dimL COIN
: For (K, 1, N, 1)
: iPart 2rand → COIN(K)
: End
: sum COIN → H
: Disp H

c) Even though it is not difficult to further edit the program to keep track of the number of heads for many repetitions of the experiment, that task is left to you.

d) Execute program RECORD for $N = 1000$. (This will take a minute or so, but wouldn't Buffon, Pearson and Kerrich have been amazed at the speed of the TI-85?) Record the number of heads obtained in the 1000 tosses of the calculator coin. How many more times would you execute the program (with $N = 1000$) to have a total of $25,000$ tosses?

e) Set your own historical record and perform the experiment. Find the relative frequency of heads for your $25,000$ tosses.

2. In the discussion of binomial probabilities, the instruction (iPart 2rand) + (iPart 2rand) + (iPart 2rand) + (iPart 2rand) is used to simulate the toss of a fair coin four times. This instruction produces the values 0, 1, 2, 3, and 4, representing the possible number of heads obtained in four tosses of the coin. Why not use the simulation step iPart 5rand that also produces the values 0, 1, 2, 3, and 4? To assist you in answering this question,

a) Edit program COINTOSS so that the simulation step : iPart 2rand + iPart 2rand + iPart 2rand + iPart 2rand → COIN(K) is the sixth line of the program. Reset the range so that xMin = 0, xMax = 5, xScl = 1 and yMin = 0. (The value of yMax is set by the program.) Execute the program for $N = 25, 50$ and 75 to simulate N repetitions of the toss of a fair coin 4 times. Carefully observe the general shape of the resulting histograms. What is your estimate of the probability of obtaining two heads?

b) Enter a new program, UNIFORM, by recalling the contents of program COINTOSS to your new program. Edit the program so that the simulation step in the sixth line is : iPart 5rand. Execute

the program for $N = 25$, 50 and 75 to simulate N repetitions of
the toss of a fair coin four times. (Use the same range parameters
as those specified in the previous part of this problem.) Carefully
observe the general shape of the resulting histograms. What is
your estimate of the probability of obtaining two heads?

c) Execute program BINOGRAF for $N = 4$ and $P = .5$. Use program CUMPROB to find the theoretical probability of obtaining
two heads in the four tosses. Is this value closer to your estimate
from part a) or part b) of this problem?

d) How do the shapes of the histograms in parts a) and b) and the
answer to c) help you answer the original question posed in this
problem? A very important theorem in statistics, the *Central
Limit Theorem*, is illustrated in this problem. (The Central Limit
Theorem is discussed in Section 3.3.)

3. Even though it is not difficult to obtain exact values of probabilities
using the programs in your TI-85, let's further explore approximations
of one distribution by another.

a) Construct a graph of the binomial distribution for $N = 50$ and
$P = .4$, and store the picture. After recording the values for the
mean and standard deviation, press $\boxed{\text{GRAPH}}$ and note the values
for xMax and yMax. Use these range values and program COMP
to overlay a graph of the Poisson distribution with $\lambda = 50(.4) = 2$.
Is the approximation of the binomial by the Poisson "good"? Why
or why not?

b) Use program NMOVLY to overlay a graph of the normal distribution on the graph of the binomial distribution for $N = 50$ and
$P = .4$. Do you think a normal approximation to this binomial
distribution is appropriate? Find, for each probability distribution, $P(15 \leq x \leq 45)$.

4. In this exploration, you will consider three simulations other than the
toss of two dice that can be performed with program RELFREQ. In
each part, *replace* the simulation step in program RELFREQ with the
indicated program line. You may find it necessary to reset yMax to
obtain a good graph.

a) Simulate the roll of one fair die N times to estimate the probability of obtaining four dots on the upturned face of the die. The
simulation is accomplished with the program step : IPart (6Rand
$+ 1$). When you are prompted for EVENT, enter 4. In response
to the PROBABILITY = prompt, enter your guess for the probability that four dots appear on the upturned face when the fair

die is rolled one time. Execute program RELFREQ for $N = 10$, 20, and 50. Do the relative frequencies you obtain approach the value you entered for the probability?

b) Simulate the birth of three children N times to estimate the probability that 2 boys and 1 girl are born. (Assume the birth of either sex is equally likely.) Consider the variable of interest, x, to be the number of boys. (If $x = 2$, the other child must be a girl.) Since the birth of either sex is equally likely, this simulation is equivalent to tossing three coins and counting, say, the number of heads. Therefore, the simulation is accomplished with the program step : IPart (2Rand) + IPart (2Rand) + IPart(2 Rand). When prompted for EVENT, enter 2. In response to the PROBABILITY = prompt, enter your guess of the probability that two boys are born. (Hint: Is this a binomial experiment? Is there a program available to obtain the probability that exactly two boys are obtained in the three births?) Execute program RELFREQ for $N = 10, 20$ and 50. Do the relative frequencies approach your theoretical probability line?

c) Simulate the toss of five coins N times to estimate the probability that exactly 3 heads result when 5 fair coins are tossed. The simulation is accomplished with : IPart 2Rand + IPart 2Rand + IPart 2Rand + IPart 2Rand + IPart 2Rand. When prompted for EVENT, enter 3. In response to the PROBABILITY = prompt, enter your guess for the probability that exactly three heads appear on the upturned faces of the five coins. (Refer to the hint given in part b) of this problem.) Execute program RELFREQ for $N = 10$, 20 and 50. Do the relative frequencies you obtain approach the value you entered for the probability?

5. Construct graphs of the Poisson distribution for $\lambda = 5$, the binomial distribution for $N = 25$, $P = .2$, and the normal distribution for $\mu = 5$ and $\sigma = 2$. For each distribution, find $P(3 \leq x \leq 6)$. Write a paragraph discussing any similarities and/or differences you observe.

3.3. Inferential Statistics

Systematic methods that allow conclusions to be drawn from data while giving an associated measure of the reliability of those conclusions are the subject of statistical inference. The process of gathering data from an experiment with chance outcomes is called *sampling*. If all possible samples of a certain size are chosen from a given population, and if, for each of those samples, a certain numerical value called the *test statistic* is computed, the value of this statistic varies from sample to sample. The *sampling distribution* of the statistic is the probability distribution of all possible values of the statistic. Conclusions made in the two major areas of *inferential statistics, estimation* and *hypothesis testing*, are based on the sampling distribution of the test statistic.

The Sampling Distribution of the Sample Mean

Many sets of data are approximately described by normal distributions, and many test statistics have sampling distributions that are close to a normal distribution. In particular, the sampling distribution of the sample mean $\bar{x}$ will approach a normal distribution as the sample size increases *regardless of the population distribution*. This result, perhaps the most important in all of statistical inference, is known as the *Central Limit Theorem*.

Let's explore the Central Limit Theorem with a program that generates S random samples, each of size N, from populations having the exponential, standard normal, or uniform distributions. The program then draws a histogram of each sample, computes the mean of each sample, and draws a histogram of the sample means computed from each of the S samples. The subroutines that generate the random values from each population are given first. Program ZEXPRN generates random values from an exponential distribution with mean λ and sets range parameters for drawing histograms of the S samples.

```
PROGRAM: ZEXPRN
:0→xMin:5L→xMax
:L/2→xScl:5→yScl
:0→yMin:N/2→yMax
:For(K,1,N,1)
:L*-ln rand→W(K)
:End
:Return
```

Program 3.18.

Program ZNMRN generates random values [9] from the standard normal distribution and sets range parameters for drawing histograms of the samples.

Program ZUFMRN generates random values from the uniform distribution on the interval [0,1] and sets range parameters for drawing histograms of the samples.

Program CLT generates and graphs S random samples of size N from exponential, normal or Poisson populations and displays the graph of the sample means.

```
PROGRAM: ZNMRN
: -3→xMin:3→xMax
:1→xScl:5→yScl
:0→yMin:N/2→yMax
:For(K,1,N,1)
:√(-2ln (rand))*cos (2πrand)→W(K)
:End
:Return
```

Program 3.19.

```
PROGRAM: ZUFMRN
:0→xMin:1→xMax
:.1→xScl:5→yScl
:0→yMin:N/3→yMax
:For(K,1,N,1)
:rand→W(K)
:End
:Return
```

Program 3.20.

The sampling distribution of the sample mean is the probability distribution of the means of *all* possible samples of size **N** chosen from a population. Program CLT draws a graph of some of those sample means. Therefore, the graph shown at the conclusion of program CLT will vary with the values you input for **N** and **S**. As the number of samples increases, the graph of the sampling distribution of the sample means given by the program will closer approximate the actual sampling distribution of the sample mean $\bar{x}$.

Before executing program CLT, press MODE and be certain that you have Radian and Func selected. What can this program tell you about the sampling distribution of the sample mean for samples randomly chosen from a population that has an exponential distribution with mean λ? Press CLT and input $N = 5$ and # samples $= $ **S** $ = 10$. When prompted, enter 1 to choose the exponential population, and then input $\lambda = 2$. Carefully observe the histograms of five data values in each of the ten samples, and notice the rightward skewness of the data. Record the displayed value of the mean of the distribution of the $\bar{x}$'s, and press ENTER to observe the

```
 PROGRAM: CLT
:ZFSTEP
:Input "# samples=",S
:N→dimL W:S→dimL M
:Input "EXP(1),NORM(2),UNF(3)",C
:If C==1
:Input "λ=",L
:For(J,1,S,1)
:If C==1
:ZEXPRN
:If C==2
:ZNMRN
:If C==3
:ZUFMRN
:ClDrw
:Hist W
:OneVar W
:x̄→M(J)
:End
:OneVar M
:Disp "mean of dist of x̄'s is",x̄
:Pause
:x̄-3Sx→xMin
:x̄+3Sx→xMax
:Sx→xScl
:.75→yMax
:-.2*yMax→yMin
:Hist M
```

Program 3.21.

graph of the means of the ten samples. (If you wish to view the individual means of the ten samples, they are stored in list **M**.) Does the graph of the distribution of the means of the ten samples infer that the distribution of the sample mean is normal? Probably not, because $N = 5$ is a small sample size. Repeat the experiment for $N = 15$, $S = 10$ and $\lambda = 2$. You should find the rightward skewness of the samples more pronounced, but again, the distribution of the sample means will not generally appear to be that of the mound-shaped normal distribution. Once more repeat the experiment, this time choosing $N = 30$, $S = 10$ and $\lambda = 2$. (You will get more consistent results using a larger number of samples, but the program will take longer to execute.) Look at the values of the mean of the distribution of the $\bar{x}$'s each time you execute the program. Do you find most of them close to λ, the mean of the exponential population? You will probably find, if you repeat this simulation several times for $N \geq 30$, that the distribution of the sample means will appear approximately mound-shaped most of the time.

Repeat the entire experiment for the exponential population with $\lambda = 10$. You should obtain similar results with the exception that each time you execute CLT, the mean of the distribution of the $\bar{x}$'s should now be close to 10. One possible graph for the *sample* with $N = 30$ and $\lambda = 10$ is shown in Figure 3.16.

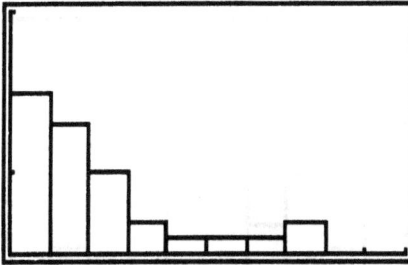

Figure 3.16. Histogram of random sample from exponential population, $\lambda = 10$.

A graph that could be obtained for the distribution of the *sample means* with mean of dist of $\bar{x}$'s $= 10.083$ when $N = 30$, $S = 15$ and $\lambda = 10$ is shown in Figure 3.17.

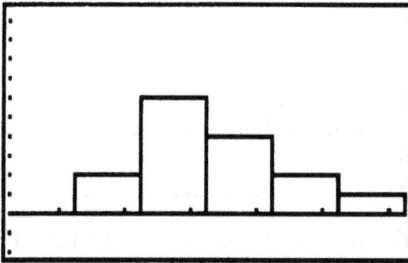

Figure 3.17. Distribution of sample means for 15 random samples of size 30 chosen from exponential population, $\lambda = 10$.

Program CLT sets the same range as program DATADISP when graphing the distribution of the sample means. According to the Central Limit Theorem and the Empirical Rule, for large N, the percentage of sample means falling in the intervals $\bar{x} \pm s_x$, $\bar{x} \pm 2s_x$, and $\bar{x} \pm 3s_x$ should be close to 68%, 95% and 99.7%, respectively. Compare the percentage of sample means falling in each of these intervals for your last execution of program CLT to the Empirical Rule percentages.

Let's perform the same experiment using the uniform distribution. The probability distribution resulting from the roll of a single die is an example of a discrete uniform distribution. The uniform distribution used in subroutine ZUFMRN is a continuous uniform distribution with mean equal to 0.5. Each number in the interval $[0, 1)$ has an equal chance of being selected for inclusion in the sample. Observe the values for the mean of the distribution of the $\overline{x}$'s each time you execute program CLT for samples of size $N = 5, 15$ and 30 drawn from this uniform population. It is suggested that you use 10 or 15 samples. One possible graph of a *sample* of size $N = 30$ is shown in Figure 3.18.

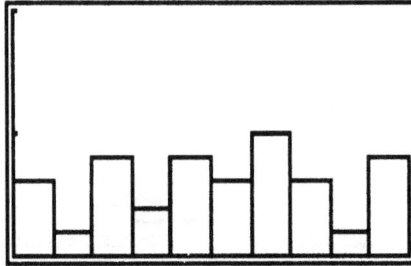

Figure 3.18. Histogram of random sample from uniform population.

A possible graph for the distribution of the *sample means* with mean of dist of $\overline{x}$'s $= 0.496$, $N = 30$ and $S = 10$ is shown in Figure 3.19.

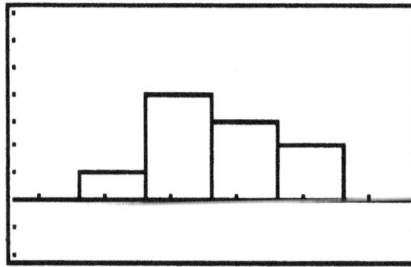

Figure 3.19. Distribution of sample means for 10 random samples
of size 30 chosen from uniform population.

Repeated execution of this simulation for larger values of N and S should show the "level" or uniform nature of the graphs of the samples and, for large N, a mound shape for the graph of the distribution of the sample means. Do you find the means of the distribution of the $\overline{x}$'s close to 0.5 no matter what sample size you use?

Now consider samples consisting of values randomly chosen from a normal population. Subroutine ZNMRN generates random values from the standard normal distribution with $\mu = 0$ and $\sigma = 1$. What value would you expect the means of the distribution of the $\bar{x}$'s to be near when you execute program CLT for this population? Run program CLT for 15 or 20 samples of sizes $N = 5, 15$ and 30 drawn from the standard normal population. Carefully observe the shape of the samples and the shape of the sampling distribution of the $\bar{x}$'s for the small sample sizes $N = 5$ and 15. A possible graph of the *sample* for $N = 15$ is given in Figure 3.20.

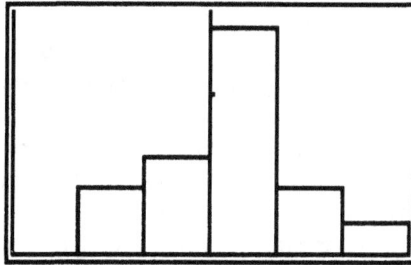

Figure 3.20. Histogram of random sample from normal population.

A possible graph for the distribution of the *sample means* with mean of dist of $\bar{x}$'s $= 0.004$, $N = 15$ and $S = 20$ is shown in Figure 3.21.

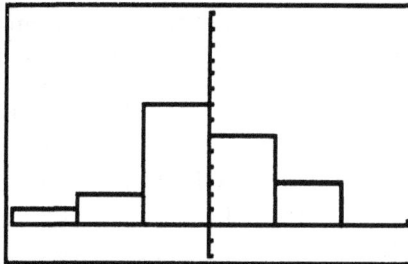

Figure 3.21. Distribution of sample means for 20 random samples of size 15 chosen from normal population.

Do you notice the approximate mound shape of the histograms of most of the samples and of the distribution of the sample means, even when the sample size is small?

Your exploration with program CLT should aid in your understanding of the following statistical theorems:

1) The mean of the sampling distribution of the sample mean *equals* the mean μ of the population from which the random samples are chosen.

2) *Regardless of the size of the sample*, the sampling distribution of the sample mean is normal when the random samples are chosen from a normal population.

3) *Regardless of the distribution of the population*, the sampling distribution of the sample mean approaches the normal distribution as the sample size increases. (The Central Limit Theorem)

Confidence Intervals and Hypothesis Tests

While statistical inference is the process of reaching conclusions about population values from evidence gathered in a sample, formal statistical reasoning is based on considering what would happen in many repetitions of the experiment that gathers that evidence. Both *interval estimation and tests of hypotheses* use measures of reliability that state what would happen if the decision were made many times under identical conditions. However, when the methods of inferential statistics are used to arrive at a conclusion, the decision is based on the results of a single sample, *not* the results of many repetitions of the experiment. You must therefore ask if the conclusion drawn from your sample is convincing or merely due to chance. The answer to this question involves the *confidence* that you have in your result, and the confidence is determined by the laws of probability and the sampling distribution of the appropriate test statistic.

It is important to note that the sampling distribution of the test statistic $z = \frac{\bar{x} - \mu}{\sigma/\sqrt{n}}$ is the standard normal distribution when σ is known and either the population is normal or the size of the random sample is large enough for the Central Limit Theorem to apply. The quantity $\sigma/\sqrt{n}$ is called the *standard error of the mean* and is the proper denominator for the z test statistic when the population is infinite or is large relative to the sample size n. When the population from which the sample is chosen is finite of size K and the sample size is large relative to the population size, the denominator of z should be corrected using the *finite population correction factor* by replacing $\frac{\sigma}{\sqrt{n}}$ with $\frac{\sigma}{\sqrt{n}}\sqrt{\frac{K-n}{K-1}}$. (The rule of thumb for when n is considered "large" relative to K varies slightly among textbooks.)

Two other test statistics that are used in one-sample inferential statistics are the t statistic for small sample inferences about a population mean and the z statistic for decisions about a binomial population proportion of successes. Recall that if σ is unknown and a random sample of size n is chosen from a population that has a normal distribution with mean μ and variance σ, the distribution of the statistic $t = \frac{\bar{x} - \mu}{s/\sqrt{n}}$ is the Student t distribution. When the sample size is *large* and σ is unknown, the sample standard deviation s estimates the population standard deviation σ. In this case, the statistic $z = \frac{\bar{x} - \mu}{s/\sqrt{n}}$ is used as the test statistic in inferences about one-sample population means. The reason for this is that for large samples, the values of the t and z distributions are close enough to be used interchangeably in most applied problems. Thus, the Central Limit Theorem can be applied so that the assumption of a normal population for use of the t statistic is not necessary.

When inferences about the proportion p of successes in a binomial population are desired, the sample proportion of successes, x/n, is used as the estimator of p. The appropriate test statistic whenever the normal distribution can be used to approximate the binomial distribution is

$$ z = \frac{\frac{x}{n} - p}{\sqrt{\frac{p(1-p)}{n}}} . $$

A *confidence-interval estimate of a parameter* (population value) consists of an interval of numbers that is predicted to include the parameter and a probability that specifies how confident you are that the parameter lies in that interval. When the probability is expressed as a percentage, the percentage is called the *confidence level*. Program 3.22, CINTVL, outputs the endpoints of large and small-sample confidence intervals for the population mean μ and for the proportion p of successes in a binomial population. When the t distribution is the sampling distribution of the estimator, program CINTVL assigns $N - 1$ as the number of degrees of freedom. When the program requests input of the confidence, you should enter the confidence level expressed as a probability between 0 and 1.

Consider this example: The quality control engineer in a plant producing cans of cola needs to estimate the mean amount of soft drink in 12 oz. cans filled by a machine in the plant. A random sample of 50 cans yields the results $\bar{x} = 11.95$ oz. and $s = 0.2$ oz. To find a 95% confidence interval for the true mean amount of cola (μ) in all cans filled by this machine, execute program CINTVL. Since the inference concerns a mean, enter 1 to choose the μ option. The population standard deviation σ is unknown, so t is the theoretical choice for the test statistic. However, if the t statistic is chosen, a test should be performed to see if the population from which this sample is drawn can be considered normally distributed. Notice that this is not necessary because of the large sample size. Why? The Central Limit Theorem can be applied, and s can be used to estimate σ. Therefore, enter 3 to choose the z test statistic, and input 11.95 for xbar, 50 for N, and .2 for s. When you are asked to input the confidence, enter .95. The program will give the endpoints of the confidence interval. The quality control engineer can be 95% certain that the population mean fill is between 11.895 and 12.005 oz.

```
PROGRAM: CINTVL
:Input "µ(1) or p(2)?",C
:If C==1
:Goto H
:Prompt x,N
:x/N→P
:√(p(1-p)/N)→D
:p→R
:Goto J
:Lbl H
:Input "z(3) or t(4)?",C
:Prompt xbar,N
:Input "σ or s=",s
:N-1→F
:xbar→R
:s/√N→D
:Lbl J
:Input "confidence=",P
:(1-P)/2→P
:ZVAL
:If C==4
:ZTNUM
:R-z*D→L
:R+z*D→U
:Disp "interval is"
:Disp L:Disp U
```

Program 3.22.

What result would be obtained if the quality control engineer uses t instead of z? Rerun program CINTVL using the same information, but this time, enter 4 to choose the t statistic. You should obtain the interval 11.893 to 12.007. Notice that the z and t intervals are almost identical.

Suppose the sample size in this problem had not been 50, but 15. A sample of size 15 is not large enough for the Central Limit Theorem to apply or for s to give a good estimate of σ, so the t test statistic is the one that should be used. (In actual practice, you must determine if the population from which the sample is chosen can be considered approximately normally distributed in order to use the t statistic.) To determine a 95% confidence interval for μ, execute program CINTVL, choosing 1 for the μ option and

4 for the t statistic. Input 11.95 for xbar, 15 for N, .2 for s, and .95 for confidence to obtain the interval (11.839, 12.061).

The interpretation of the numerical value of the confidence in an interval estimate is often misunderstood. For instance, in the example given above, it is *incorrect* to say that the probability the true mean fill falls in the interval (11.839, 12.061) is .95. The population parameter μ is constant and does not have sampling variability. The probability the true mean fill falls in this interval is either 0 or 1 depending on whether or not μ is between the two endpoints of the interval! Let's explore this important concept. Suppose you could generate many interval estimates from a population with a known mean. You could then examine the intervals and determine the proportion (relative frequency) of those intervals that actually contain the population mean. The proportion of intervals containing μ would be your estimate of the confidence that μ is in a randomly chosen interval.

Any simulation that generates random values from a distribution with a known mean can be used, so choose subroutine ZUFMRN to generate a list of random values from the uniform population with mean $\mu = 0.5$. The values generated by this subroutine are stored in list W. Choose a sample size of 30 with the keystrokes 30 $\boxed{\text{STO>}}$ N. Press $\boxed{\text{ZUFMRN}}$ and when the Done message appears, calculate the mean and standard deviation of the sample with $\boxed{\text{OneVar}}$ W. Execute program CINTVL, input 1 for the μ option and 3 for the z statistic (30 and over is considered a large sample). Enter the value of xbar with the keystrokes $\boxed{\text{STAT}}$ $\boxed{\text{VARS}}$ $\boxed{\bar{x}}$, input $N = 30$, and when requested for the value of s, press $\boxed{\text{STAT}}$ $\boxed{\text{VARS}}$ $\boxed{s_x}$. Enter some value for the confidence, say .90. Record whether or not the value 0.5 is between the two endpoints of the displayed interval. Repeat this process 9 more times for a total of 10 intervals. One possible result of this experiment is given in Table V.

Table V. 90% confidence interval simulation results.

Interval determined by program CINTVL	Does the interval contain 0.5?
(.4121, .5793)	Yes
(.4184, .5829)	Yes
(.5608, .6934)	No
(.4914, .6566)	Yes
(.3958, .5983)	Yes
(.4440, .6171)	Yes
(.4055, .5954)	Yes
(.4525, .6328)	Yes
(.4584, .6351)	Yes
(.4393, .6237)	Yes

The percentage of intervals containing $\mu = 0.5$ is $9/10$ indicating 90% confidence. In how many of these intervals did you find the true mean 0.5? You may or may not find exactly 90%, but if you repeat this experiment several more times for increasingly larger values of N, the percentage of intervals containing 0.5 should approach 90%. The proper interpretation of the value of the *confidence* in a p% confidence interval estimate is that if we select many random samples of the same size under similar conditions, and if we calculate a confidence interval for each of these samples, in p% of the cases, the population mean will lie within that interval. Repeat this experiment using different confidence levels and generate your samples from the exponential population with mean λ (program ZEXPRN) and the normal population with mean 0 (program ZNMRN).

How would you obtain a 93% confidence interval estimate of the proportion of all students at your school who prefer diet cola to regular cola? Suppose you choose a random sample of 200 students and find that 83 prefer diet cola, 100 prefer regular cola, and 17 have no opinion. Execute program CINTVL, choose 2 for the p option, input x, the number of successes in your sample, 83, $N = 200$, and the confidence .93 to obtain the interval $(.3519, .4781)$. What about the validity of this result? First, the conditions of a binomial experiment should hold true. Second, the normal approximation to the binomial distribution should be "good." You could visually check the adequacy of the approximation by constructing a graph of the binomial distribution for $N = 200$ and $P = 83/200$ using program BINOGRAF and overlay the normal distribution with program NMOVLY. However, since N is so large, this may take a while. Another approach is to recall that about 99.7% of the normal distribution falls within three standard deviations of its mean. If the approximation of the binomial by the normal is "good," a 99.7% confidence interval for the population proportion should fall within the interval $[0, 1]$ that contains all possible values of the estimator x/n. Execute program CINTVL using option 2 and input $x = 83$, $N = 200$, and confidence $= .997$. The resulting interval, $(.3116, .5184)$, is contained within the interval $[0, 1]$ indicating that the approximation is good.

Another type of statistical inference is a test of significance. This inference, usually called a *hypothesis test*, assesses the evidence provided by the data in regard to a claim or statement about a population parameter. The statement being tested is called the *null hypothesis*, and the hypothesis testing procedure is designed to determine the strength of the evidence *against* the null hypothesis. The null hypothesis is denoted by the symbol H_o and the statement we suspect is true instead of H_o is called the *alternative or research hypothesis* H_a. Once the hypotheses are determined, a random sample is chosen, and the value of the appropriate test statistic is computed. If the observed outcome is unlikely when the null hypothesis is true but more probable when the alternative hypothesis is true, that outcome is evidence against H_o in favor of H_a, and the decision is to reject H_o. If it is probable that the observed outcome can occur when the null hypothesis is true, the decision is to not reject H_o. The conclusion "do not reject H_o" indicates that the evidence gathered in the sample does not provide sufficient evidence to support the alternative hypothesis.

There are three choices for the form of any alternative hypothesis. Consider these choices for a significance test of the population mean. If you are primarily interested in deciding whether μ differs from a specified value, say m, H_a is $\mu \neq m$. This is called a *two-tail test*. When you are to decide if a population mean is less than a specified value m, H_a is $\mu < m$, and when you hope to support the claim that a population mean is greater than a specified value m, H_a is $\mu > m$. These tests of significance are called *one-tail tests*.

Because this type of inference is based on sampling, there is always a chance that an error will be made in the decision. A *type I error* is made when the decision is to reject the null hypothesis if H_o is actually true. The probability of a type I error is denoted by α and is called the *level of significance* of the test. The value of α is usually predetermined by the experimenter before any sample results are obtained and represents the area under the sampling distribution of the test statistic corresponding to the set of values of the test statistic that lead to rejection of the null hypothesis. This region of values is called the *rejection region*, and the *critical value(s)* is the value of the test statistic that separates the rejection region from the "do not reject" region. A *type II error* is made when the decision is to accept the null hypothesis if H_o is actually false. The probability of type II error is denoted by β.

The probability of obtaining an outcome at least as far from what is expected if H_o is true is called the *p-value* for the hypothesis test. The smaller the p-value, the stronger is the evidence provided by the data against H_o. Therefore, the null hypothesis is rejected for any choice of α that is greater than the p-value.

Program 3.24, HYPTST, calculates the value of the appropriate test statistic for a hypothesis test of a single population mean or proportion. The program requests the value of α, the level of significance, displays the critical value, the value of the test statistic, and the p-value for the test. For two-tailed tests of hypothesis, the program displays only the critical value falling in the same tail as the test statistic. The conditions for use of the appropriate formula are the same as those given in the discussion of the corresponding confidence interval estimates. Tail areas used in determining p-values for the Student t distribution are calculated using an approximating formula in subroutine ZTAREA, Program 3.23.

```
PROGRAM: ZTAREA
:(8F+1)(F  ln  (R²/F+1))^.5/(8F+3)→z
:ZAREA
:Return
```

Program 3.23.

Test the hypothesis $H_o : \mu = 8$ against the alternative $H_a : \mu > 8$ using the following information from a random sample of size 36: the sample mean is 9.35 and the sample variance is 20.79. The level of significance of the test, α, is chosen to be 0.05. Execute program HYPTST with inputs of 1 for the μ option, xbar $= 9.35$, 8 for μ, the hypothesized value of the population mean, $N = 36$, $\sigma = \sqrt{20.79}$, 3 for the z test statistic, 1 for the one-tail option, and .05 for α. The displayed critical value, 1.645, is the value to the *right* of which the null hypothesis should be rejected using a .05 significance level. (Since the alternative hypothesis is supported for values of the test statistic significantly larger than 8, the null hypothesis is rejected for values in the right-tail of the sampling distribution of the test statistic.) The value of the test statistic is $z = 1.7765$, a value that falls in the rejection region since it is larger than the critical value. Thus, the decision based on the predetermined value of α is "reject H_o." How much does this decision depend on the choice of α? The p-value for the test is .0378. The decision to reject H_o will be made for any $\alpha > .0378$.

Let's consider another application of these ideas. The probability of rolling a sum of 7 or 11 for the toss of a pair of fair dice is 8/36. Suppose you wish to test, at the .08 level of significance, the hypothesis that a pair of dice is fair. You roll them 50 times and obtain a 7 or 11 ten times. Is it appropriate to use the normal approximation to the binomial distribution in this situation? That is, does 99.7% of the normal curve fall within the interval $[0, 1]$, the limits on the proportion of successes in the binomial population? Execute program CINTVL with inputs of 2 for the proportion option, $x = 10$, $N = 50$, and .997 for confidence to obtain the interval (.0321, .3679). Since this interval is within the interval $[0, 1]$, the normal approximation to the binomial distribution is appropriate and the results of the hypothesis test should be valid. To see if the dice are fair, test $H_o : p = 8/36$ versus $H_a : p \neq 8/36$ using program HYPTST. Enter 2 to choose the proportion option, input $x = 10$, $N = 50$, 8/36 for the

```
 PROGRAM: HYPTST
:Input "μ(1) or p(2)?",C
:If C==1:Goto L
:If C==2:Then
:Prompt x,N,P
:(x/N-P)√N/√(P(1-P))→R
:End:Goto H
:Lbl L
:Prompt xbar,μ,N
:Input "σ or s=",s
:N-1→F
:(xbar-μ)√N/s→R
:Input "z(3) or t(4)?",C
:Lbl H
:Input "1 tail or 2?",T
:Input "α=",P
:If T==1:Then
:ZVAL
:If C==4
:ZTNUM
:End
:If T==2:Then
:P/2→P
:ZVAL
:If C==4
:ZTNUM
:End
:If R<0: -z→z
:Disp "critical value=",z
:Disp "test statistic=",R
:R→z
:If C≠4
:ZAREA
:If C==4:ZTAREA
:If T==2:AREA*2→AREA
:Disp "p value",AREA
```

Program 3.24.

hypothesized value of p, option 2 for the two-tail test, and $\alpha = .08$. Since the test result $z = -.3780$ is not within the rejection region $z < -1.7508$ or $z > 1.7508$, the decision is made that there is *not* enough evidence, using a level of significance of .08, to reject the null hypothesis. (Notice that for a two-tail test, program HYPTST gives only the critical value falling on the same side of the mean as the test statistic.) The p-value .7055 shows that the decision "do not reject H_o" is highly significant since the same conclusion would be reached for *any* reasonable choice of α.

Let's now consider a *paired difference test*. When the elements of two populations are matched or paired by design, outcomes are compared within each matched pair. A common situation calling for a paired comparison experiment is a before-and-after study on the same subjects. A random sample is chosen from the population of pairs, each pair is assigned a single value, the difference in the two individual data values, and a t-test is used with the data consisting of the differences in the sampled pairs. The condition under which the result of this test is valid is that the population of paired differences is approximately normally distributed. Consider an example.

The pulse rates of a random sample of eleven patients before and after being given a certain tranquilizer are listed in Table VI.

Table VI. Patient pulse rates.

Person	Pulse Rate Before Tranquilizer	Pulse Rate After Tranquilizer
1	81	77
2	80	79
3	78	75
4	79	80
5	84	79
6	80	74
7	83	79
8	82	79
9	75	76
10	88	83
11	81	80

Using $\alpha = .01$ and the paired data above, can you conclude that the use of the tranquilizer significantly reduces pulse rate? Rather than risk making an arithmetic mistake in obtaining the differences, let's let the TI-85 do the subtraction. Enter the two sets of data as lists and store them under the names BF (pulse rate before taking the tranquilizer) and AFT (pulse rate after taking the tranquilizer). To obtain the data of differences, enter BF−AFT and store the result as list DIFF. Let's call the mean of the population of pulse rates before using the tranquilizer μ_1 and the mean of the population of pulse rates after using the tranquilizer μ_2. To see if the use of the tranquilizer reduces the pulse rate, test the hypothesis $H_o : \mu_1 - \mu_2 = 0$ versus the alternative hypothesis $H_a : \mu_1 - \mu_2 > 0$.

Let's digress for a moment and view a scatter plot of BF (x) versus AFT (y) to see the location of the data points in relation to the line $y = x$. If the points appear close to the line, you would probably believe that $x = y$ ($\mu_1 - \mu_2 = 0$). If most of the data points are above the line, you would suspect that $x < y$ ($\mu_1 - \mu_2 < 0$), and if the majority of the data points are below the line, you would suspect that $x > y$ ($\mu_1 - \mu_2 > 0$). Press $\boxed{\text{GRAPH}}$ $\boxed{y(x) =}$ and enter x in the $y1$ location. Set the range so that all data values can clearly be seen, say a $[73, 90]$ by $[73, 85]$ viewing window with xScl = 0 and yScl = 0. Draw the scatter plot with $\boxed{\text{Scatter}}$ $\boxed{\ (\ }$ $\boxed{\text{BF}}$ $\boxed{\ ,\ }$ $\boxed{\text{AFT}}$ $\boxed{\ )\ }$ $\boxed{\text{ENTER}}$. Press $\boxed{\text{CLEAR}}$ to eliminate the menu from the bottom of the screen for the best view.

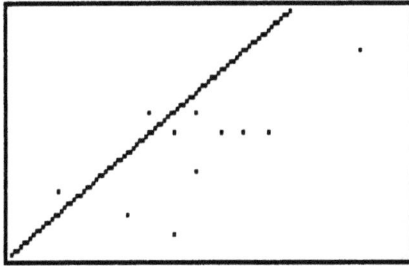

Figure 3.22. Scatter plot of before tranquilizer versus
after tranquilizer pulse rates.

Notice that most of the data points are below the line, indicating that "pulse rate before tranquilizer" – "pulse rate after tranquilizer" is positive most of the time.

To perform the hypothesis test, calculate the one-variable statistics with $\boxed{\text{OneVar}}$ $\boxed{\text{DIFF}}$ and execute program HYPTST. Choose 1 for the mean option, use the STAT VARS menu to enter $\bar{x}$ for xbar, enter 0 for μ, the hypothesized difference in the two means, and input 11 for N, the number of matched pairs. Use the STAT VARS menu to enter s_x for s, and choose 4 for the t statistic option. Since this is a one-tail test, enter 1 at the next prompt. Input the level of significance $\alpha = .01$, and obtain the results: critical value = 2.7664, $t = 3.75$, and p-value = .0019. Therefore, H_o is rejected, and you conclude that the pulse rate is lower after using the tranquilizer. Notice that you would reach the same conclusion for any $\alpha > .0019$. Also note that program CINTVL gives a 98% confidence for the population mean difference, $\mu_1 - \mu_2$, as $(.7154, 4.7392)$. The hypothesized

difference, 0, is not in this interval of positive values, supporting the belief
that $\mu_1 - \mu_2 > 0$.

Realize that these conclusions are based on the assumption that the
population of differences is normally distributed. One way to assess the
adequacy of the normal model is to construct a *normal quantile plot*, also
called a *normal probability plot*. This graph compares the sampled population distribution to the normal distribution by plotting their percentiles
against one another. If the distributions are nearly the same, the data
values will fall close to a straight line. If the distribution is negatively (leftward) skewed, the smallest observations fall distinctly to the left of a line
drawn through the main body of the points. Positive (rightward) skewness
or large outliers cause the largest observations to fall distinctly to the right
of a line drawn through the smaller data values. [5] A plot that is bent
down on the left and bent up on the right means that the data has longer
tails than the normal distribution.

```
PROGRAM: NQPLT
:ZFSTEP
:Input "data list?",D
:sortA D
:Ans→D
:D(1)-1→xMin
:D(N)+1→xMax
:0→xScl:0→yScl
: -3→yMin:3→yMax
:N→dimL Q
:For(K,1,N,1)
:(K-.375)/(N+.25)→P
:If P≤.5:Then
:ZVAL
: -z→Q(K):End
:If P>.5:Then
:1-P→P
:ZVAL
:z→Q(K)
:End:End
:Scatter (D,Q)
```

Program 3.25.

The construction of a normal quantile plot is usually done with computer software, but you can use Program 3.25 to easily draw the plot. Use program NQPLT to construct a normal probability plot for the data list DIFF.

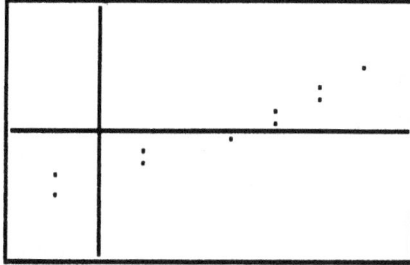

Figure 3.23. Normal quantile plot for DIFF.

The graph in Figure 3.23 indicates that the overall pattern of the points is roughly that of a straight line. Thus, the results of the paired difference hypothesis test and confidence interval are valid. (The stacking or granularity in the graph is because some integer values are repeated.)

Programming the formulas for two-sample confidence intervals and hypothesis tests is not difficult. Recall programs CINTVL and HYPTST, and edit each by replacing the steps that store into R with the appropriate formulas.

Explorations with Inferential Statistics

1. Use program COINTOSS to simulate the toss of a coin $N = 25$ times and record the number of tails obtained. Find a 95% confidence interval for the true proportion of tails when your "calculator coin" is tossed. Repeat the experiment for $N = 50$ and $N = 100$ tosses. How does the increasing sample size affect the width of the confidence interval?

2. Use program COINTOSS to simulate the toss of a coin $N = 50$ times and record the number of tails obtained. Find an 86% confidence interval and a 95% confidence interval for the true proportion of tails when your "calculator coin" is tossed. How does the confidence level affect the width of the interval?

3. According to the National Research Council, the suggested weight for a 6'2" male 19 to 34 years old is 148 to 195 pounds. [6] As reported in the January 9, 1992 edition of *USA Today*, Table VII gives the weights of all 6'2" males listed on the Washington Redskins, Detroit Lions, Denver Broncos, and Buffalo Bills rosters.

Table VII. Weights of 6'2" football players on Redskins, Lions, Broncos, and Bills rosters.

202	223	210	196	230	260	245
251	255	234	276	205	282	240
179	226	281	243	213	189	260
197	237	256	208	235	235	225
190	238	270	286	230	245	

a) Assuming this data represents a random sample of the weights of 6'2" professional football players (it may or may not), find a 90% confidence interval estimate for the mean weight of the population of professional football players. How does this interval compare with the suggested weight interval for all males between ages 19 and 34?

b) Using this data, test the hypothesis that the mean weight of a 6'2" professional football player is more than 1.2 times the average suggested weight of all 6'2" males. Use a .05 level of significance and interpret the p-value for the test. Discuss the validity of your conclusion.

4. Refer to the BF and AFT lists given in Table VI. Verify that the mean of the difference data stored in list DIFF equals the difference $\overline{x}_{BF} - \overline{x}_{AFT}$. Does a similar relationship hold for the standard deviations? That is, is the standard deviation of the difference data in DIFF equal to the difference $s_{BF} - s_{AFT}$?

5. Recall that the strength of the evidence provided against the null hypothesis in a test of significance is measured by the p-value. When the p-value is smaller than α, the data is *statistically significant* and H_o is rejected. Choosing $\alpha = .10$, for instance, requires that the data give evidence against H_o so strong that a false rejection of the null hypothesis would happen no more than 10% of the time for samples chosen randomly from the population of interest. We can explore this idea for a small number of repetitions and conjecture what would happen if the experiment were repeated a large number of times.

a) Choose a sample size with the keystrokes 15 STO> N, and execute program ZNMRN to generate a list, W, of 15 values chosen at random from the standard normal distribution with $\mu = 0$ and $\sigma = 1$. Before executing program HYPTST, you could find the sample mean and sample standard deviation with the keystrokes OneVar ALPHA **W** ENTER STAT VARS $\overline{x}$ and s_x. You may, however, find it helpful to enter Program 3.26, SIGCK. This program will generate the random values, calculate the summary statistics, and call the hypothesis testing program.

```
 PROGRAM: SIGCK
:Prompt N
:ZNMRN
:OneVar W
:x̄→B:Sx→S
:HYPTST
```

Program 3.26.

To repeatedly test $H_o : \mu = 0$ against the alternative that μ differs from 0, press [SIGCK], enter 15 for N, and 1 for the μ option. Enter [ALPHA] B when prompted for $\bar{x}$, 0 for the null hypothesis value of μ, 15 for N, [ALPHA] S when prompted for s. Since the sample is chosen from a normal population and s does not approximate σ due to the small sample size, the t-test statistic should be used. Next enter 2 for the two-tail option and .10 for α. Observe the p-value, and record whether or not you reject the null hypothesis. Press [ENTER] to again execute program SIGCK, and repeat the experiment for nine more times. In how many of the ten hypothesis tests would you *expect* to reject the null hypothesis? In how many of the ten hypothesis tests *did* you reject H_o? Since α, the probability that the null hypothesis is rejected when it is actually true, is approximated by the proportion of times you reject H_o, what is your estimate of α?

b) Replace the lines :Prompt N and :ZNMRN in program SIGCK with :Prompt N, λ and :ZEXPRN. Choose a value for $\lambda = \mu$, and repeat the experiment described in part a) of this problem for $N = 30$ and $\alpha = .05$ to test the hypothesis $H_o : \mu = \lambda$ against the alternative $H_a : \mu \neq \lambda$. Repeat this experiment 10 times to determine your estimate of α.

3.4. Regression and Analysis of Variance

Many statistical problems are concerned with the relation, if such a relation exists, between two or more variables. We usually wish to know if the variables of interest are related, and if so, what is the nature of the relationship? If an appropriate mathematical model can be found, how can information about one of the variables be used to predict another?

Simple Linear Regression

A possible relationship between two variables is sometimes visually identified by looking at a scatter diagram of the data points. When a scatter plot suggests that the dependence of the *response variable y* on the *explanatory variable x* is in the form of a straight line, the *linear regression* model is appropriate. Consider a method for finding an appropriate linear model for the relation, if it exists, between x and y, and use the equation $y = a + bx$ with y-intercept a and slope b as the equation of the *regression line*. There are many different lines that could be drawn "through" the data points. To find the linear equation that is the *best-fitting line* for the data, the *least squares criterion* is used.

The basic idea in linear regression is to construct the line so that the sum of the squares of the vertical distances from the data points to the line is as small as possible. The vertical distance between each data point and the corresponding point on the line, $y_{data} - y_{line}$, is called the *deviation* or *residual*. Therefore, if we denote a data point by (x, y) and the linear model by $y = a + bx$, we must find the coefficients a and b that minimize

$$\sum_{\text{all } x} \left(y_{data} - y_{line}\right)^2 .$$

Techniques in calculus are applied to give the values of a and b that minimize this *sum of the squares of the deviations*. Even though your TI-85 will easily compute these values, it is important to understand the process of the least-squares technique and the conditions necessary for its application if you intend to use the regression line for predictions. Let's explore with Program 3.27, LSLINE, to see what is actually happening when the least squares criterion is used.

```
PROGRAM: LSLINE
:FnOff
:0→A:0→B:1→C
:y1=A+B x
:Input "XLIST?",xStat
:Input "YLIST?",yStat
:dimL xStat→N
:min(xStat)-2→xMin
:max(xStat)+2→xMax
:min(yStat)-2→yMin
:max(yStat)+2→yMax
:0→xScl:0→yScl
:N→dimL RES
```

```
:Lbl A1
:Scatter
:Pause
:Input "slope=",B
:Input "y intercept=",A
:1→K:0→x:0→S
:Lbl V
:xStat(K)→x
:y1→Y
:yStat(K)-Y→RES(K)
:(yStat(K)-Y)²+S→S
:Line(xStat(K),yStat(K),x,Y)
:K+1→K
:If K≤N
:Goto V
:Pause
:Disp "SSD=",S
:Pause
:If C==2
:Goto W
:Input "try again? Y(1) N(2) ",C
:If C==1
:Goto A1
:LinR
:ShwSt
:Pause
:DrawF RegEq
:Pause
:a→A:b→B
:1→K:0→x:0→S
:Goto V
:Lbl W
:Disp "RESIDUAL PLOT"
:Pause
:FnOff
:min(RES)-2→yMin
:max(RES)+2→yMax
:0→xScl:0→yScl
:Scatter (xStat,RES)
```

Program 3.27

Program LSLINE will first draw a scatter plot for data that has been entered in an xlist and a ylist. The program requests that you find the *y*-intercept and the slope of the best-fitting line that minimize the sum of the squares of the deviations (residuals) for the two-variable data. The line you fit through the data is drawn, the residuals are shown as vertical line segments on the graph, and the sum of squares of the deviations of the points from the line is displayed as the quantity SSD.

Let's investigate the information obtained from program LSLINE with data collected by the University of Michigan under a series of grants from the National Institute on Drug Abuse. As reported in a study titled "Monitoring the Future" [6], the percent of college students one to four years beyond high school who used alcohol and/or marijuana in the last twelve months of the indicated year is listed in Table VIII.

Table VIII. Percent of college students using alcohol and/or marijuana

Drug	1985	1986	1987	1988	1989	1990
Alcohol	92.0	91.5	90.9	89.6	89.6	89.0
Marijuana	41.7	40.9	37.0	34.6	33.6	29.4

Enter this data in the TI-85 as the lists ALCH and MARJ. Do these data indicate a relationship between alcohol and marijuana use among college students? If so, can the relationship be modeled with a straight line? Execute program LSLINE with XLIST = ALCH (explanatory variable) and YLIST = MARJ (response variable). What information does the scatter plot displayed by the program reveal? It appears that as the percentage of students who use alcohol increases, the percentage of students who use marijuana increases. The scatter diagram does not rule out the possibility of a linear model.

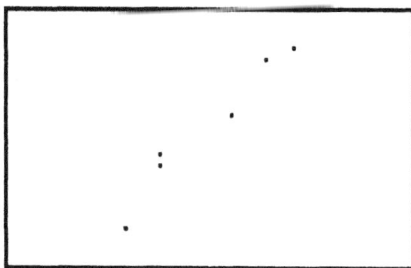

Figure 3.24. Scatter plot of ALCH versus MARJ.

Press ENTER to continue the program. You are next requested to enter a guess for the slope of the regression line. A suggestion for obtaining the initial estimate is to look at any two data points, say $(92, 41.7)$ and

(89, 29.4), and compute the slope as $\frac{29.4-41.7}{89-92} \approx 4.1$. Enter an initial guess of -330 for the y-intercept of the least squares line and press ENTER. (The value of the y-intercept can be roughly estimated with the cross-hair cursor.) Notice that the line is above the data points and that SSD = 132.6938. Try to make SSD smaller. You can move the line "down" by decreasing the y-intercept. Press ENTER and input 1 at the try again? prompt. You are shown your previous guess for the line (with the residuals shown as vertical line segments) graphed on the scatter plot. ENTER allows input of your next guess for the slope and y-intercept. Let's try slope = 4.1 and y-intercept = -337. View the line, and notice that SSD has been reduced to 42.2538. See if changing the slope can further reduce the sum of the squares of the deviations. Enter 1 at the try again? prompt, and after viewing your previous estimate of the regression line, press ENTER and input slope = 3.8, y-intercept = -307. The resulting line seems to fit the data much better, doesn't it? Notice that SSD is now 7.3752. Is this as small as SSD can be? Enter 2 at the try again? prompt, and the *least* squares coefficients, y-intercept = $a = -303.5484$ and slope = $b = 3.7569$, are calculated and displayed. Press ENTER and you will see the least-squares regression line overdrawn on the scatter plot containing your last estimate. ENTER displays the least squares line with the residuals for the best fit. Pressing ENTER once more shows the minimum SSD is 6.1647.

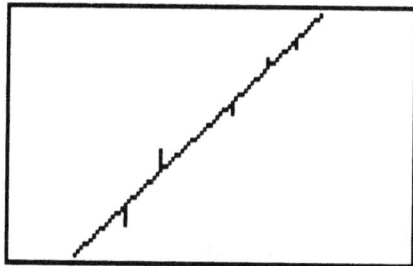

Figure 3.25. Least squares line with residuals for linear regression of ALCH vs MARJ.

The question now becomes "Is a line the appropriate model for this data?" If so, you have found it to be $y = -303.5484 + 3.7569x$. If not, another model should be explored. There are several conditions that must hold true in order that the linear regression analysis be valid. Recall that residual = $y_{\text{data}} - y_{\text{line}}$ = observed value − predicted value. The residuals calculated from the least squares line have the properties that the sum, and therefore the mean, of the residuals is zero, and the variance of the residuals is the smallest possible for any *line* that is drawn through the data points. Examination of a scatter plot of the residuals, plotted against the values of the explanatory variable, should not show any unusual patterns or systematic departures from the fitted line (represented by $y = 0$) if the least squares model is appropriate. Press $\boxed{\text{ENTER}}$, and program LSLINE will draw a scatter plot of the alcohol use percentage versus the residuals for the least squares line.

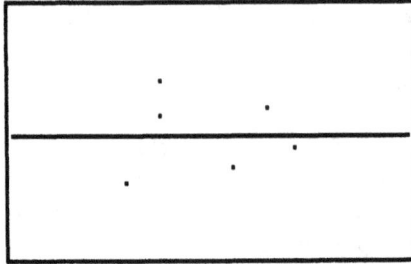

Figure 3.26. Scatter plot of residuals for linear
regression of ALCH vs MARJ.

An examination of this residual plot reveals no unusual values or obvious patterns that would cause you to question the validity of the linear model.

A measure of the strength of the linear relationship between x and y is the *correlation coefficient*. The symbol for the sample correlation coefficient is r, and correlation coefficients will always range between -1 and 1. The closer r is to 1 or -1, the closer the points come to falling on a straight line. If $r = 0$, there is no *linear* relationship between x and y. The value of r is displayed by program LSLINE with the least squares values of the slope and y intercept. It can also be accessed by pressing $\boxed{\text{STAT}}$ $\boxed{\text{CALC}}$, entering the xlist and ylist names, and pressing $\boxed{\text{LINR}}$. For the regression of MARJ on ALCH, $r = .9712$. Since this value is close to 1, you suspect that a strong positive linear relation exists between the percentage of students using alcohol and the percentage of students using marijuana.

When interpreting the correlation coefficient, several words of warning are necessary. The correlation coefficient r measures the *strength* of the linear relationship between two variables. It does *not* mean that one is *causing* the other. A strong correlation between x and y is often due to other variables. There are factors other than use of alcohol that would contribute to the reason a student would use marijuana.

When values of r close to zero are found, it is tempting to conclude that there is little or no relationship between the variables under consideration. However, remember that the correlation coefficient measures only the *linear relationship* between the values. If r is close to zero, there could still be a relationship, possibly strong, between the variables, but it would be non-linear. The correlation coefficient measures the strength of a relationship;

it does *not* prove that such a relationship exists or does not exist.

An analysis of the residuals is very important in determining whether or not a condition violation is severe enough to invalidate the use of the population regression model $y = A + Bx + \epsilon$. The residuals are estimates of the population error term ϵ, a random quantity that accounts for all of the variables that are not part of the model. Conditions necessary for the validity of the linear regression model are 1) ϵ is a random variable with mean of zero, 2) ϵ is a normally distributed random variable, 3) the variance of ϵ is the same for all values of x, and 4) the values of ϵ are independent.

To see if ϵ can be considered a normally distributed random variable, execute program NQPLT, input $N = 6$, and enter RES at the data list? prompt. Notice that the normal probability plot is fairly straight indicating that the assumption of a normally distributed error term seems reasonable.

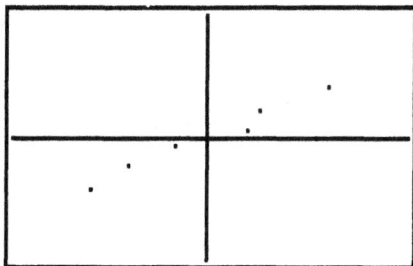

Figure 3.27. Normal probability plot of residuals for linear regression of ALCH vs MARJ.

There are several patterns statisticians look for in a residual plot. If the condition that the variance of ϵ is the same for all values of x is valid, the residual plot should show a horizontal band of points randomly spread about the line $y = 0$. Refer to Figure 3.26 and notice that this is the case in the residual plot for the regression of ALC versus MARJ.

If the variability of ϵ is not constant, a "fan-shaped" residual plot is observed, and a condition known as *heteroscedasticity* exists. For instance, consider the data xlist = $\{.5, 1.5, 2, 3, 4, 5\}$ and ylist = $\{1, 1, 2.5, 1, 5, 2\}$. Execute program LSLINE, and input these lists at the XLIST? and YLIST? prompts. Input some estimates of the slope and y-intercept, and then choose the no (2) option to observe the least squares line. Is the line a good fit to the data? The residual plot shows the variability is greater for larger values of x, indicating an inconsistent variance of the error term. Therefore, a linear model is not appropriate for this data.

An obvious pattern in the residuals indicates that the assumption of independence of the error terms is not true. Consider the data xlist = $\{1.5, 2, 3, 4, 5, 6\}$ and ylist = $\{4, 2, 1, 1.5, 2, 4\}$. Execute program LSLINE

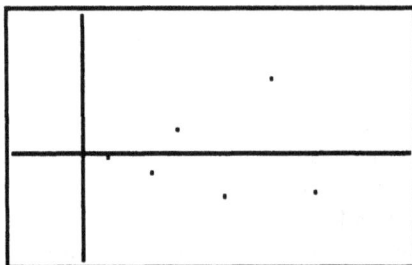

Figure 3.28. Residual plot showing nonconstant variance of ϵ.

and input these lists at the XLIST? and YLIST? prompts. Input some estimate for the slope and y-intercept, and then choose the no (2) option to observe the least squares line. The scatter diagram of the data indicates a nonlinear relationship, and the obvious pattern in the residual plot shows that the values of ϵ are not independent. The linear model is therefore not adequate; perhaps a multiple regression model, one in which more than one variable is used in the prediction of another, should be used.

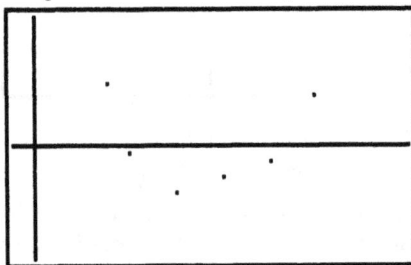

Figure 3.29. Residual plot showing values of ϵ are not independent.

Once you have checked to see that the underlying assumptions concerning the use of the linear model are appropriate and examined the strength of the relationship with the correlation coefficient, a hypothesis test on the slope B of the population regression line $y = A + Bx$ should be performed. The results of this test allow you to draw a conclusion as to the significance of the relationship between the exploratory and response variables. The specifics of the test are not covered in this manual.

Return to the example in which you found the regression line with ALC as the explanatory variable and MARJ as the response variable. Suppose you wished to predict the percentage of students using marijuana for an alcohol use percentage equal to 90. Assuming that all conditions have been checked and the linear model is appropriate, press $\boxed{\text{STAT}}$ $\boxed{\text{CALC}}$, enter ALC as the xlist Name and MARJ as the ylist Name. Press $\boxed{\text{ENTER}}$, and choose the linear model with $\boxed{\text{LINR}}$. Press $\boxed{\text{FCST}}$, enter 90 for x, press $\boxed{\text{ENTER}}$ and then $\boxed{\text{SOLVE}}$ to observe $y = 34.5720$.

When you predict y-values for x-values that are within the range of data in your scatter plot, you are using a process called *interpolation*. Predicting y-values for x-values that are well beyond the range of the data is called *extrapolation*. You do not know how the data will behave outside the range you have plotted. Extrapolation will very often lead to incorrect predictions. For example, press $\boxed{\blacktriangle}$ to position the cursor over the x-value in the FCST SOLVE display screen, enter 25 for x, and solve for the predicted y from the current least squares equation. The value -209.6261 certainly makes no sense as a prediction for the percent of students using marijuana!

Enter the list YEAR with the command seq $(A, A, 1985, 1990, 1)$. Execute program LSLINE twice with YEAR as the explanatory variable and ALCH and MARJ as response variables. If you do not wish to explore finding the least squares line with the program, choose 2 for the try again? option, and observe the regression line, regression coefficients and correlation coefficient. How is the percentage of students using alcohol and marijuana changing over time? Do the residual plots over time yield further information?

Other Regression Models

When the model $y = Ax + B$ is not appropriate, it may be that the data can be transformed so that the transformed data fits a linear model. If so, there are six other regression models you can obtain from the TI-85: LNR, EXPR, PWRR, P2REG, P3REG, and P4REG. Explore these models with the Table IX data [6] on women in the labor force as reported by the Bureau of the Census and the Bureau of Labor Statistics.

Table IX. Percentage of women in the labor force.

Year	% of female population aged 16 and over in the labor force
1900	18.8
1910	21.5
1920	21.4
1930	22.0
1940	25.4
1950	33.9
1960	37.8
1970	43.4
1980	51.6
1988	56.6
1989	57.5
1990	57.5

Input the two columns in Table IX as lists and store them to the names YRWOM and WOM, respectively. Execute program LSLINE with XLIST = YRWOM and YLIST = WOM. Notice that the scatter plot does not indicate a linear trend for the data even though $r = .9700$. Continue with the program, and notice the obvious pattern in the residual graph. Let's see if there is a better model. Execute program LSLINE to set the range parameters and draw the scatter plot of year versus percentage women in the labor force. After the scatter diagram appears on the screen, press $\boxed{\text{ON}}$ and $\boxed{\text{QUIT}}$ to exit the program. Press $\boxed{\text{STAT}}$ $\boxed{\text{CALC}}$, enter YRWOM as the xlist Name, WOM as the ylist Name and press $\boxed{\text{ENTER}}$. The menu at the bottom of the display screen shows the other regression models available in your TI-85. Let's explore each of these models and see if any seem to visually fit this data.

The logarithmic model is $y = a + b(\ell n x)$. That is, the transformed data used to determine the regression equation is $(\ell n x, y)$. Press $\boxed{\text{LNR}}$ to view the logarithmic model coefficients. Note that $r = .9685$. Press $\boxed{\text{MORE}}$ to access the remainder of the menu and $\boxed{\text{STREG}}$. This command allows you to store the current regression equation to the equation variable you specify for graphing. Enter $y2$ at the Name= prompt. (You must use the lowercase letter y.)

The exponential model is $y = ab^x$. The transformed data used to determine the regression equation is $(x, \ell n y)$. Press $\boxed{\text{MORE}}$ $\boxed{\text{EXPR}}$ to view the exponential model coefficients. Note that $r = .9847$. Press $\boxed{\text{STREG}}$ and enter $y3$ at the Name= prompt.

The power regression model is $y = ax^b$. The transformed data used to determine the regression equation is $(\ell n x, \ell n y)$. Press $\boxed{\text{MORE}}$ $\boxed{\text{PWRR}}$ to view the power model coefficients. Note that $r = .9840$. Press $\boxed{\text{STREG}}$ and enter $y4$ at the Name= prompt.

For comparison purposes, press $\boxed{\text{LINR}}$ to view the linear model coefficients. Press $\boxed{\text{STREG}}$ and enter $y1$ at the Name= prompt. Press $\boxed{\text{GRAPH}}$ and $\boxed{y(x) =}$ to observe the four regression models in the function graphing list. View the graphs of these models on the scatter diagram by entering $\boxed{\text{QUIT}}$ $\boxed{\text{Scatter}}$ $\boxed{(}$ $\boxed{\text{YRWOM}}$ $\boxed{,}$ $\boxed{\text{WOM}}$ $\boxed{)}$. The closeness of the correlation coefficients for the linear and logarithmic models and for the exponential and power models is reflected in the graph. Also notice that the highest correlation coefficient is for the exponential model whose graph seems to "best fit" the data.

Is there a better model? The regression models P2REG, P3REG, and P4REG are used when the relationship between the x and y is *curvilinear*. The polynomial models $y = ax^2 + bx + c$ (P2REG), $y = ax^3 + bx^2 + cx + d$ (P3REG) and $y = ax^4 + bx^3 + cx^2 + dx + e$ (P4REG) use quadratic, cubic, and quartic regression techniques to find the coefficients for the best fit.

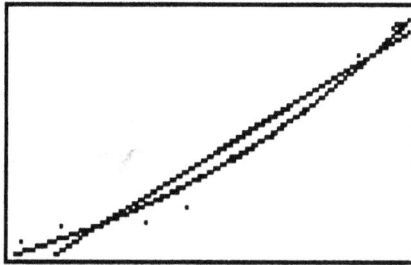

Figure 3.30. Linear, logarithmic, exponential and power
models for regression of YRWOM versus WOM.

The coefficients, in descending order by powers of x, are returned in the list PRegC that is accessed in the STAT VARS menu once the regression has been performed with $\boxed{\text{STAT}}$ $\boxed{\text{CALC}}$ followed by the indicated curvilinear model key. Do any of these models better fit the women in the labor force data? Press $\boxed{\text{STAT}}$ $\boxed{\text{CALC}}$ $\boxed{\text{ENTER}}$ $\boxed{\text{ENTER}}$ $\boxed{\text{MORE}}$ and $\boxed{\text{P2REG}}$. Use $\boxed{\text{STREG}}$ to enter $y1$ at the Name= prompt. Press $\boxed{\text{P3REG}}$, and use $\boxed{\text{STREG}}$ to store the cubic equation in $y2$. Repeat the procedure to store the $\boxed{\text{P4REG}}$ equation in $y3$. Use $\boxed{\text{GRAPH}}$ $\boxed{\text{RANGE}}$ to check that the range parameters are those set by program LSLINE, $[1898, 1992]$ by $[16.8, 59.5]$. Press $\boxed{y(x) =}$ and delete $y4$. Press $\boxed{\text{QUIT}}$ to exit the range, and draw the graph of the function list and the scatter plot with the instruction Scatter (YRWOM, WOM). It appears that of these three models, the quadratic model gives the best fit. (To verify this, you may need to graph each individual regression model on the scatter plot.)

While you have these graphs on the screen, zoom out twice by pressing $\boxed{\text{ZOOM}}$ $\boxed{\text{ZOUT}}$ $\boxed{\text{ENTER}}$ $\boxed{\text{ENTER}}$. Note the closeness of the quadratic and cubic models *for the range of the original data* in the zoomed-out viewing window. Also notice why you did not see the graph of the equation obtained with P4REG when viewing the graphs in the viewing window set by program LSLINE. There are not enough points in this data for the quartic regression model to give a fit to the data. Note the dangers of extrapolation if predictions are made using an inappropriate model outside the range of the original viewing window set by program LSLINE.

Does the exponential model or the quadratic model give a better fit for this data? You be the judge! Return to the STAT CALC menu and enter the exponential model in $y2$. Press $\boxed{\text{GRAPH}}$ $\boxed{y(x) =}$, delete $y3$, set the range parameters to view the scatter diagram in the $[1898, 1992]$ by $[16.8, 59.5]$ viewing window, and graph the exponential model in $y1$ and the quadratic model in $y2$ with the instruction Scatter (YRWOM, WOM).

Remember that even if a model seems appropriate, you must be careful about using it to make predictions until you can answer the question "How reliable is the prediction?"

Analysis of Variance

Many statistical applications concern comparing the means of two or more populations. The procedure for comparing these means involves analysis of the variation in the sample data. When comparing two population means, the sources of variability are the difference between the sample means and the variability within the two samples chosen from those populations. When using analysis of variance procedures for comparing the means of three or more populations, variability is measured and allocated among its sources. The process of planning the experiment to collect the data, the *experimental design*, determines the proper analysis of variance, abbreviated ANOVA, procedure. This exploration considers the one-way ANOVA procedure for a completely randomized design.

The hypotheses to be tested are $H_o : \mu_1 = \mu_2 = \mu_3 = \cdots = \mu_k$ versus H_a: Not all the means are equal (that is, at least two of the means $\mu_1, \mu_2, \mu_3, \ldots, \mu_k$ are unequal) where μ_i is the mean of the ith population and k is the number of populations. In one-way ANOVA, the variation between the sample means is measured by a weighted average of the squared deviations about the mean of the combined sample data, $SSTR/(k-1)$. The measure of variation within the samples is the pooled estimate of the assumed common population variance and is denoted by $SSE/(n-k)$, where n is the total number of data values from all the samples. An F test statistic equal to $SSTR/(k-1) \div SSE/(n-k)$ and its p-value are used to determine if the alternative hypothesis is statistically significant. If the null hypothesis is true, the numerator and denominator of the F statistic should be approximately the same. Therefore, large values of F indicate the null hypothesis of equal population means should be rejected.

The ANOVA procedure is based on the assumptions that the randomly selected samples are independent of one another, each of the populations is normally distributed, and the variances of the populations are equal. Provided the number of values in each sample is not too small, program NQPLT can be used with each sample to see if the data look reasonably normal. Residual analysis similar to that used in linear regression can be employed to check the validity of the other two conditions.

Program 3.29, ANOVA, performs the lengthy computations of the one-way analysis of variance procedure. Program 3.28 gives subroutine ZFAREA that calculates areas associated with the F probability distribution using an approximating formula [1] based on the standard normal distribution.

```
 PROGRAM: ZFAREA
:9N→N:9D→D
:((2/D-1)F^(1/3)+(1-2/N))/
: √((2/D)F²^(1/3)+2/N)→z
:ZAREA
:Return
```

Program 3.28.

```
 PROGRAM: ANOVA
:Input "number samples?",K
:0→T:0→A:0→B:0→E:1→C
:Disp "sample information?"
:Lbl L
:Prompt N,xbar,s
:N+T→T
:xbar→x
:N x+A→A
:N x²+B→B
:(N-1)s²+E→E
:IS>(C,K)
:Goto L
:B-A²/T→SSTR
:T-K→D
:K-1→N
:SSTR*D/(E*N)→F
:Disp "test statistic=",F
:ZFAREA
:Disp "p value=",AREA
```

Program 3.29.

Let's explore the ANOVA procedure with an example. Suppose ten questions on a standardized statistics examination are graded by each of four teaching assistants in an experiment to assess differences in grading techniques. The questions are graded by each teaching assistant in random order, and each question is scored using a grade ranging from 0 (no credit) to 10 (full credit). Do the four assistants differ in terms of mean scores assigned to the tests at the .05 level of significance?

	Teaching Assistant			
	Bob	Carol	Ted	Alice
number of questions graded	10	10	10	10
mean score	78	75	72	76
standard deviation of scores	12.35	14.62	10.34	15.80

Execute program ANOVA and input 4 for the number of samples. Next enter the sample information for each of the four teaching assistants. The F test statistic value is displayed as .3459. Since the p-value = .2056 is not smaller than $\alpha = .05$, the evidence is not strong enough to reject the null hypothesis of equal population means.

If you are given the sample data instead of the summary statistics, enter the data as k lists and compute the summary statistics for each sample with OneVar before executing program ANOVA.

Explorations with Regression and Analysis of Variance

1. The data in Table X, reported by the National Center for Health Statistics, lists years of life expected at birth. [10]

   a) Construct a scatter plot for the year versus (all) male life expectancy at birth. The year in which the data is recorded should be the explanatory variable. What information does the graph for the male life expectancy reveal? It appears that as time progresses, male life expectancy increases. The general pattern seems almost linear, but a closer examination reveals that there is a larger increase than you might expect between the years 1940 and 1950. There is also interesting behavior in the '60's and '70's. Notice that between 1960 and 1965 there was very little growth in years of life expected at birth for males. What historical events would account for these two deviations from a general straight line pattern for this data?

   b) Look now at the scatter plot for the year in which the data was recorded versus (all) female life expectancy. The overall pattern does not seem to be as linear as the data for males. Does it appear that there is more than one linear pattern? Between 1982 and 1989 there has been very little increase in female life expectancy. Discuss possible reasons for this behavior of the data.

Table X. Years of life expected at birth.

Year	All Males	All Females	White Males	White Females	Black & Other Males	Black & Other Females
1920	53.6	54.6	54.4	55.6	45.5	45.2
1930	58.1	61.6	59.7	63.5	47.3	49.2
1940	60.8	65.2	62.1	66.6	51.5	54.9
1950	65.6	71.1	66.5	72.2	59.1	62.9
1960	66.6	73.1	67.4	74.1	61.1	66.3
1965	66.8	73.7	67.6	74.7	61.1	67.4
1970	67.1	74.7	68.0	75.6	61.3	69.4
1971	67.4	75.0	68.3	75.8	61.6	69.8
1972	67.4	75.1	68.3	75.9	61.5	70.1
1973	67.6	75.3	68.5	76.1	62.0	70.3
1974	68.2	75.9	69.0	76.7	62.9	71.3
1975	68.8	76.6	69.5	77.3	63.7	72.4
1976	69.1	76.8	69.9	77.5	64.2	72.7
1977	69.2	77.2	70.2	77.9	64.7	73.2
1979	70.0	77.8	70.8	78.4	65.4	74.1
1980	70.0	77.5	70.7	78.1	65.3	73.6
1981	70.4	77.8	71.1	78.4	66.1	74.4
1982	70.9	78.1	71.5	78.7	66.8	75.0
1983	71.0	78.1	71.7	78.7	67.2	74.3
1984	71.2	78.2	71.8	78.7	67.4	75.0
1985	71.2	78.2	71.9	78.7	67.2	75.0
1986	71.3	78.3	72.0	78.8	67.2	75.1
1987	71.5	78.4	72.2	78.9	67.3	75.2
1988	71.4	78.3	72.1	78.9	67.4	75.5
1989	71.8	78.5	72.6	79.1	67.5	75.7

If you are a male:

c) Construct a scatter plot of year (x) versus white male life expectancy (y). Discuss any pattern or deviations from a pattern that you observe. Store the graph. Now construct a scatter plot of year (x) versus black and other male life expectancy (y). Discuss any pattern or deviations from a pattern that you see.

d) Recall the scatter diagram for white male life expectancy so that it is overdrawn on the scatter diagram for black and other male life expectancy. (If it is difficult to distinguish the data points, use the command xyline instead of Scatter.) Do you see any pattern relating white male life expectancy to black and other male life expectancy? If so, are there any significant deviations from that pattern? Comments?

If you are a female,
Answer parts c) and d) with the word *female* substituted for *male*.

  e) Use program LSLINE to find the regression line for the data for your sex and race. Carefully observe the residual graph. Discuss what the residual graph tells you about the adequacy of the linear model.

  f) Find the best-fitting regression model your calculator has to offer for the data for your sex and race. Predict the life expectancy in the year 1992. Predict the life expectancy in the year 2005. Do the values seem reasonable?

2. Are state general sales and use taxes randomly distributed by region or do the regions differ more than random sorting would suggest? Perform an analysis of variance using the data [6] in Table XI to see if the mean state general sales and use tax differs across the five geographical regions. (Local and county taxes are not included.)

Table XI. State general sales and use taxes (% rate).

Northeast		Southeast		Central		Southwest		West	
CT	8	AL	4	IL	6.25	AZ	5	AK	0
DE	0	FL	6	IN	5	AR	4.5	CA	4.75
ME	5	GA	4	IA	4	CO	3	HI	4
MA	5	KY	6	KS	4.25	NM	5	ID	5
NH	0	LA	4	MI	4	OK	4.5	MT	0
NJ	7	MD	5	MN	6	TX	6.25	NV	0
NY	4	MS	6	MO	4.225			OR	0
PA	5	NC	3	NE	5			UT	5
RI	7	SC	5	ND	5			WA	6
VT	5	TN	5.5	OH	5			WY	3
WV	6	VA	3.5	SD	4				
				WI	5				

3.5. Programming Notes

1. Before you enter the programs in this chapter, you will find it helpful to create a custom menu of frequently-used symbols. From the home screen, press [CATALOG] and move the arrow to the item you wish to copy to your custom menu. You can find the item by scrolling with ▼ or by typing the first letter of the name of the item to quickly access the names beginning with that letter. Once you have located

the desired instruction or function, press $\boxed{\text{CUSTM}}$ followed by $\boxed{\text{F1}}$ to copy the instruction or function to the custom menu in location F1. Find the next instruction or function, and press $\boxed{\text{F2}}$ to copy it to the second location in your custom menu, etc. Once the first five custom menu positions are filled, $\boxed{\text{MORE}}$ will access the next five. You can enter 15 custom menu items. The following instructions and functions are suggested for inclusion in your custom menu:

Input	Disp	Goto	Lbl	"
rand	iPart	Hist	If	>
FnOff	Scatter	OneVar	ClDrw	dimL

To copy any of these items to a program line or the home screen, press $\boxed{\text{CUSTOM}}$ followed by the F key where the item is stored.

2. Programs are entered and edited in the PRGM EDIT application. Each line of a program must begin with a colon : that is supplied by the TI-85 when you press $\boxed{\text{ENTER}}$. Programs are executed from the PRGM NAMES application.

3. When typing words or names consisting of several letters, you can set the TI-85 to upper-case alpha lock with the keystrokes $\boxed{\text{ALPHA}}$ $\boxed{\text{ALPHA}}$. Pressing $\boxed{\text{ALPHA}}$ unlocks the alpha cursor. If you wish to type in words consisting of lower-case letters, activate the lower-case alpha lock with $\boxed{\text{2nd}}$ $\boxed{\text{ALPHA}}$ $\boxed{\text{ALPHA}}$ and unlock with $\boxed{\text{ALPHA}}$ $\boxed{\text{ALPHA}}$.

4. The upper-case alpha lock is automatically active following $\boxed{\text{STO>}}$, $\boxed{\text{STO>}}$ $\boxed{\text{dimL}}$, $\boxed{\text{STAT}}$ $\boxed{\text{EDIT}}$, $\boxed{\text{PRGM}}$ $\boxed{\text{EDIT}}$, $\boxed{\text{CATALOG}}$, $\boxed{\text{ : }}$, $\boxed{\text{STPIC}}$, and $\boxed{\text{RCPIC}}$. It is always a good idea to look carefully at each program line as you press the calculator keys to be certain the proper symbols appear on the display screen.

5. The question mark character $\boxed{\text{ ? }}$ key and the apostrophe $\boxed{\text{ ' }}$ key are found in the $\boxed{\text{CHAR}}$ $\boxed{\text{MISC}}$ menu. The symbols μ, σ and λ are located in the $\boxed{\text{CHAR}}$ $\boxed{\text{GREEK}}$ menu. When you wish to insert a space between words in a displayed statement, use the space character, ⊔ ($\boxed{\text{2nd}}$ $\boxed{\text{(-)}}$).

6. To use a program as a subroutine in another program, enter the name of the program on the indicated line with $\boxed{\text{PRGM}}$ followed by the menu key giving the subroutine name.

7. Other than user-defined names, some upper and lower-case symbols reference special instructions in the TI-85. For instance, N represents any value stored in the N memory location whereas n represents the sample size when calculating one-variable statistical results. If your

program will not execute properly, check that you have used the proper upper-case and lower-case symbols.

8. Do not confuse the equals sign on the keyboard ($\boxed{\text{APLHA}}$ $\boxed{=}$) with the relational operator $\boxed{= =}$ found in $\boxed{\text{TEST}}$. The equals sign $\boxed{=}$ is used to assign values to variables or equations while the relational operator $\boxed{= =}$ is used to test a condition.

9. Whenever one program is similar to another, you can type in the new program name in the PRGM EDIT mode and then recall the contents of the similar program with $\boxed{\text{RCL}}$ followed by the existing program name. Edit the steps to those for the new program.

10. When one programming line is similar to another, position the cursor on the line you have already entered and press the program edit menu key $\boxed{\text{DELc}}$. Press $\boxed{\text{UNDEL}}$ to reinsert that line, move to the new program line, and press $\boxed{\text{UNDEL}}$ again to copy the line to the new position. Edit the copied line to the proper symbols.

11. If the same group of symbols appears on a single program line, you can copy a block of symbols within a line using $\boxed{\text{UNDEL}}$. For instance, to type in

: iPart 2rand + iPart 2rand + iPart 2rand + iPart 2rand + iPart 2rand

enter iPart 2rand + and press $\boxed{\text{DELc}}$ followed by $\boxed{\text{UNDEL}}$. Move the cursor to the end of the simulation line and once more press UNDEL. Repeat this process three more times. Press $\boxed{\blacktriangleright}$ $\boxed{\text{DEL}}$ to eliminate the extra + symbol at the end of the line and to eliminate the blank lines.

References

1. Derenzo, S., *Mathematics of Computation*, Vol. 31, No. 137, January 1977.

2. Fetta, I., *Explorations in Algebra, Precalculus and Statistics: A Manual for the TI-81 Graphing Calculator*, HBJ/Saunders, Philadelphia, PA, 1992.

3. Fetta, I., *Calculator Enhancement for Introductory Statistics: A Manual of Applications using the Sharp EL-5200, HP-28S and HP48S Graphing Calculators*, HBJ/Saunders, San Diego, CA, 1992.

4. Moore, D., *Statistical Concepts and Controversies*, 3rd edition, W. H. Freeman & Company, New York, 1991.

5. Moore, D. and G. McCabe, *Introduction to the Practice of Statistics*, W. H. Freeman & Company, New York, 1989.

6. *The 1992 Information Please Almanac, Atlas and Yearbook*, 45th edition, Houghton Mifflin Company, Boston, 1991.

7. Weiss, N. and M. Hassett, *Introductory Statistics*, 3rd edition, Addison-Wesley Publishing Company, Reading, MA, 1991.

8. Triola, M., *Elementary Statistics*, 4th edition, Benjamin Cummings Publishing Company, Inc., Redwood City, CA, 1989.

9. Wickes, W. C., *HP-28 Insights, Principles and Programming of the HP-28C/S*, Larken Publications, Corvallis, Oregon, 1988.

10. *World Almanac and Book of Facts*, Press Publishing Company, New York, 1990.

4 Explorations in Calculus I

John Harvey
University of Wisconsin–Madison

John Kenelly
Clemson University

For several years now, changes in the initial calculus course have been advocated. Most of the changes proposed are intended to help students better understand the concepts of calculus, to improve student problem solving, and to assist students in attacking and solving "real-world" or applications problems – the kinds of problems that many of them will encounter in their college majors and in their professions whether they be engineers, accountants, biologists, physicists, economists, or whatever. To make the changes proposed it is necessary that the paper and pencil skills and techniques we have taught for so long be de-emphasized or eliminated from the calculus curriculum. With the introduction of powerful, hand-held calculating tools like the TI-85, it seems possible to realize these goals.

This chapter focuses on the elementary concepts of calculus: limits and continuity, interpretations of the derivative and the integral, applications of the derivative and integral, and numeric integration. Chapter 5 considers topics that can be learned once the basic material of the calculus has been learned. We hope that this chapter and Chapter 5 will help both students and teachers to see ways in which learning and teaching calculus can be enriched by appropriate, constant use of the TI-85.

Chapter 4 is divided into seven sections. Section 4.1, Limits and Continuity, is an exploration of the ways in which the graphing capabilities of the TI-85 can be used to help find the limit of a function at a point and, having found it, to suggest ways to deductively prove that the limit exists and has a particular value. The "Squeeze Theorem" is used and is demonstrated graphically here.

Section 4.2 explores rates and rates of rates; that is, the first and second derivative of functions and the interpretations of those derivatives.

Explorations with the Texas Instruments TI-85
John G. Harvey and John W. Kenelly (eds.), pp. 135–188
©1993 by Academic Press, Inc.
ISBN: 0-12-329070-8

Two problems are explored in this section: cyclic rates and sequences of numbers. The first problem shows ways to use the graphing and derivative capabilities of the TI-85 to discover trends. The second problem looks at the ways in which calculus can be used to explore sequences of numbers.

Section 4.3 demonstrates ways in which the TI-85 can be used to analyze the shape of a curve. In this section many of the functions in the TI-85's **CALC** and **GRAPH MATH** menus are discussed and demonstrated; these functions include ones that find the (local) extrema, the inflection points, and the zeros of a function and that will graph the tangent line to the curve and find the derivative of a function at a point.

The next section, Section 4.4, uses the TI-85 functions discussed in the Section 4.3 to further explore derivatives and important features of graphs. Topics discussed in this section are: the development of a "good graph," a generalized related rates problem, and an optimization problem.

In Sections 4.5 and 4.6 the integral of a function is interpreted as an average and an area. The integral as the average value of a function is discussed in Program 4.1 – a guessing game that shows the graph of a function and asks you to estimate the total area bounded by it and the x-axis. Contrasts between the integral as an average and an area are discussed while determining the area of the windows that cover the end of an airport terminal. These two sections conclude with programs that produce the lower- and upper-averages, the trapezoid estimate, the midpoint estimate, and the Simpson Rule estimate.

The concluding section, Section 4.7, discusses ways in which both the graphing and numeric integration capabilities of the TI-85 can be used to find the area bounded between two curves and to find volumes of revolution. In this section, Program 4.5 evaluates an improper integral by finding the value of the integral at a finite sequence of points and estimating the limit.

Each section concludes with some explorations that should further help you to understand ways in which the TI-85 can be a good tool while solving calculus problems. However, we urge you to develop and explore your own functions and to use the strategies we develop here to solve you own calculus problems.

4.1. Limits and Continuity

The informal definitions that typically have been used for limits and continuity are phrased in terms of the graph of the function. That is, we have interpreted

$$\lim_{x \to a} f(x) = L$$

in this way: As x approaches a, the graph of $f(x)$ approaches the graph of the line $y = L$. Similarly, we have interpreted continuity as the ability to traverse the graph of f without lifting our pencil from the graph. Thus, the TI-85 can be useful in helping us to determine whether or not a function has a limit at a point and whether or not a function is continuous on some interval. To explore this we will first consider the function

$$f(x) = \sin\frac{1}{x}.$$

This function is not defined and has no limit at $x = 0$. Graphing $\sin\frac{1}{x}$ using a succession of smaller and smaller viewing rectangles can help us to see that there is no limit at $x = 0$. Begin by pressing $\boxed{\text{RANGE}}$ $\boxed{y(x)=}$ and entering the function $\sin\frac{1}{x}$ into the function list. Set the **RANGE** values as: **xMin** $= -\pi$, **xMax** $= \pi$, **xScl** $= \pi/4$, **yMin** $= -1.5$, **yMax** $= 1.5$, **yScl** $= .5$ and press $\boxed{\text{GRAPH}}$; the graph that should appear is shown in Figure 4.1.

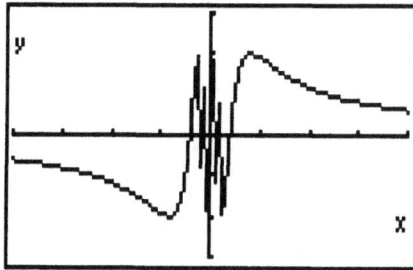

Figure 4.1. Graph of $f(x) = \sin\left(\frac{1}{x}\right)$
in $[-\pi, \pi] \times [-1.5, 1.5]$.

This graph shows that $\sin(1/x)$ is changing rapidly in the interval from $-\pi/4$ to $\pi/4$, but it does not show that the limit at $x = 0$ does not exist. Perhaps all we need to do is to zoom-in on the graph of f, and then we will see that the limit exists. Since the *sine* function always has values between -1 and 1, we want to keep the y-range values constant while making the x-range smaller and smaller. To do this easily we will set the **ZFACT** values so that each time the x-range is one-half its last value and the y-range values remain unchanged; these keystrokes on the TI-85 will halve the x-range and keep the y-range constant: $\boxed{\text{ZOOM}}$ $\boxed{\text{MORE}}$ $\boxed{\text{MORE}}$ $\boxed{\text{ZFACT}}$ **xFact** $= 2$, **yFact** $= 1$. Having done this press $\boxed{\text{ZOOM}}$ $\boxed{\text{ZIN}}$; the TI-85 screen should be as shown in Figure 4.2.

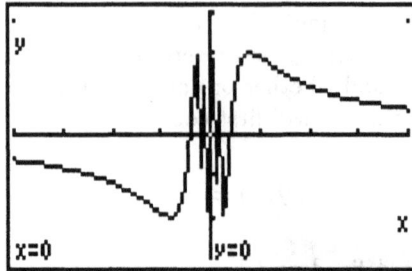

Figure 4.2. Before the first zoom-in.

Now press [ENTER], and the graph of $\sin(1/x)$ in the viewing rectangle $[-\pi/2, \pi/2] \times [-1.5, 1.5]$ will appear. Pressing [ENTER] again will graph the function in the interval $[-\pi/4, \pi/4]$. Each time you press [ENTER] a new graph will appear; each time the interval will be half the length it was before. Figures 4.3, 4.4, and 4.5 show the graph of $\sin(1/x)$ in the intervals $[-\pi/4, \pi/4]$, $[-\pi/8, \pi/8]$, and $[-\pi/16, \pi/16]$, respectively. Continuing to zoom-in will reveal that the graph of $\sin(1/x)$ looks slightly different each

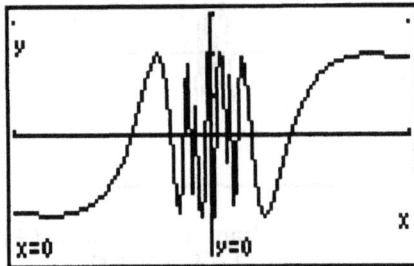

Figure 4.3. $f(x) = \sin(1/x)$ in the
interval $[-\pi/4, \pi/4]$.

time "away from zero" but that near $x = 0$, the graph changes often and appears to be somewhat irregular and chaotic. A part of this irregularity is attributable to the TI-85 since it plots and connects a finite number of points. But the calculator behaves in this way because, in some sense, plotting $\sin(1/x)$ in any interval around zero is compressing into that interval the graph of the function as $|x| \to \infty$ as a result $\sin(1/x)$ can have no limit at $x = 0$ because the function oscillates between $y = \pm 1$. Another way to see this would be to press [TRACE] and then to trace the points that were plotted. If you do this, be sure to look at the values of y as you change the values of x by pressing either the left or right cursor keys.

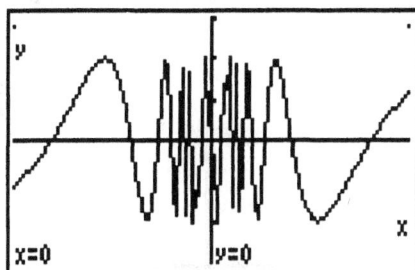

Figure 4.4. $f(x) = \sin(1/x)$ in the
interval $[-\pi/8, \pi/8]$.

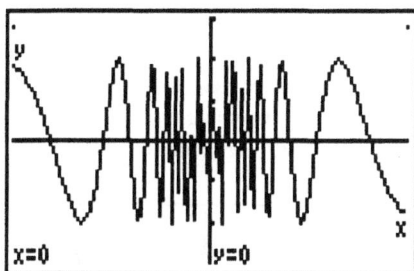

Figure 4.5. $f(x) = \sin(1/x)$ in the
interval $[-\pi/16, \pi/16]$.

Functions can fluctuate a great deal near a point and still have a limit
there. Graphs of a function and of functions that bound it can help us (a)
to see that the limit exists at a point and to find it and (b) to find ways
to prove that the limit exists there. As an example, we will explore the
function

$$f(x) = x \sin \frac{1}{x}.$$

This function, like $\sin(1/x)$, is not defined at $x = 0$, but unlike $\sin(1/x)$ it
has a limit of 0 at $x = 0$; that is,

$$\lim_{x \to 0} x \sin \frac{1}{x} = 0.$$

Begin by entering this function into the function list and graphing it in the
viewing rectangle $[-\pi/2, \pi/2] \times [-\pi/2, \pi/2]$; the graph in Figure 4.6 is the
one that should appear when $\boxed{\text{GRAPH}}$ is pressed.

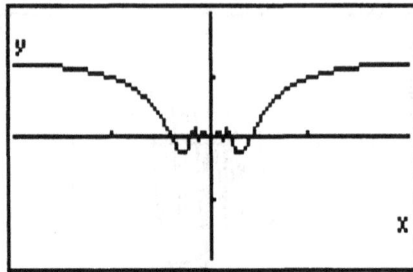

Figure 4.6. Graph of $f(x) = x\sin(1/x)$
in the interval $[-\pi/2, \pi/2]$.

To see that the limit of $x\sin(1/x)$ is probably 0 at $x = 0$, we will use the zoom-in technique employed with $\sin(1/x)$, except that this time we will decrease both the x- and the y-ranges by a factor of 2; that is, each time we will halve the length of both intervals. In order to do this, set **xFact** $= 2$ and **yFact** $= 2$. Figure 4.7 shows the result of zooming-in until $x\sin(1/x)$ is graphed in the interval $[-\pi/16, \pi/16]$.

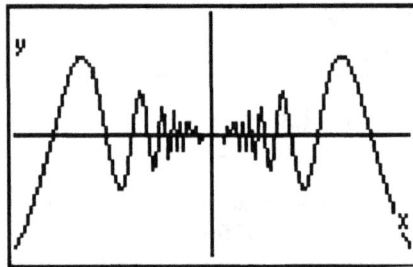

Figure 4.7. Graph of $f(x) = x\sin(1/x)$
in the interval $[-\pi/16, \pi/16]$.

Each time we zoom-in the new graph looks very much like the previous one, especially near the origin; we can look at the previous graph by pressing [QUIT] [ZPREV] and return to the present graph by pressing [ZPREV] again.

The function $f(x) = x\sin(1/x)$ is bounded by the lines $y = \pm x$. It is easy to show this. Return to the $\mathbf{y(x)}=$ menu and put both x and $-x$ on the function list. Having done that press [EXIT] [MORE] [FORMT] and place the TI-85 in the **SimulG** mode. Now when [GRAPH] is pressed, you will see our original function "bouncing" back and forth between the two lines. Since both $y = x$ and $y = -x$ have a limit of 0 at $x = 0$ and since these functions bound $x\sin(1/x)$, we have proved that the limit of our function at

$x = 0$ is 0 using the result that most calculus textbooks call "The Squeeze Theorem."

This example also shows how we can rephrase the usual definition of the limit of a function at a point; that definition is usually phrased in this way:

> **Definition.** The limit of a function f at a is L if and only if given any $\epsilon > 0$, there exists $\delta > 0$ such that $|f(x) - L| < \epsilon$ whenever $|x - a| < \delta$.

Equivalent expressions for $|f(x) - L| < \epsilon$ and $|x - a| < \delta$ are $L - \epsilon < f(x) < L + \epsilon$ and $a - \delta < x < a + \delta$, respectively. If **yMin** $= L - \epsilon$ and **yMax** $= L + \epsilon$, then δ is a positive number such that for **xMin** $= a - \delta$ and **xMax** $= a + \delta$, each point of $(x, f(x))$ is in the rectangle $[a - \delta, a + \delta] \times [L - \epsilon, L + \epsilon]$. Thus, for the function $f(x) = x \sin(1/x)$, for each $\epsilon > 0$, we have shown that $\delta = \epsilon$ will satisfy the required condition. This kind of exploration will work equally well for other functions as the next example shows.

The usual way of computing the derivative of the *sine* is first to find the value of the derivative of the *sine* at 0. To compute the derivative at 0, we must compute:

$$\lim_{x \to 0} \frac{\sin x}{x}.$$

Begin this exploration by placing $\sin(x)/x$ on the function list and setting the **RANGE** values to: **xMin** $= -\pi/2$, **xMax** $= \pi/2$, **yMin** $= .5$, and **yMax** $= 1.5$. Before pressing $\boxed{\text{GRAPH}}$ turn **AxesOff** in the **FORMT** menu. When you press $\boxed{\text{GRAPH}}$, you should see the graph shown in Figure 4.8.

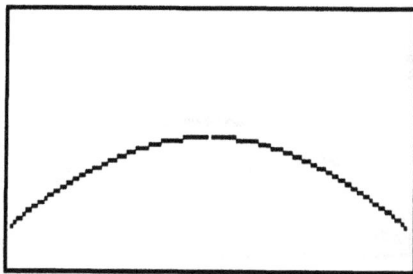

Figure 4.8. Graph of $f(x) = \frac{\sin x}{x}$
in $[-\pi/2, \pi/2) \times [.5, 1.5]$.

Since the x- and y-axes have been "turned off," this graph clearly shows that $\sin(x)/x$ is not defined at $x = 0$ since the point in the center of the screen has coordinates $(0, 0)$.

Our function, $\sin(x)/x$ is bounded above by $y = 1$; to see this, enter $y = 1$ into the function list and press $\boxed{\text{GRAPH}}$ again. This graph also shows that $\sin(x)/x$ appears to be tangent to $y = 1$ at $x = 0$; thus, we hypothesize that the derivative of *sine* is 1 at $x = 0$. In order to show that our hypothesis is true, we need determine for each $\epsilon > 0$ a $\delta > 0$ so that when we graph $\sin(x)/x$ in the viewing rectangle $[-\delta, \delta] \times [1 - \epsilon, 1 + \epsilon]$ all of the points $(x, f(x))$ are graphed. Figure 4.8 shows that if $\epsilon = 0.5$, then $\delta = \pi/2$ satisfies the condition. Table 4.1 gives a set of values for δ and ϵ that determine appropriate viewing rectangles.

Table 4.1. Values of δ and ϵ for
$$f(x) = \sin(x)/x.$$

ϵ	δ
.5	$\pi/2$
.125	$\pi/4$
.03125	$\pi/8$
.008359375	$\pi/15$

Studying this table will show that each time a current ϵ is divided by 4 to obtain the ϵ in the next row of the table, the corresponding δ is divided by 2 to obtain the δ in the next row. It is easy to see that these pairs (ϵ, δ) satisfy the "rectangle conditions" we have discussed by looking at the graphs of $\sin(x)/x$ in the specified viewing rectangles. This sequence of graphs can best be viewed by setting **xFact** = 2 and **yFact** = 4 and then using **ZIN** to look at successive graphs starting with $\epsilon = .5$ and $\delta = \pi/2$; each of these graphs should be like the one shown in Figure 4.9.

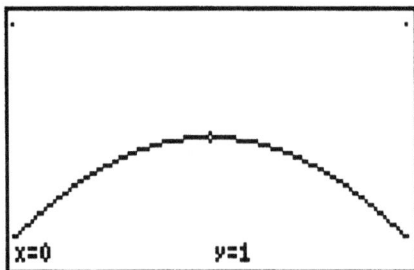

Figure 4.9. Graph of $\sin(x)/x$ in one of the $\epsilon - \delta$ rectangles.

These three explorations show that limits and continuity can be explored using graphing techniques. The next section of this chapter uses the TI-85 to explore derivatives and their interpretations.

Explorations with Limits and Continuity

1. Use graphs and the Squeeze Theorem to find L where:

$$L = \lim_{x \to 0} \frac{x^2 \cos 2x}{\sin^2 2x}.$$

   Then for $\epsilon = 0.1, 0.01$, and 0.001, find δ so that when x is in $[\mathbf{xMin} = -\delta, \mathbf{xMax} = \delta]$ then the value of the function at x is in $[\mathbf{yMin} = L - \epsilon, \mathbf{yMax} = L + \epsilon]$.

2. For $0 < x < \pi/2$, $\cos x < \frac{\sin x}{x} < \frac{1}{\cos x}$. For yMin $= .95$ and yMax $= 1.05$, find $\delta > 0$ so that when xMin $= -\delta$ and xMax $= \delta$, then all three of the functions have all of their values in the interval [yMin, yMax]. With **YFACT**= 10, find an **XFACT** so that repeated "zooming-in" at $(0,1)$ produces graphs that are, as nearly as possible, the same.

3. Use graphical techniques to find:

$$\lim_{x \to 0} \frac{\tan 3x}{\sqrt{|x|}}.$$

4. Let $f(x) = 6x^5 - 15x^4 + 30x^3 - 48x^2 + 36x - 9$; f has real zeros at $x = 1$ and $x = \frac{1}{2}$. For each of these zeros, find the slope of the tangent line to the curve at that point by using the **TANLN** command (included in the **GRAPH MATH** menu). Use zoom-in procedures to estimate the value of the slope by zooming-in at each zero until the graph of the curve looks like a straight line and then by using **TRACE** to find two points on the curve that you use to estimate the slope.

4.2. Rates and Rates of Rates

Calculus is the mathematics of *change*. From the very beginnings of calculus the measurement of *rates* and *rates of rates* has been central to the study of change. The derivative of a function measures the rate, and the second derivative measures the rate of the rate. Both of these numbers describe the dynamics of the function in the immediate neighborhood of the domain value that is being investigated. Since the **der1** command on the TI-85 evaluates the derivative and the **der2** command evaluates the second derivative, powerful calculators, like the TI-85, have created an exciting new era for studying rates and rates of rates.

In calculus courses in the past, the initial study of rates has focused on simple functions; for example, on the cycles of $\sin x$. Here we will use the power of the TI-85 to move beyond examples that yield to paper-and-pencil techniques and study sophisticated functions with an ease that was not possible before the development of the graphing calculator.

Cyclic Rates

This exploration is about a rate problem whose function is $f(x) =$ $\sin(\cos x)$. This exploration can be easily adapted to any other function by placing that function on the $y(x)=$ list as **y1**. The different rate functions, defined in **y2** and **y3**, remain unchanged and are used to explore the qualitative and quantitative aspects of the change that are present at different input values of the function stored as **y1**.

In particular, the dynamics of $f(x) = \sin(\cos x)$ at $x = 15$ are explored as follows. First, put the function $f(x) = \sin(\cos x)$ on the function list as **y1**. Then put the first derivative of **y1** in the **y2** position by pressing $\boxed{\text{CALC}}$ $\boxed{\text{der1}}$ $\boxed{\text{y}}$ $\boxed{\text{1}}$ $\boxed{\text{,}}$ $\boxed{\text{x}}$ $\boxed{\text{)}}$. A similar sequence puts the second derivative, der2, in the **y3** position on the function list; that is, press $\boxed{\text{CALC}}$ $\boxed{\text{der2}}$ $\boxed{\text{y}}$ $\boxed{\text{1}}$ $\boxed{\text{,}}$ $\boxed{\text{x}}$ $\boxed{\text{)}}$. Now store 15 as the value of x. Then find the values **y1** and of the first and second derivatives of **y1** by recalling **y1**, **y2**, and **y3** on the home screen of the TI-85. (The value of **yn** at the current value of x is found by pressing $\boxed{\text{ALPHA}}$ $\boxed{\text{y}}$ $\boxed{\text{n}}$ $\boxed{\text{ENTER}}$.) The values of **y1**, **y2**, and **y3** at $x = 15$ are: (a) $y1 = -.688\cdots$, (b) $y2 = -.471\cdots$, and (c) $y3 = .842\cdots$. Thus, at $x = 15$, the function $f(x) = \sin(\cos x)$ has a value of $-0.688\cdots$, a rate of $-0.471\cdots$ and a rate of its rate of $0.842\cdots$. These three values are the essence of the function's behavior in the immediate neighborhood of the point $(15, -.688\cdots)$. We see that the output of the function at $x = 15$ is negative, that the output would be expected to decrease with an increased input, and that this decrease would be less severe in the near term. In technical terms, at $x = 15$ we have negative output, a negative derivative (i.e., a downward slope), and a positive second derivative (i.e., the function is concave upward). These three values give, respectively, *position* (**y1**), *direction* (**y2**), and *trend* (**y3**) at any point. In colloquial terms, these three values are where you are "*at,*" where you are "*going,*" and where you are "*fixing to be going.*"

If we set $x = 100$, we have an entirely different situation, since then $y1 = 0.759\cdots$, $y2 = 0.329\cdots$, and $y3 = -0.755\cdots$. That is, we have positive output with an expected *increase* when the input is increased and an anticipated deterioration of the amount of that increase in the near term. That is, at $x = 100$, the graph will be positive, has an upward slope, and is concave downward.

If this process represented a business cycle, at 15 the output and its marginal change would be disappointing, but we could be optimistic about future improvement. At 100, a totally different state exists; the process is producing positive results that are increasing in the near term but there is cause for future concern – the nice state of affairs will deteriorate very shortly and corrective action should be applied.

This exploration shows us why many business analysts spend large amounts of time looking at local differences and differences of those differences in their attempts to analyze streams of numerical data. These business analysts are trying to determine estimates of the derivative and the second derivative. By necessity their investigations must deal with discrete sequences of data, and we will see in the next topic how calculus techniques apply to those sequences. However, the study of rates of functions like we have just described is an excellent way to understand the nature of changes in a process and the changes in those changes; that is, the immediate direction and the long term trend of the process.

Sequences of Numbers

This part of Section 4.2 will define and investigate changes in a sequence of numbers. The purpose of this exploration is to explain how you would analyze a sequence to estimate the rates of change in those numbers and the changes in those rates. In a "true application," the sequence would be observed numerical values reported over equally spaced time intervals (e.g., daily sales reports). However in this illustration, we will use the values generated by a known function, $\sin x$ and thus, see the problem in a wider context. In this more complete setting, we should fully understand both the nature of the observations and the conclusions that we reach from our analysis. After this exploration you will be prepared to enter any sequence of equally spaced data items and to apply differencing techniques to the resulting lists in order to investigate changes and trend behaviors. These first and second differences we will investigate approximate the first and second derivative, and as such, they estimate the *direction* and *trend* of the information supplied by the data.

First set the TI-85 so that it displays only two decimal digits by pressing $\boxed{\text{MODE}}$ $\boxed{\blacktriangledown}$ $\boxed{\blacktriangleright}$ $\boxed{\blacktriangleright}$ $\boxed{\blacktriangleright}$ $\boxed{\text{ENTER}}$. Now we will construct two copies of a list that contain 63 entries; i.e., the values of the $\sin x$ at increments of 0.1 over the interval $[0, 2\pi]$. The list is constructed by pressing $\boxed{\text{LIST}}$ $\boxed{\text{OPS}}$ $\boxed{\text{MORE}}$ $\boxed{\text{seq}}$ $\boxed{\text{SIN}}$ $\boxed{\text{x}}$ $\boxed{\text{,}}$ $\boxed{\text{x}}$ $\boxed{\text{,}}$ $\boxed{\text{0}}$ $\boxed{\text{,}}$ $\boxed{\text{2}}$ $\boxed{\pi}$ $\boxed{\text{,}}$ $\boxed{\text{.}}$ $\boxed{\text{1}}$ $\boxed{\text{)}}$ $\boxed{\text{ENTER}}$;the resulting list is $\{0.00\ .10\ .20\ .30\ .3\ \cdots\ -.37\ -.28\ -.18\ -.08\}$. Store duplicate copies of the list under the names SINE and NSINE; these names reflect "sine value" and "next sine value."

In order for the second copy, NSINE, to be true to its name, we have to delete one value at the front of the list. Arithmetic operations require that the list operands be of the same length; so, we must also delete one entry from the end of SINE. When the two lists are edited so that they are both 62 entries long we will, using subtraction, be able to create a list of the differences between adjacent entries in the original SINE list. To delete the first entry from NSINE use these keystrokes: $\boxed{\text{LIST}}$ $\boxed{\text{EDIT}}$ $\boxed{\text{NSINE}}$ $\boxed{\text{ENTER}}$

DELi . A similar set of keystrokes will delete the 63rd entry of SINE except that before pressing DELi we will need to cursor down to that entry, e63. The resulting lists are:

$$SINE = \{0.00 \ .10 \ .20 \ .30 \ .3 \cdots -.46 \ -.37 \ -.28 \ -.18\}$$

and

$$NSINE = \{.10 \ .20 \ .30 \ .39 \ .48 \cdots -.37 \ -.28 \ -.18 \ -.08\}.$$

Now we can use SINE and NSINE to create a list that is the difference of successive terms; we create this new list by pressing LIST   NAMES   SINE   −   NSINE   ENTER . Store duplicate copies of this list under the names ΔSINE and NΔSINE. (The symbol Δ is found in **CHAR GREEK**.)

The two new lists, ΔSINE and NΔSINE, are supposed to represent the *rate* of change in the values, but they currently reflect only the changes in the data. We must divide the differences in the sequence values by the differences, 0.1, in the x-values to have the rate; we can accomplish this by a simple arithmetic operation on each of the two lists. To alter the ΔSINE list press LIST   NAMES ΔSINE   /   .   1   ENTER ; store the resulting list in ΔSINE. The list NΔSINE is altered in the same way.

Just as before, to calculate the second differences we must remove the first term in the NΔSINE list and the last term of ΔSINE list. Next we form and store the list of second differences under the name $\Delta\Delta$SINE by taking the differences of the terms in ΔSINE and NΔSINE and dividing each of those differences by 0.1, the increment in the domain variable; the keystrokes used to develop and alter ΔSINE are used here with the appropriate changes in the names of the lists used.

Now we are prepared to investigate the sequence of differences and the sequence of second differences. We want to look at both the values of these differences and the graphs of those values. We look first at the values at particular positions in the difference sequences; individual values in a list may be found by evaluating the list at an index (e.g., SINE(8)). We want to make a table of values for the sequences by using these evaluation commands. In general, to find the value of the nth term of a list L press: LIST   NAMES   L   (n   )   ENTER . Using the appropriate list names, these keystrokes were used to develop Table 4.2; the entries in this table are the values of the lists SINE, ΔSINE, and $\Delta\Delta$SINE for $n = 8, 24, 40,$ and 56.

Table 4.2 describes the process at list positions $8, 24, 40,$ and 56. At position 8, the output is positive, the immediately future direction is up, but the future is declining. At position 24, we have compounded pessimism (i.e., the positive output is on the decline and will decline even more in the future). At position 40, the output is negative, the immediate margin is negative, but there is hope for a future upturn. At position 56, the negative output is overshadowed by the increasing slope and even further improvement is suggested by the positive second difference.

Table 4.2. Values of SINE(n), ΔSINE(n), and $\Delta\Delta$SINE(n) for $n = 8, 24, 40$, and 56.

List	ENTRY			
	8	24	40	56
SINE	.64	.75	-.69	-,71
ΔSINE	.73	-.70	-.69	.74
$\Delta\Delta$SINE	-.72	-.67	.76	.63

Since this sequence comes from the well known *sine* function, we can graph the function and relate our analysis of the differences to points on the sine curve. The graph of the *sine* function in Figure 4.10 was generated using these **RANGE** settings: **xMin = 0, xMax = 6.5, xScl = .8, yMin = −1, yMax = 1, yScl = 1**. Table 4.3 shows the value of the f, f', and f'' and of SINE, ΔSINE, and $\Delta\Delta$SINE at the list positions $8, 24, 40$,

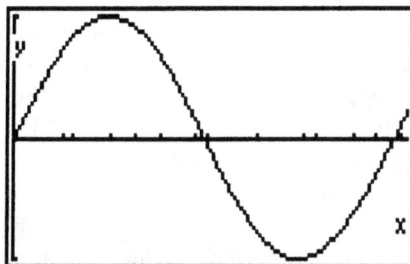

Figure 4.10. Graph of $f(x) = \sin x$
in the rectangle $[0, 6.5] \times [-1, 1]$.

and 56. We see that the ΔSINE and $\Delta\Delta$SINE approximations of the first and second derivative values, respectively, of $\sin x$ are only crude approximations of those values but that they correctly reflect the qualitative behavior of the changes and the changes in the change in the underlying set of data values.

Since real applications of this technique will involve data for which the underlying function is unknown, we need to see how to graph lists of numerical values. We will graph these values by going to the **STAT** menu and using **SCAT** after first deleting the last element in the list SINE so that it is the same length as the lists ΔSINE and $\Delta\Delta$SINE. When the last entry of SINE is deleted, all three of the lists will have 61 entries in them.

Table 4.3. Comparisons of the values of f, f', and f
and SINE, ΔSINE, and $\Delta\Delta$SINE.

	ENTRY			
x-value	8	24	40	56
x-value	.70	2.30	3.90	5.50
$\sin x$	.64	.75	-.69	-.71
SINE	.64	.75	-.69	-.71
sign	+	+	−	−
$\cos x$	.76	-.67	-.73	.71
ΔSINE	.73	-.70	-.69	.74
slope	up	down	down	up
$-\sin x$	-.64	-.75	.69	.71
$\Delta\Delta$SINE	-.72	-.67	.76	.63
concavity	down	down	up	up

In order to get scatter plot graphs of all three lists, we must first create
a list, called ENTRY, of the position numbers in the list. This list will serve
as our domain variable in each of the graphs. The list ENTRY is created
with the command $\text{seq}(x, x, 1, 61, 1)$. Next, we set the range to reflect the
list values: **xMin = 0, xMax = 62, xScl = 10, yMin = −1, yMax = 1,
yScl = 1**. Then we graph the list values by making them the *xlist* and the
ylist in statistical scatter plots; this sequence of keystrokes will graph the
SINE list: $\boxed{\text{STAT}}$ $\boxed{\text{CALC}}$ (xlist Name=) $\boxed{\text{ENTRY}}$ $\boxed{\text{ENTER}}$ (ylist Name=)
$\boxed{\text{SINE}}$ $\boxed{\text{ENTER}}$ (x=ENTRY, y=SINE) $\boxed{\text{LINR}}$ $\boxed{\text{DRAW}}$ $\boxed{\text{SCAT}}$. Draw scatter
plots of the ΔSINE and $\Delta\Delta$SINE lists by placing the names of these lists,
one at a time, in the ylist and repeating the sequence of keystrokes just
outlined. Figure 4.11 shows the result of graphing each of the lists.

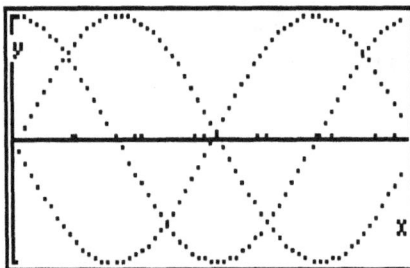

Figure 4.11. Graph of sequence, rate, and rate of rate values for $\sin x$.

Section 4.2 has shown how to study rates and rates of rates using the TI-85. The next section will demonstrate the ways in which the TI-85 can be used to find the local maximum and minimum values, inflection points, and zeros of a function and to draw the tangent line to a function of a point.

Explorations in Rates and Rates of Rates

1. Use the techniques in the part "Cyclic Rates" to investigate the local behavior of the following functions at the values of 15 and 100:

   (a) $\cos(\sin x)$

   (b) $\tan(\sin x)$

   (c) $\cos(\tan x)$

2. Make a list to examine the sequence of values of the $\cos x$ function in steps of 0.1 over the interval $[0, 2\pi]$. In the same manner as in "Changes in Sequences of Numbers" construct a chart of the changes in the rates of change and the changes in those rates of change. Graph the $\cos(x)$ function and its first and second derivatives. Confirm that the incremental analysis identifies the qualitative behavior of the changes in the function.

3. Use a list to form a sequence of values of the logistic curve

$$f(x) = \frac{1}{1 + e^{-x}}$$

in increments of 0.1 over the interval $[-1, 1]$ and confirm that an inflection point exists at $x = 0$. At the inflection point the graph of f shifts from being concave upward to being concave downward. The logistic curve is very important since it models the growth and fade of processes that grow exponentially until they start to exhaust their resources. Many products are thought to have sales patterns that parallel this model. When the product is in its initial exponential growth phase, the product manager should focus the firm's resources on production. When the growth "S's-out" into the logarithmic part of the curve, then the firm's resources are better utilized in the creation of demand; i.e., advertising. Missing this change in the demand for the product will result in excess production and an expensive growth in inventory. Tracking the daily sales data with the techniques in this section would be a important quantitative way for a manager to analyze sales data.

4. Determine the pattern of differences and second differences in quadratic, cubic, and exponential expressions. Then, classify the following data sequences as quadratic, cubic, or exponential. For a data sequence

to be quadratic, it must be a constant times a variable squared (e.g., kx^2), and for it to be cubic, it must a constant times a variable cubed (e.g., kx^3). For the data to be exponential, it must be a constant times a number raised to a power that is a constant times a variable (e.g., $k2^{cx}$).

List A

1.0	13.3
1.7	14.2
2.4	15.3
3.1	16.4
3.8	17.6
4.5	18.8

List B

1.5	415
2.0	984
2.5	1,922
3.0	3,321
3.5	5,274
4.0	7,872

List C

1.7	35,677
2.0	49,380
2.3	65,305
2.6	83,452
2.9	103,821
3.2	126,412

4.3. Curve Sketching Insights

The special capabilities of the TI-85 can be used to analyze the shape of a curve. We all know that the reason for looking at the graph of a function is to describe the behavior of that function; that is, we want to know when the function (a) is increasing and decreasing, (b) reaches a maximum or minimum, and (c) shifts from being concave up or down into being of the other concavity. These questions can be answered by single keystrokes on the TI-85; the ease of obtaining answers to these questions does not place any less importance on knowing what these answers say about the nature of the function.

Graphically Observing Behavior

To give our suggestions some substance, we will look at the graph of the *sine* curve in the **ZTRIG** viewing window (i.e., the "trig" viewing window). First, set **y1=SIN**x; then, press ZOOM MORE ZTRIG. Now, simply touching the **GRAPH** key gives us the graph of the *sine* function in the rectangle $[-8.246\cdots, 8.246\cdots] \times [-4, 4]$ shown in Figure 4.12.

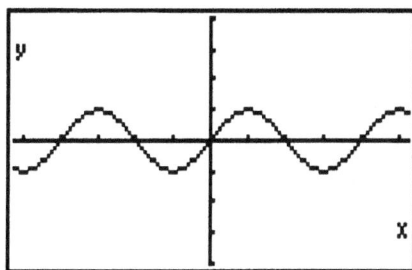

Figure 4.12. Graph of $f(x) = \sin x$.

We tend to underestimate the importance of the *sine* curve, but it happens to be one of the most important curves that we will ever study. The *sine* curve represents the "pure form" of all the cycles of nature; it is the graphical realization of how many things respond to external stimuli. For example, when you strike a tuning fork, or any other object, the object vibrates, and the wave form of the vibration is the *sine* curve. Just as different forces make economic markets go "up and down" their behavior is in its "pure" form the *sine* wave. So the question remains, "How do those cycles change?"

We can look at the overall nature of these changes by using the TI-85 to graph the *sine* curve and its first derivative on the same axes. On the TI-85 this is done by setting **y2=der1(y1,x)**; with the cursor flashing on the **y2=** line of the function list press 2nd CALC der1 y 1 , x to do this. Now, graph both **y1** and **y2**, and both the *sine* function and its derivative, the *cosine* function, are graphed as shown in Figure 4.13. In Figure 4.13, notice the interplay of the graphs of the two functions in the interval $[0, 2\pi]$. The *sine* curve starts at the origin (i.e., $(0,0)$) and rises with a decreasing rate of increase (i.e., the values of the derivative of sine decrease) until it "caps out" at $(\pi/2, 1)$; it continues down with an increasing rate of decrease (i.e., the *cosine* continues to decrease) until it gets to $(\pi, 0)$. As the *sine* curve crosses the horizontal axis at $(\pi, 0)$, it goes through an inflection point and proceeds to fall, but with a diminishing rate of decrease until it "bottoms out" at the point $(3\pi/2, -1)$; from that point it rises to 0 at $x = 2\pi$ where the cycle starts all over again.

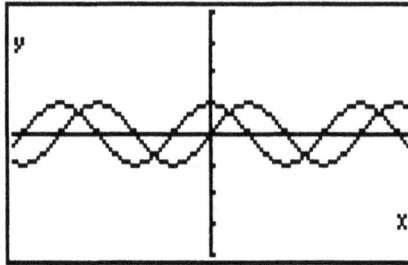

Figure 4.13. Graphs of sin x and its first derivative
in the rectangle $[-8.246 \cdots, 8.246 \cdots] \times [-4, 4]$.

A critically important observation needs to be made at this point. When the *sine* curve is going through one of its inflection points, its first derivative curve, the *cosine*, is going through one of its "cap" points; this is why business analysts look at "rate" curves. By looking at rate curves these analysts can determine when the underlying process has shifted (i.e., that the process has fundamentally changed the rate of its increase or decrease). When a shift takes place in the "cycles," the rate curve reaches a (local) maximum or minimum value. Since "cap" points are far easier to recognize than inflection points, analysts look for "cap" points in the rate curves in order to find the "shift" points in the basic process. In the language of calculus, business analysts are trying to find the inflection points in the basic curve by looking for the local maximum and local minimum points on the basic curve. We can do the same with a TI-85 and, by doing so, develop a strong intuition about the behavior of cycles.

Finding Values

Now, we will use the powers of the TI-85 to find maximum and minimum values of a function easily. To do this let's graph the function $f(x) = .1(x^3 - 25x + 2)$ in the standard rectangle (i.e., $[-10, 10] \times [-10, 10]$). Set $y1 = .1(x^3 - 25x + 2)$ and then, press $\boxed{\text{ZOOM}}$ $\boxed{\text{ZSTD}}$; the resulting graph is shown in Figure 4.14. Next, we want to define an additional interval that will control our exploration. Our additional interval is the [**UPPER, LOWER**] interval; these commands are in the **MATH** menu. Use the $\boxed{\blacktriangleright}$ and $\boxed{\blacktriangleleft}$ cursor keys to place the cursor close to $x = -7$ and then press $\boxed{\text{MATH}}$ $\boxed{\text{LOWER}}$ $\boxed{\text{ENTER}}$. A triangle pointing to the right will appear at the top of the screen. Now place the cursor close to $x = 5$ and $\boxed{\text{UPPER}}$ $\boxed{\text{ENTER}}$; a triangle pointing to the left will appear at the top of the screen; at this point the screen should be like that one shown in Figure 4.15. The left pointing triangle along with the earlier right pointing triangle mark the

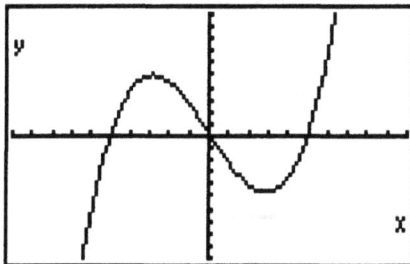

Figure 4.14. Graph of $f(x) = .1(x^3 - 25x + 2)$
in the standard rectangle.

interval of application of several very important functions – in particular,
the **FMIN, FMAX, ROOT**, and **INFLC** keys.

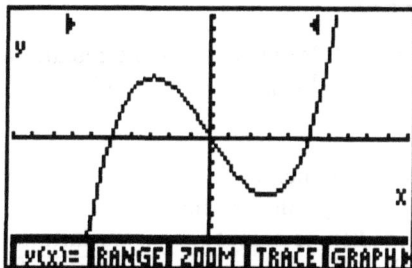

Figure 4.15. Graph of $f(x) = .1(x^3 - 25x + 2)$
with the additional interval shown.

    We will now illustrate the use of the **FMAX** and **FMIN** keys. With
the graph showing, press $\boxed{\text{MATH}}$ $\boxed{\text{MORE}}$ $\boxed{\text{FMAX}}$ $\boxed{\text{ENTER}}$. After a very brief
moment, the cursor moves to the peak of the curve in the interval $[-7, 5]$,
and the coordinates $x = -2.8898\cdots$ and $y = 5.0112\cdots$ are displayed at
the bottom of the screen as in Figure 4.16. The calculator has found the
maximum point of the graph in this interval and has displayed its coordinates. Repeat this exercise using the **FMIN** and **ENTER** keys. The
cursor moves to the minimum on the curve in $[-7, 5]$, and the calculator
displays the coordinates $x = 2.9038\cdots$ and $y = -4.6109\cdots$ of that point.
The **INFLC** key will locate the inflection point(s), and the **ROOT** key will
locate the intersections of the curve with the horizontal axis. To use the
ROOT key effectively, the interval determined by the "pointing triangles"
(cf. Figure 4.15) should include only one zero of the function. When the
pointing triangles interval contains only one zero, press the **ROOT** key,

trace along the curve until you are close to the x-intersection, and then press the **ENTER** key. Defining the pointing triangles interval this way gives the calculator a sensible starting point in its numerical search for the zero and a single goal for it to achieve. If you give the root finder two or more zeros in its search interval, it can become confused. If you do not give it a starting point in the vicinity of the root, then it will be inefficient and may be unable to isolate the zero.

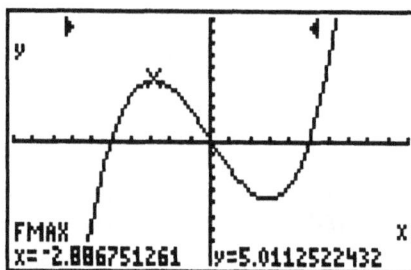

Figure 4.16. Graph showing the maximum value
of $f(x) = .1(x^3 - 25x + 2)$ in the interval $[-7, 5]$.

Tangent Lines

Perhaps one of the most helpful TI-85 keys in the study of the dynamics of a curve is the key that graphs the line tangent to a function at a point; this key, the **TANLN** key, is easy to use. To initiate drawing a tangent line at a point use the keystroke sequence: GRAPH MORE MATH MORE MORE TANLN . Then, move the cursor to a point on the curve where you want to draw a tangent line and press ENTER . There you have it – the tangent line at the point requested! Figure 4.17 shows a tangent line to $f(x) = .1(x^3 - 25x + 2)$ at $x = 3.9682 \cdots$.

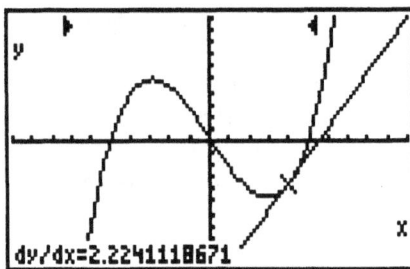

Figure 4.17. Graph of the tangent line to
$f(x) = .1(x^3 - 25x + 2)$ at $x = 3.682539683$.

With this facility and the knowledge that the tangent line indicates the local behavior of the curve, we can generate tremendous insights into the behavior of the graph. For example, for the function $f(x) = .1(x^3 - 25x + 2)$, we see that the slopes of the tangent lines are positive at all of the points on the graph whose x-coordinates are less than -2.8; thus, for these values of x the function is increasing. For the values of x between -2.8 and 2.8, the tangent lines all have negative slope, and this indicates a decreasing curve. Beyond $x = 2.8$ the slope again is positive, and the function is increasing.

Section 4.3 has shown how the TI-85 can be used to study the behavior of functions. Section 4.4 continues that exploration and focuses principally on finding the maximum and minimum values of a function.

Explorations with Maximum and Minimum Values and Tangent Lines

1. Show that the inflection points of a *cosine* curve are directly above the maximum and minimum points of the *sine* curve.

2. Use the **FMAX** key to find the local maximum of a function (take $f(x) = .1(x^3 - 25x + 2)$ in this section as one possibility) and then, draw the **TANLN** on the graph to show that the tangent line at a maximum point on a curve is horizontal. Do the same with **FMIN** and illustrate that the tangent line at a local minimum is horizontal.

3. Investigate the inflection point on the function

$$f(x) = \frac{1}{1 + e^{-x}}.$$

4.4. Maxima and Minima with the TI-85

In this section we will explore ways in which the TI-85 can be used to determine the relative maxima and minima of functions and their inflection points. Both the graphing and the computational power of the calculator will be used.

A Good Graph

If asked why they were asked to learn how to differentiate, many former calculus students might respond that they learned to differentiate so that they could find the maximum and minimum values of functions and thus, so that they could sketch good graphs of functions. If this were the only reason for finding derivatives, this part of calculus might disappear since with the TI-85, finding the zeros and the maximum and minimum values of a function can easily be approximated and examined. For example, let us explore this fifth-degree polynomial function using the TI-85:

$$f(x) = 6x^5 - 15x^4 + 30x^3 - 48x^2 + 36x - 9\,.$$

First, place the polynomial on the function list as **y1**, and graph it by pressing $\boxed{\text{ZOOM}}$ $\boxed{\text{ZSTD}}$. A portion of the graph of f will appear on the screen; however, the graph is not a very good one. It looks like a better graph might emerge if the xMin and xMax values were changed to, say, -1 and 3, respectively; the resulting graph, shown in Figure 4.18, has the characteristic shape of the graph of a fifth-degree polynomial. But how do

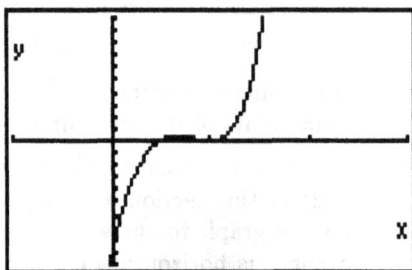

Figure 4.18. Graph of f in the
interval $[-1, 3]$.

we know that we are looking at a "good" graph of f? One criterion for a good graph is that all of the real zeros of f are shown. Using the TI-85, we can obtain approximations of both the real and the complex zeros of a polynomial function by using the **POLY** function. In order to do this for our polynomial, we must first **EXIT** the graphing screen and then press $\boxed{\text{2nd}}$ $\boxed{\text{POLY}}$. Now you are ready to enter the following values: order $= 5$, $a5 = 6$, $a4 = -15$, $a3 = 30$, $a2 = -48$, $a1 = 36$, $a0 = -9$. Once these values have been entered, press $\boxed{\text{SOLVE}}$, and the TI-85 will approximate the zeros of f. Rounded to five decimal places, the zeros of the polynomial are:

$$x_1 = 1.73205i$$
$$x_2 = -1.73205i$$
$$x_3 = 1 \qquad\qquad \text{(a double zero)}$$
$$x_4 = 1/2$$

Thus, in the interval $[-1, 3]$ all of the real zeros of f are shown.

Another criterion for a good graph is that the values of all of the relative extrema (i.e., the relative maxima and minima) are shown. Another look at our graph in the rectangle $[-1, 3] \times [-10, 10]$ shows us that the relative extrema occur somewhere on the interval $[0, 1.5]$ and have relatively small y values. Thus, we look more closely at that portion of the graph of f by changing the rectangle to $[.25, 1.25] \times [-.5, .5]$ and setting $\mathbf{xScl} = .25$ and $\mathbf{yScl} = .25$; the resulting graph is shown in Figure 4.19. To approximate the x- and y-coordinates of the relative maximum of f, press $\boxed{\text{MORE}}$ $\boxed{\text{MATH}}$ $\boxed{\text{MORE}}$ $\boxed{\text{FMAX}}$, move the cursor to the neighborhood of the relative maximum, and press $\boxed{\text{ENTER}}$. Figure 4.19 shows the approximate position and coordinates of the relative maximum of f. Figure 4.20 shows the approximate position and coordinates of the relative minimum of f; this graph was produced by pressing $\boxed{\text{EXIT}}$ $\boxed{\text{FMIN}}$ and moving to the neighborhood of the relative minimum before pressing $\boxed{\text{ENTER}}$. Just how good are these approximations? Is, for example, the relative minimum of f at $x = 1$ or at

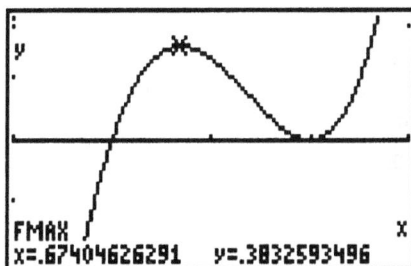

Figure 4.19. Graph of f showing the relative maximum and its coordinates.

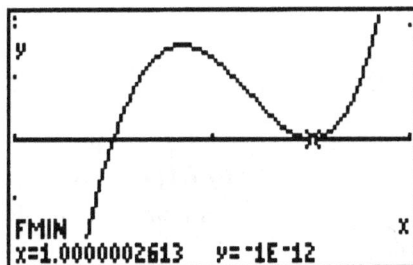

Figure 4.20. Graph of f showing the relative minimum and its coordinates.

$x = 1.0000002613$? The first derivative can help us to determine how good the TI-85 approximations are. The first derivative of f is

$$f'(x) = 30x^4 - 60x^3 + 90x^2 - 96x + 36.$$

By direct substitution it is easy to see that $f'(1) = 0$; we can also use the TI-85 to compute the value. First, enter f' as y2 on the function list. Then, from the home screen, enter these keystrokes: 1 $\boxed{\text{STO▶}}$ x $\boxed{\text{ENTER}}$ $\boxed{\text{2nd}}$ $\boxed{\text{ALPHA}}$ y2 $\boxed{\text{ENTER}}$. Since $f'(1) = 0$ there is a relative minimum of f at $x = 1$ and not at $x = 1.0000002613$ ($f'(1.0000002613) \simeq 0.000006271209$).

It is not as easy to approximate the x-coordinate of the relative maximum of f. However, the **POLY** function approximates the other real zero of f' as $x = 0.674046641091$. This value of x is close to $\frac{2}{3}$, and so, it may be that $\frac{2}{3}$ is the x-coordinate of the relative maximum. But $f'(\frac{2}{3}) \simeq 0.148148148149$ or $\frac{4}{27}$. (After computing the value of $f'(\frac{2}{3})$, press $\boxed{\text{2nd}}$ $\boxed{\text{MATH}}$ $\boxed{\text{MISC}}$ $\boxed{\text{▶Frac}}$ $\boxed{\text{ENTER}}$ to see that $f'(\frac{2}{3}) = \frac{4}{27}$.)

The last feature of f that we will examine is the position of its inflection point. There will be only one inflection point since f has only one relative maximum and one relative minimum; the single inflection point will occur somewhere between the x-coordinates of the relative maximum and relative minimum (i.e., in the interval $[.674046641091, 1]$). There are two ways to locate the x-coordinate of the inflection point. One way would be to graph f in the rectangle $[.75, 1] \times [-.25, .75]$ and then to press $\boxed{\text{MORE}}$ $\boxed{\text{MATH}}$ $\boxed{\text{MORE}}$ $\boxed{\text{INFLC}}$ $\boxed{\text{ENTER}}$. Figure 4.21 shows the x- and y-coordinates and the location of the inflection point of f.

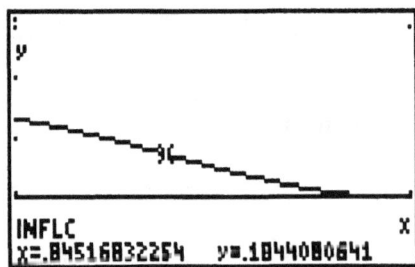

Figure 4.21. The inflection point of f
in the rectangle $[.75, 1] \times [-.25, .75]$.

Another way to find the inflection point of f would be to find the real zeros of f''. The second derivative of f is

$$f''(x) = 120x^3 - 180x^2 + 180x - 96.$$

Use the **POLY** function to find the zeros of f''.

This exploration shows that finding and using the derivatives of a function is still important but that many of the computational procedures we have previously used to help us distinguish relevant features of a function are less important than they were. The next exploration shows how we can use the computational power of the TI-85 to explore problems more completely.

A Precarious Ladder

One of the problems that always seems to appear in introductory calculus textbooks can be called the "sliding ladder problem"; it is a related rates problem that asks the student to find the rate at which one end of the ladder is sliding when the distance of the foot of the ladder from a wall, the length of the ladder, and the velocity of the other end of the ladder is known. Here is a generalization of this problem that can be explored with the TI-85.

A 13-foot ladder is leaning against a wall so that the foot of the ladder is 1 foot from the wall. A gust of wind causes the ladder to begin sliding down the wall; the motion of the end of the ladder resting against the wall is described by:

$$y = -16t^2 + .05t + \sqrt{168}.$$

Describe the velocity and the acceleration of the end of the ladder that is resting on the ground. In particular, does the end of the ladder resting on the ground have a maximum velocity? If it does, where does this maximum occur?

Since $x^2 + y^2 = 13$, we know that

$$2x\frac{dx}{dt} + 2y\frac{dy}{dt} = 0.$$

Thus, the velocity of the foot of the ladder is given by

$$\frac{dx}{dt} = -\frac{y}{x}\frac{dy}{dt},$$

and the acceleration of the foot is described by

$$\frac{d^2x}{dt^2} = -\frac{y}{x}\left(\frac{d^2y}{dt^2}\right) - \frac{dy}{dt}\left(\frac{x\frac{dy}{dt} - y\frac{dx}{dt}}{x^2}\right).$$

To explore this problem we must first place the TI-85 in the **Param** mode; to do this press $\boxed{\text{2nd}}$ $\boxed{\text{MODE}}$, press $\boxed{\blacktriangledown}$ four times, press $\boxed{\blacktriangleright}$ three times, and press $\boxed{\text{ENTER}}$. Now press $\boxed{\text{GRAPH}}$ and $\boxed{\text{E(t)=}}$ and enter these functions into the parametric function list:

xt1=t
yt1=−16t²+.05t+√168

xt2=t

yt2=$\sqrt{(13^2\text{-yt1}^2)}$

xt3=t

yt3=(yt1/yt2)*der1(yt1,t)

xt4=t

yt4=(-yt1/yt2)der2(yt1,t)-der1(yt1,t)*
$$((\text{yt2*der1(yt1,t)-yt1*der1(yt2,t)})/\text{yt2}^2).$$

The first of these functions represents the position of the top of the ladder as it slides down the wall; the second function represents the distance of the base of the ladder from the wall. The third function is the velocity of the base of the ladder as it moves away from the wall. And, the last function represents the acceleration of the base of the ladder. Figure 4.22 shows the graphs of the first and second functions using the **RANGE** values: **tMin = 0, tMax = .9106, tStep = .01, xMin = 0, xMax = .9106, xScl = .1, yMin = 0, yMax = 15, yScl = 1**. The graph of the velocity of

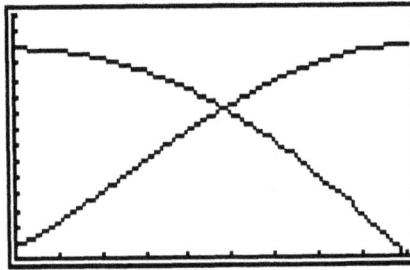

Figure 4.22. Graph of the position of the
top and base of the ladder as functions of time.

the foot of the ladder as a function of time is shown in Figure 4.23, and the graph of the acceleration of the ladder is shown in Figure 4.24. The y intervals in Figures 4.23 and 4.24 are $[0, 20]$ and $[-75, 425]$, respectively.

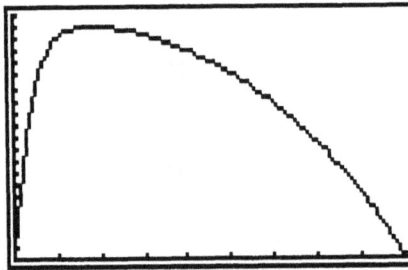

Figure 4.23. Graph of the velocity of the foot of the ladder.

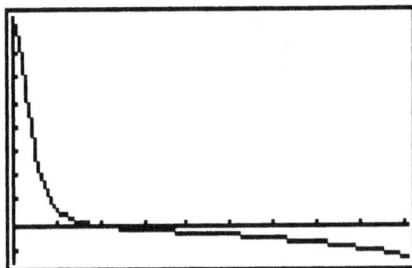

Figure 4.24. Graph of the acceleration of the foot of
the ladder.

From these graphs we see that the foot of the ladder has a maximum
velocity. We can find the t-coordinate of the velocity using the **SOLVER**.
To find that coordinate press 2nd SOLVER yt4 ENTER ; then, set **exp**
equal to zero, place the cursor by **t=**, and press ENTER . After a few
seconds the TI-85 will compute that the t-coordinate is 0.1858452983459
when the acceleration of the foot of the ladder is zero. To find the maximum
velocity, we can use almost the same procedure except that this time we set
exp equal to yt3 and place the cursor after **exp** before pressing ENTER .
This time the value of the velocity at $t = 0.1858452983459$ is computed;
that value is 19.042212703812.

Getting the Most Out of a Log

Also among the problems that always appear in introductory calculus
books are problems in which you are asked to maximize or minimize a
function. Here is one of those problems.

A round log that has a radius of 1 foot is going to be cut into five
pieces. One of those pieces will be a square beam whose sides
are $\sqrt{2}$. The other four pieces will each be identical boards; each
one will be cut from one of the four pieces of the log that results
when the square beam is cut out of the center of the log. The
end of the beam and of one of the four boards is shown in Figure
4.25. We want to cut the boards so that the cross-sectional area
of the each of them is a maximum. If, we let x be the width of
the board, y be one-half the height of the board, and θ be the
angle as shown in Figure 4.25, then it is easily seen that

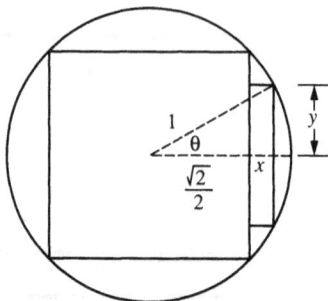

Figure 4.25. Cross-sectional picture of the beam and one
of the boards to be cut from the circular log.

$$x + \frac{\sqrt{2}}{2} = \cos\theta$$

$$y = \sin\theta$$

and thus, that the cross-sectional area of the log is:

$$A(x) = 2x\sqrt{1 - \left(x + \frac{\sqrt{2}}{2}\right)^2}$$

where $0 \le x \le 1 - \sqrt{2}/2 \simeq 0.29289\cdots$. In order to maximize A in this interval, we will find its first derivative, find where $A'(x) = 0$, and then determine whether the maximum of A occurs at one the points where $A'(x) = 0$
or at one of the end points. The first derivative of A is:

$$A'(x) = \sqrt{2}\sqrt{-2x^2 - 2\sqrt{2}x + 1} - \frac{2x(\sqrt{2}x + 1)}{\sqrt{-2x^2 - 2\sqrt{2}x + 1}}.$$

Undoubtedly, we could find the values of x at which A' is zero using paper
and pencil, but it would be easier to graph both of these functions so that
we see where A' is zero and, at the same time, see that A achieves its
maximum there. Figure 4.26 pictures the graphs of both A and A' in the
interval $[0, .3]$. To find the x-intercept of A' we will use $\boxed{\textbf{ROOT}}$; to do
this press $\boxed{\texttt{GRAPH}}$ $\boxed{\texttt{MORE}}$ $\boxed{\texttt{MATH}}$ $\boxed{\texttt{ROOT}}$. Then, if necessary, use the $\boxed{\blacktriangle}$ or
$\boxed{\blacktriangledown}$ cursor key to place the cursor on the graph of A' and the $\boxed{\blacktriangleright}$ and $\boxed{\blacktriangleleft}$
cursor keys to place the cursor near the zero of A' before pressing $\boxed{\texttt{ENTER}}$.
The x-intercept of A' is approximately 0.1985389.

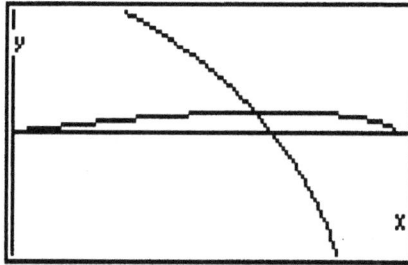

Figure 4.26. Graphs of A and A'
in the rectangle $[0, .3] \times [-1, 1]$.

Another way to find the zero of A' and the value of A at that zero would be to use the **SOLVER** as we did while looking for the maximum value of the velocity function in the preceding exploration. If you use the **SOLVER** to find the zero of A' and the maximum value of A, first find the x-intercept of A' and then the value of A.

These three explorations have shown some ways in which both the graphing and computational power of the TI-85 can be used to investigate problems which involve the first and second derivatives of functions. We begin our study of integration and the solution of integration problems in the next section.

Explorations with Maxima and Minima Problems

1. Generate a third-, a fourth-, a fifth-, and a sixth-degree polynomial function with real coefficients. Use the techniques discussed in "A Good Graph" to analyze the behavior of each of these functions.

2. Suppose that in "A Precarious Ladder" the equation of the motion of the end of the ladder resting against the wall is of the form:

$$y = -16t^2 + bt + \sqrt{168}$$

where b is parameter taking on small values such as 0.01, 0.02, and 0.03 and somewhat larger values such as 0.1, 0.2, and 0.3. Explore the motion of the ladder for different values of the parameter b. For a given, positive real number is it possible to find a value of b so that the given real number is the maximum velocity of the ladder as it slides down the wall? If it is not possible to find a value of b for each positive, real number within what interval is it possible?

3. Suppose that, instead of a circular log in "Getting the Most Out of a Log," we have an elliptical log whose cross-section is given by the equation:

$$\frac{x^2}{16} + \frac{y^2}{9} = 1.$$

What is the maximum area of a beam with rectangular cross-section that can be cut from this log? What are the maximum areas of the boards that can be cut from the four pieces that remain after the beam of maximum rectangular cross-section has been cut from the log?

4.5. The Definite Integral as an Average

The definite integral of a function over an interval $[a, b]$ is the average value of the function over that interval multiplied by the length of the interval; this suggests that integration is the "opposite" of differentiation in a way other than that stated as the Fundamental Theorem of Calculus [1, p. 268]. Whereas differentiation can be thought of as the measurement of change, integration can be thought of as the measurement of "anti-change"; that is, as the value that would remain when all the change is removed. In the case of area, the definite integral value of a non-negative function is the constant (i.e., non-changing) value that must be maintained to create the same amount of area over the given interval. In the case of variable forces or variable masses, the integral represents the constant value that achieves the same effect as do the variable quantities.

To explore these points, let's imagine that there is a new game in town in which players estimate the height of a fluid in a tank. Two *non-mixing* fluids fill a thin tank that has a transparent side. The red fluid floats on top of the green fluid. The tank is rocked from side-to-side until the boundary between the two fluids is a wave. While the fluids are moving, contestants must estimate how high the top of the lower green fluid will be when the waves subside and the boundary between the two liquids becomes a horizontal line. The TI-85 user can "practice" making these estimates and thus, can be better prepared to win the game when it is played at the local fair.

All kinds of wave patterns can be created on the TI-85 screen using these directions. First, enter into your TI-85 a set of range values that makes the viewing window into a rectangle two units high and one unit wide: **xMin = 0, xMax = 1, xScl = 0, yMin = −1, yMax = 1, yScl = 0.** Then turn the display of both the $x-$ and $y-$ axes off by pressing [GRAPH] [FORMT] [▼] [▼] [▼] [▼] [▼] [▶] [ENTER]; the *AxesOff* statement will be highlighted after you press **ENTER**. (The *beginners version* of the game allows the players to compete with the *AxesOn*, but we will describe and use the advanced version here.) Finally, we need to place a function on the function list that generates values between -1 and 1 in the interval $[0, 1]$. We can do this by taking the *sine* of any function whose domain includes the interval $[0, 1]$. For this exploration we will use $f(x) = \sin e^{2x}$; we will set **y1**$= \sin e\wedge(2x)$. After graphing **y1** and pressing **CLEAR** you should see the graph of the function shown in Figure 4.27.

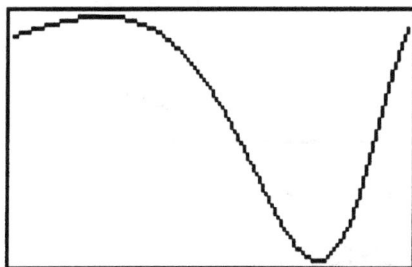

Figure 4.27. Graph of $f(x) = \sin e^{2x}$
in the rectangle $[0, 1] \times [-1, 1]$.

Now, to play the game we need to guess where the top of the lower fluid will be when the fluids stop moving and the separating surfaces are horizontal. Remember that the y-coordinates range from -1 to 1. Let's guess that the top of the lower fluid will be at $y = 0.1$ when the fluids stop moving — remember that the y-coordinate of the center of the screen is 0. To determine how good our guess is, we need to find the answer; that answer is, as we have already said, the average value of the function (i.e., the definite integral of f on $[0, 1]$). To compute this integral we will write

$$\text{fnInt}(y1, x, 0, 1)$$

on the TI-85's home screen by pressing $\boxed{\text{2nd}}$ $\boxed{\text{CALC}}$ $\boxed{\text{fnInt}}$ y1, x, 0, 1 $\boxed{\text{ENTER}}$. After a few moments the calculator will display the answer: 0.27546758687, and thus, we know that

$$\int_0^1 \sin e^{2x} dx = 0.27546758587 \,.$$

Now, toggle between the graphic screen and the home screen using the **GRAPH**, **CLEAR**, and **EXIT** keys so as to practice your geometric estimation skills.

Additional plays of the game using other functions are easy when you use the **GRAPH**, **y(x) =**, and **ENTRY** keys. Change **y1**$=$ to $\sin \ln 2x$. Graph this function and guess what the average value is (e.g., $-.3$). Then see what the answer is by going from the graph screen to the home screen by computing the integral. The **fnInt**$(y1, x, 0, 1)$ command stays the same, and the **ENTER** key causes the calculator to evaluate the integral with the new **y1** function. The answer for **y1**$=$ sin ln $2x$ is -0.065138809206. The graphs of $f(x) = \sin \ln(2x)$ and of $y = -0.065138809206$ are shown in Figure 4.28.

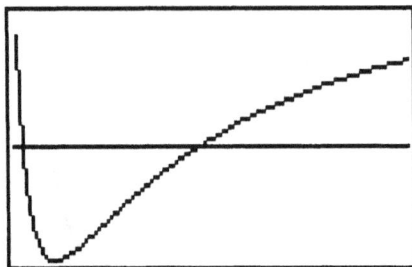

Figure 4.28. Graphs of $f(x) = \sin \ln(2x)$ and $y = -0.065138809206$
in the rectangle $[0, 1] \times [-1, 1]$.

It is important to recognize the definite integral as an average because this concept applies more easily during the early, theoretical development of the calculus. For example, in contrast to the usual approach where the integral is introduced as an area function, in averages you do not have to digress to worry about where the graph of the function crosses the horizontal axes — and talk about strange objects like *negative area*! The definite integral is simply a number, and the calculator readily calculates it for you. Likewise, as you will see in the next section, averages are naturally thought of in terms of "sums" divided by "n;" this is the natural way to introduce Riemann Sums.

The average value game can be programmed on the TI-85. The program IVAG generates a "random" wave in the rectangle $[0, 1] \times [-1, 1]$; prompts the player for an estimate of the average value of the function; graphs the player's guess, the average value calculated by the TI-85, and the wave simultaneously; and finally, prints the values of the estimated and calculated values. Notice that (a) **tol** has been set fairly large so as to speed up the game and (b) the user must press the **ENTER** key both after the graph of the "random" wave is displayed and after the simultaneous graphs of the wave, the player's estimate, and the actual value are displayed.

This section has explored interpretations of the definite integral. Section 4.6 will investigate several ways of approximating definite integrals.

```
PROGRAM: IVAG
:AxesOff
:LabelOff
:Func
:Fix 2
:ZU=tol
:0→xMin
:1→xMax
:0→xScl
:-1→yMin
:1→yMax
:0→yScl
:rand→B
:FUNCTION=sin (30*B*cos x)
:y1=FUNCTION
:FnOff 2,3
:DispG
:Pause
:Disp
:Input "Estimate Average Value",ESTIMATE
:y2=ESTIMATE
:fnInt(y1,x,0,1)→ACTUAL
:y3=ACTUAL
:DispG
:Pause
:Disp "ESTIMATE",ESTIMATE
:Disp "ACTUAL",ACTUAL
:Disp
```

Program 4.1.

Exploring the Integral as an Average

1. Play the average guessing game with the following functions:

    (a) $\sin(15 \cos x)$

    (b) $\sin(15 \tan x)$

    (c) $\sin(\log x)$

2. What changes in the calculation of the average value of the function over the integral would you have to make when the rectangle

in the average guessing game is changed to $[0, 2] \times [-1, 1]$ and the **fnInt**$(y1, x, 0, 2)$ value is calculated? What changes would you have to make for the interval [A,B] and a calculated definite integral value **fnInt**$(y1, x, A, B)$ where A and B represent a pair of specific numbers?

4.6. Numerical Integration

Now that calculators readily calculate the values of definite integrals, the study of calculus will shift away from the study of integration techniques used to find the values of definite integrals to an analysis of the quality and use of definite integrals computed electronically. The study of the use and interpretation of the values of definite integrals; that is, mathematical modeling, is beyond the scope of this section, but we do encourage you to pursue this topic elsewhere. However, the understanding of how calculator and computer definite integral routines arrive at values for definite integrals is the topic of this section; by understanding these techniques, you should have a better grasp of the numbers that are produced by these routines.

In the simplest terms, the definite integral is calculated by the TI-85 when the **fnInt** command is executed; the **fnInt** command is found in the **CALC** menu. When the **fnInt** command is executed, the TI-85 attempts to calculate the value of the definite integral within the present tolerance setting of the machine; this tolerance setting is called **tol**. The value of **tol** can be changed by pressing the **TOLER** key; the default setting is **tol**= $1E-5$. In addition to calculating the value of the definite integral, the **fnInt** command generates a value **fnIntErr** that indicates the possible solution error. After computing the value of a definite integral using **fnInt**, to see the value of **fnIntErr** press $\boxed{\text{VARS}}$, $\boxed{\text{ALL}}$, and the $\boxed{\text{PAGE}\downarrow}$, $\boxed{\blacktriangle}$, and $\boxed{\blacktriangledown}$ keys until the cursor is positioned beside **fnIntErr** before pressing $\boxed{\text{ENTER}}$. The details of the precise numerical methods used in the TI-85 to compute the values of definite integrals is proprietary to Texas Instruments Incorporated, but the general nature of numerical integration is not. Understanding this theory makes the user able to make better judgments about the quality of numerical results.

First, we will examine the general nature of integral approximations; these approximations are analogous to computing the average of a list of numbers. If we wanted to calculate the average value of a list of numbers, we would simply add the numbers and divide by n, the length of our list. When we want to determine the average value of a function, we sample its value at n points and "average them out" (i.e., add them up and divide by n). Conceivably, when the size of the sample, n, is very large the calculated average would closely approximate the average value of the function. But would it? Notice that we are allowing any given function value to contribute

$\frac{1}{n^{\text{th}}}$ of the answer; so, we should add some requirement that the sampled values are "representative"; that is, that they are evenly distributed over the interval. Otherwise, we could load up on values in a region where the function values are not representative and miss important sections where the function values are more typical.

Airport Windows

To give our discussion a context, let's assume that we are waiting in a new airport terminal between flights and that, out of curiosity, we try to estimate the cross sectional area in the two identical ends of the building and with that guess, at the total volume of the building. The ends of this modern building are window walls; the grid of windows at each end of the terminal matches that shown in Figure 4.29.

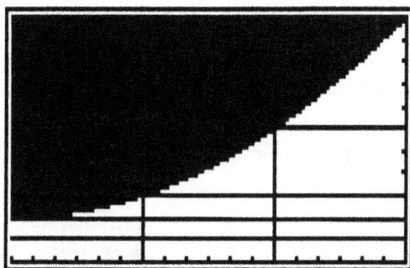

Figure 4.29. Airport terminal window wall.

A graph like that shown in Figure 4.29 can be produced on your TI-85 by setting the range values to: **xMin** = −3, **xMax** = 0, **xScl** = 3/18, **yMin** = 0, **yMax** = 11, and **yScl** = 1. The function list is: **y1**= $(x+3)^2+2$, **y2**= 1, **y3**= $2(x > -3)$, **y4**= $3(x > -2)$, and **y5**= $6(x > -1)$. The Boolean commands, $(x > -3)$, $(x > -2)$, and $(x > -1)$, in the **y3**, **y4**, and **y5** members of the function list graph the values of 2, 3, and 6 when the domain variable x exceeds −3, −2 and −1 respectively; for example, to insert the command $(x > -3)$ into **y3** press these keys $\boxed{(}$, $\boxed{x}$ $\boxed{\text{TEST}}$ $\boxed{>}$ $\boxed{-}$ $\boxed{3}$ $\boxed{)}$. The remainder of the graph is completed by executing **Shade**($y1, 11$); Shade is found by pressing **GRAPH** and then **DRAW**. The **CLEAR** key will remove all the menus in the graph viewing window and give you the unrestricted view shown in Figure 4.29.

By counting the courses of tiles on the front (right) edge of the building, we determine that the three lower rows of window panes are one unit high, the fourth row is three units high and the top curved pane is five units high at the front edge and totally curved on the back (left) edge. Again, by counting some tiles on the wall surface of the building, we determine that

the sections are six units wide on the bottom edge. We quickly determine
that the area of the window section is assuredly larger than 11×6 square
units and certainly less that 20×6 square units. Notice that these crude
estimates are determined by counting the eleven 6×1 rectangles that are
totally contained in the window (see Figure 4.30) and comparing that with
an upper bound determined by the twenty 6×1 rectangles that would
totally enclose the window (see Figure 4.31).

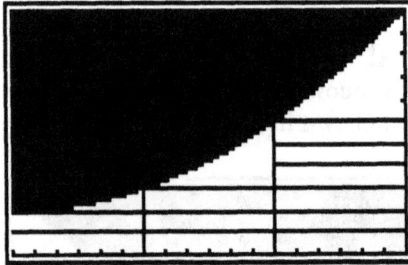

Figure 4.30. Inscribed rectangles in airport terminal window.

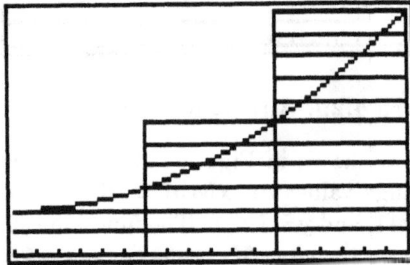

Figure 4.31. Circumscribed rectangles over airport terminal window.

We are looking for the average height of the windows over the length
of the end wall of the airport. By looking at Figure 4.30 we see that the
left most section is at least 2 units tall; the middle section is at least 3 units
tall; and the right most section is at least 6 units tall. The average of those
heights is $\frac{2+3+6}{3} = \frac{11}{3}$ units. When we look at Figure 4.31 we see that the
left most panel is at most 3 units; the middle, at most 6 units; and the
right, at most 11 units. The average value of the upper bound values is:
$\frac{3+6+11}{3} = \frac{20}{3}$ units. These two average values are pictured in Figure 4.32.

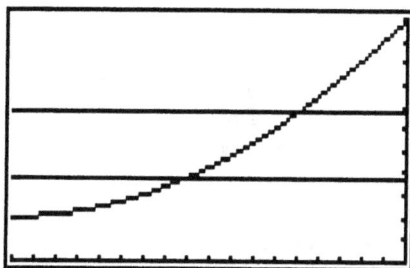

Figure 4.32. The lower and upper average heights
of the airport terminal windows.

We can interpret these average calculations in two ways. One way is to assume that the height is a composition of the representative values and we just "average" them out (i.e., that we add them up and divide by their count) as we did in the last paragraph. Since the lower average height was $\frac{11}{3}$, a lower bound on the area of the window is $\frac{11}{3} \times 3 = 11$ square units. Since the upper average height is $\frac{20}{3}$, an upper bound on the area of the window is $\frac{20}{3} \times 3 = 11$ square units. You might wonder why we don't simply take all four of the known values and average them out. That would give $\frac{2+3+6+11}{4} = \frac{11}{2}$ as an average height. Even though this "total average" might be even closer to the actual average height of the windows, we would not know whether it was a "high" or a "low" estimate. This "total average height," 5.5, is much closer to the actual average height (i.e., 5). But the difference between the total average height and the pair of heights consisting of the left average height and the right average height is that this pair of average heights catches the actual value between them. Thus, starting with the pair of average heights, we can keep improving on the upper and lower bounds to achieve better and better estimates of the actual average height. In most applications, it is just as important to know the "quality" as it is to know a "quantity" of our approximations.

The other interpretation of the average calculations is that we have bounds on the height of each of the three sections of the wall, and we use these bounds as estimates of the areas in the individual sections. Thus, we have a lower bound on the area of $2 \times 1 + 3 \times 1 + 6 \times 1 = 11$ and an upper bound on the area of $3 \times 1 + 6 \times 1 + 11 \times 1 = 20$. Why is this different? In the previous paragraph we were using the heights to arrive at averages; in this paragraph, we are focusing on area. When we deal with a non-negative function as we are here, these two interpretations hardly differ. When we deal with functions that cross the x-axis, then the two concepts diverge.

The "average" interpretation remains natural — you still just add things up and divide by n. However, the area interpretation becomes "forced," and we have to make up artificial items like "negative area" to maintain consistency.

We will explore only the average value interpretation here. Whenever we calculate the average value of a non-negative function, the area interpretation follows easily by observing that the average of that function over the identified interval will "sweep out" an area equal to the average height multiplied by the length of the interval. It is very important to remember that when we calculate a numeric integral with the TI-85, the value that it returns is the average value of the function over the specified interval multiplied by the length of that interval. Thus, when $\mathbf{y1}= (x + 3)^2 + 2$, $\mathbf{fnInt}(y1, x, -3, 0) = 15$ — the average value, 5, of the function $\mathbf{y1}$ over the interval $[-3, 0]$ times the length of the interval, 3.

The instruction $\mathbf{fnInt}(x - 1, x, 0, 3)$ returns the value 1.5; this is the average value, 0.5, of the function $x - 1$ over the interval $[0, 3]$ multiplied by the length, 3, of that interval. Within the interval $[0, 1]$ the function takes on negative values (see Figure 4.33), and those values enter the calculation of the average as negative numbers. To calculate the area trapped between the line and the x-axis in the interval $[0, 3]$, the problem must be broken into two parts; that is, $\mathbf{fnInt}(x - 1, x, 0, 1) = -.5$ and $\mathbf{fnInt}(x - 1, x, 1, 3) = 2$. This decomposition of the problem gives us a total area of 2.5 square units.

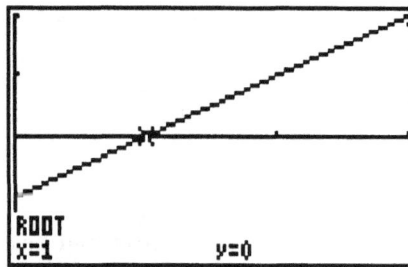

Figure 4.33. Graph of $f(x) = x - 1$
in the rectangle $[0, 3] \times [-2, 2]$.

The TI-85 is an especially powerful tool in helping us to solve integration problems. With it we can easily use graphics, symbolic expressions, and numeric values to attack area and average problems with ease and understanding; that is, we can routinely graph the functions, study the "picture," and process the symbolic expressions to arrive at numeric results. For example, in average calculations, we can easily judge the reasonableness of the answer by using our imagination and the graph to estimate where the "fluid will settle out" as we did in Section 4.5. In area calculations, we can graph the curve and quickly estimate the x-intercepts using the **TRACE** command. Or, we can take our time and use the **SOLVER** to find the x-coordinates of the intercepts very accurately. No matter which approach we take to finding the x-intercepts, once we know them it is a simple matter to use the **fnInt** command and the appropriate intervals to find the area.

```
PROGRAM: TRAP
:Disp "A"
:Input A
:Disp "B"
:Input B
:Disp "N"
:Input N
:(B-A)/N→H
:A→x
:0→S
:y1→L
:For(K,1,N-1,1)
:x+H→x
:y1+S→S
:End
:B→x
:y1→R
:(L+S)/N→L
:(R+S)/N→R
:Disp "LEFT AVERAGE"
:Disp L
:Disp "RIGHT AVERAGE"
:Disp R
:(L+R)/2→T
:Disp "TRAPEZOID AVG"
:Disp T
```

Program 4.2

Now, let's return to the function, $f(x) = (x+3)^2+2$, that represents the windows at the ends of the airport terminal. We have illustrated, over the interval $[-3, 0]$, that with a crude selection of three representative values, we can find a lower estimate of the window's average height of $11/3$ and an upper estimate of the average height of $20/3$. Program 4.2 calculates the left and right averages of a function stored in the function list as **y1** over the interval $[A, B]$. The user is prompted for the number of subdivisions, N, and for the coordinates of the endpoints, A and B, of the interval. The program also computes and reports the trapezoid estimate that is discussed next.

The Trapezoid Estimate

For an increasing function f, the left average will underestimate and the right average will overestimate the average value of the function — by roughly the same amount. This means that the left and right averages have the average of f trapped between them in the interval of length $(f(B) - f(A)) \times H$ where $f(B)$ is the value of f at the right endpoint of the interval, $f(A)$ is the value at the right endpoint, and H is the width of each of the N subintervals used to make the estimates of height. Thus, for increasing functions:

$$\text{left average} < \text{function average} < \text{right average}$$

and

$$\text{right average} - \text{left average} = (f(B) - f(A))\frac{B - A}{N}.$$

The fact that the left and right averages have about the same error and that those errors are respectively under and over estimates suggests very strongly that a better estimate would be the average of the two estimates. The average of the left and right estimates is called the *trapezoid estimate*. The trapezoid name is derived from the fact that the same estimation formula results from the summation of the areas in the inscribed trapezoids in the region under discussion. The inscribed trapezoids in our airport window example would be 15.5, that is;

$$\left(\frac{2+3}{2} \times 1\right) + \left(\frac{3+6}{2} \times 1\right) + \left(\frac{6+11}{2} \times 1\right).$$

The Midpoint Estimate

A different approach and another attractive estimation scheme is to use the midpoints of the intervals. This is called the *midpoint estimate*; it uses the values of the function at the centers of the intervals in the subdivisions. In our airport window example, we can not easily "count the blocks" and determine the heights of the windows at the points $(-2.5, 0)$, $(-1.5, 0)$ and $(-.5, 0)$. But if we were able to easily determine the heights above those points, the values would be 2.25, 4.25 and 8.25. The average of these three values is 4.92 — a value that is remarkably close to the actual average 5.0!

Program 4.2 can be modified easily to create a new program, Program 4.3, that calculates the average value of a function using the midpoint value.

```
PROGRAM: MIDPOINT
:Disp "A"
:Input A
:Disp "B"
:Input B
:Disp "N"
:Input N
:(B-A)/N→H
:(A+H/2)→x
:y1→S
:For(K,1,N-1,1)
:x+H→x
:y1+S→S
:S/N→M
:End
:Disp "MIDPOINT AVERAGE"
:Disp M
```

Program 4.3.

If you repeatedly apply the midpoint average program and the trapezoid average program to a function on a fixed interval, you will discover that one formula will always overestimate and the other will always underestimate. For concave down functions, the trapezoid rule underestimates and the midpoint rule overestimates. For concave up functions, the trapezoid formula overestimates and the midpoint rule underestimates. The error associated with the trapezoid rule is usually about twice the size of the errror associated with the midpoint rule. The observation that the error in the different rules are opposite in sign and in a ratio of 2 to 1 suggests that a better estimate would be the weighted average of the two; that is, that we should double the midpoint estimate (M), add that to the trapezoid estimate (T), and divide by three, (i.e., $\frac{2M+T}{3}$). This estimate is called Simpson's Rule; it is an amazingly accurate device for calculating approximations to the integral of a function. In Program 4.4 we have combined our two programs to create a new program that covers all of the integration rules.

```
 PROGRAM: ALL
:Disp "A"
:Input A
:Disp "B"
:Input B
:Disp "N"
:Input N
:(B-A)/N→H
:A→x
:0→S
:y1→L
:For(K,1,N-1,1)
:x+H→x
:y1+S→S
:End
:B→x
:y1→R
:(L+S)/N→L
:(R+S)/N→R
:Disp "LEFT AVERAGE"
:Disp L
:Disp "RIGHT AVERAGE"
:Disp R
:(L+R)/2→T
:Disp "TRAPEZOID AVG"
:Disp T
:A+H/2→x
:y1→S
:For(K,1,N-1,1)
:x+H→x
:S+y1→S
:End
:S/N→M
:Pause
:Disp "MIDPOINT AVERAGE"
:Disp M
:(2M+T)/3→P
:Disp "SIMPSON AVERAGE"
:Disp P
```

Program 4.4

When you run Program 4.4 using the function that represents the airport windows, you will find that the average values of the function $f(x) = (x + 3)^2 + 2$ over the interval $[-3, 0]$ are as shown in Table 4.3:

Table 4.3. Average values of $f(x) = (x + 3)^2 + 2$
using each of the estimation rules.

Averaging Rule	$N = 3$	$N = 6$
LEFT	3.66666666667	4.29166666667
RIGHT	6.66666666667	5.79166666667
TRAPEZOID	5.16666666667	5.04166666667
MIDPOINT	4.91666666666	4.97916666667
SIMPSON	5	5

This section has introduced you to ways of approximating definite integrals using five different techniques or rules. Section 4.7 explores the use of the TI-85 in calculating planar areas and volumes of revolution.

Explorations in Numerical Integration

1. Apply the estimations of this section to a hypothetical set of windows at the end of an airport terminal whose roof has a concave downward shape; that is, the shape is described by the function is $f(x) = -x^2 + 11$.

2. Apply the programs of this section to estimate the average value of the $1/x$ over the interval $[1, 11]$. How many subdivisions do you need in the left and right schemes to achieve the same level of accuracy that you have with Simpson's Rule when you use five intervals? Why is the average of the $1/x$ over the interval equal to $\ln(11)/10$?

4.7. Areas and Volumes[1]

As we discussed in Section 4.5, the value of the definite integral is frequently interpreted as the area of a region of the plane. In that section we also discussed the fact that when the function being used is not always nonnegative then we must carefully determine where that function is negative and positive so as to find the area enclosed by the function and the x-axis. In this section we will continue our discussion of the ways in which the TI-85 can help us to measure the area of a region bounded by two functions.

---

[1] Throughout this section be sure that $\mathbf{tol} = 1E - 13$ in order to be certain that you arrive at the same numeric answers as we do.

And we will extend that discussion to include ways in which we can use the
TI-85 to help us find volumes of revolution.

The Area Between Two Curves

What general procedure do we use to find the area in the plane bounded
by two functions? First, in order to understand the problem better we draw
a graph of the two functions, and having done that, we shade in the area we
want to measure (i.e., the region over which we plan to integrate). Next,
we find the relevant points of intersection of the two curves. Finally, we
write down and solve the definite integrals that will measure the areas we
seek. The TI-85 can help us in each step of this procedure as we will show
you. To do this we will solve the problem stated in Figure 4.34.

Let $f(x) = x^2 + 1$, and let $g(x) = x^3 - 4x + 2$.
Find the area of the region(s) bounded by f and g.

Figure 4.34. Area problem.

Following the general procedure we outlined above, we begin by graphing both functions in a rectangle that shows us the relative extrema, the
zeros, and the points of intersection of the two functions. Figure 4.35 shows
the graphs of f and g in the rectangle $[-3, 3] \times [-4, 9]$. This figure shows
us two regions bounded by our functions, but are there others? And, if

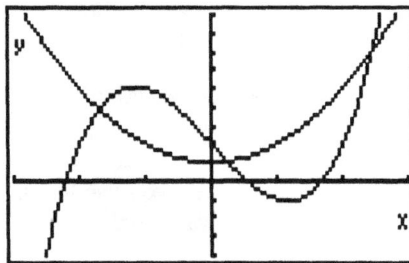

Figure 4.35. Graph of f and g
in the rectangle $[-3, 3] \times [-4, 9]$.

there are other regions, is it intended that we find their areas as well? To
try to answer these questions, let's return to the original problem. It asked
us to find the area of the region(s) bounded between the two functions. The
key phrase "bounded between" means that we are looking for those points
in the plane having y-coordinates less or equal to $f(g)$ and greater than
or equal to the $g(f)$. The TI-85 can help us to find the regions that may
be of concern to us; we will use the **SHADE** command. Having entered
$y1 = x^2 + 1$ and $y2 = x^3 - 4x + 2$ into our calculators, to shade all of the region

above **y1** and below **y2**, we key-in **SHADE**$(y1, y2)$ and press **ENTER**. Now, we enter **SHADE**$(y2, y1)$ to shade all of the region above **y2** and below **y1**; this is most easily done on the home screen by pressing ENTRY and editing the previous statement, **SHADE**$(y1, y2)$ when it appears on the screen. When **SHADE**$(y2, y1)$ has been executed, your graphics screen should look like that one shown in Figure 4.36. Only a moment's thought is needed for us to see that the shaded areas on the far left and the far right of the screen in Figure 4.36 are not bounded and have infinite area. Thus,

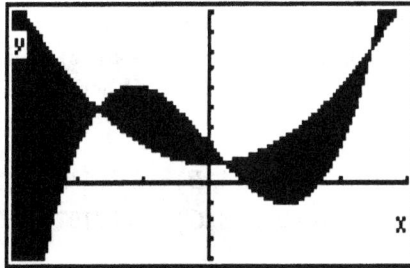

Figure 4.36. Areas bounded by f and g.

the regions that we want to measure are the two of finite area that are in the middle of the screen. The left region of finite area is bounded above by **y2** and below by **y1**; the right region of finite area, below by **y2** and above by **y1**. Thus, if the points of intersection of $f(= $ **y1**$)$ and $g(= $ **y2**$)$ are A, B, and C where $A < B < C$, then the area we seek is defined by:

$$\text{Area} = \int_A^B (g - f) + \int_B^C (f - g),$$

or

$$\text{Area} = \int_A^B (x^3 - x^2 - 4x + 1)dx + \int_B^C (-x^3 + x^2 + 4x - 1)dx \qquad (4.1)$$

Thus, our next step is to find the points of intersection A, B, and C. We will do this using the **ISECT** command. To find A, from the graphics screen we press MATH MORE ISECT and move the cursor along either **y1** or **y2** until we are near to the left-most point of intersection of the two curves; when we are in the vicinity of that point, we press **ENTER** *twice*. When the TI-85 completes its computations it tells us that $x = -1.699628148$. For a reason that will be explained very soon, let's store the value of x showing on the screen as A. To do this, exit the graphics screen for the home screen and press x STO▶ A. Now return to the graphics screen to find B and store it and finally, to find C and store it. When you finish, the values of A, B, and C should be:

$$A = -1.699628148$$
$$B = 0.23912327826$$
$$C = 2.46050487 .$$

Now that we know A, B, and C, we could calculate the symbolic integrals of the our functions and substitute in A, B, and C as we always have. However, we will use the **fnInt** command; the value of the first of the two integrals in equation 4.1 is **fnInt**$(y2 - y1, x, A, B)$, and the value of the second one is **fnInt**$(y1 - y2, x, B, C)$. When these two commands are executed we get that:

$$\textbf{fnInt}(y2 - y1, x, A, B) = 3.87533198628$$
$$\textbf{fnInt}(y1 - y2, x, B, C) = 5.57107063271$$

and thus, that area = 9.44640261899.

A Different Area Problem

Let's continue to explore finding the area between two curves by looking at a poorly defined problem: Find the area between $f(x) = \sin x$ and $g(x) = \cos 2x$. As usual, we graph both of the functions. We let **y1**$= \sin x$ and **y2**$= \cos 2x$ and graph these functions in the rectangle $[-3\pi, 3\pi] \times [-1.5, 1.5]$ with **xScl**$= \pi/2$ and **yScl**$= .5$; the resulting graph is show in Figure 4.37. This graph shows us that there are (probably) an infinite number of regions of finite area bounded by these graphs; it also

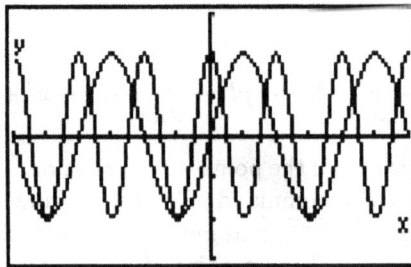

Figure 4.37. Graphs of $\sin x$ and $\cos 2x$
in the rectangle $[-3\pi, 3\pi] \times [-1.5, 1.5]$.

shows us that some of the regions are shaped like an "exclamation point," that some are shaped like "a rabbit's left ear," and that some are shaped like "a rabbit's right ear." From looking at the graph we can speculate that all of the "exclamation points" have the same area, that all of the "left ears" have the same area, that all of the "right ears" have the same area, and that "left" and "right" ears have the same area. Since both $\sin x$ and $\cos 2x$ are periodic functions, it is easily proved that all of the "exclamation points" are congruent to each other and so, have the same area, but what is that area? The same argument will prove that all of the "left ears" are congruent to each other and that all of the "right ears" are congruent to each other. So, the only remaining questions that we need to explore are: (a) What is the area of an exclamation point? and (b) Is the area of a "left ear" equal to the area of a "right ear"?

In order to answer these questions, we will re-graph the functions in the rectangle $[-4, 3] \times [-1.5, 1.5]$; this graph, shown in Figure 4.38, gives us a better picture of all three of the regions that we need to find in order

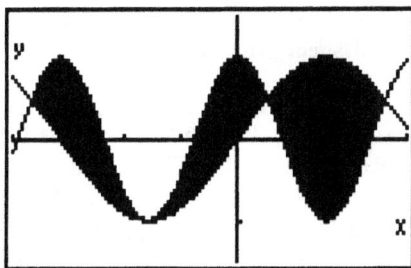

Figure 4.38. Graphs of a left ear, a right ear, and an exclamation point in the rectangle $[-4, 3] \times [-1.5, 1.5]$

to answer our remaining two questions. To calculate the areas, we need to know the upper and lower limits for the definite integrals we will compute; these limits are the points of intersection A, B, C, and D, where A and B are the lower and upper limits for the left-most "left ear" shown in Figure 4.38; B and C are the limits for the "right ear"; and C and D are the limits for the "exclamation point." Using **ISECT** we find that:

$$A = -3.665191429$$
$$B = -1.570796327$$
$$C = 0.5235987756$$
$$D = 2.617993878\,.$$

In order to express these as multiples of π, we do the following. For A, with $x = -3.665191429$ (i.e., after finding the left-most point of intersection), we return to the home screen, press $\boxed{x}$ $\boxed{\div}$ $\boxed{\pi}$ $\boxed{\text{ENTER}}$ $\boxed{\text{MATH}}$ $\boxed{\text{MISC}}$ $\boxed{\text{MORE}}$

⯈Frac ENTER , and discover that $A = \frac{-7\pi}{6}$. Subsequently, using the same procedure, we find that:

$$B = -\frac{\pi}{2}$$

$$C = \frac{\pi}{6}$$

$$D = \frac{5\pi}{6}.$$

Thus, the area of a "left ear" is $\mathbf{fnInt}(y2-y1, x, A, B) = 1.29903810568$; the area of a "right ear" (i.e., $\mathbf{fnInt}(y2-y1, x, B, C)$) equals the area of a "left ear"; and the area of an "exclamation point" (i.e., $\mathbf{fnInt}(y1-y2, x, C, D)$) is equal to the sum of the areas of a "left ear" and a "right ear."

The Volume of a Conical sheet

We continue our exploration of the use of the TI-85 by finding the volume of a thin conical sheet; that is, by solving the problem stated in Figure 4.39.

A thin conical sheet of altitude 2.4 inches is being formed by pouring metal into a mold. The outside of the conical sheet is the surface generated by rotating $f(x) = 0.75x$ around the x-axis in the interval $[0, 2.4]$. The inside of the conical sheet is the surface generated by rotating $g(x) = 0.75(x - 0.1)$ around the x-axis in the interval $[.1, 2.4]$. How much metal is needed to make each shell?

Figure 4.39. Conical sheet problem.

We will begin to solve this problem, as we usually do, by graphing both of the functions. Since the altitude of the outside surface of the cone is 2.4 inches and since the maximum value of f will be 1.8 inches at $x = 2.4$, we graph the functions in the rectangle $[0, 2.4] \times [-1.8, 1.8]$. So that we can graph g in the interval $[0, 2.4]$, we will modify this function by redefining g in this way:

$$g(x) = 0(x < .1) + .75(x - .1)(x \geq .1)(x \leq 2.4) + 0(x > 2.4). \qquad (4.2)$$

Now the function g has a domain that includes $[0, 2.4]$. When we set **y1** and **y2** to be f and g, respectively, the graph of our functions is as shown in Figure 4.40. To show the graph of the cross section of the cone we set **y3**$= -y1$ and **y4**$= -y2$; the resulting graph is shown in Figure 4.41.

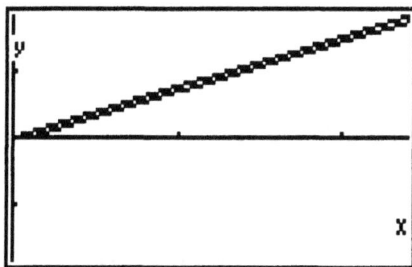

Figure 4.40. Graph of f and g in the
rectangle $[0, 2.4] \times [-1.8, 1.8]$.

(If you want to see the tip of the sheet better, re-graph that part of the
sheet using the rectangle $[0, .4] \times [-.3, .3]$.) This view of the tip of the sheet
shows you the effect of redefining g in Equation 4.2 and better helps you
to visualize the "washers" that result when a strip of width Δx that is

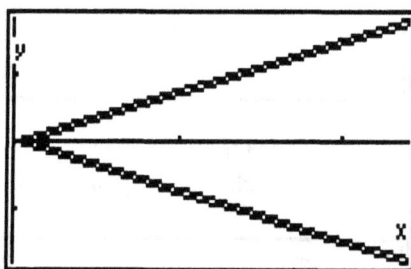

Figure 4.41. Cross-section of the conical sheet.

bounded by f and g is revolved about the x-axis; the volume ΔV of that
strip is:

$$\Delta V = \pi(y1^2 - y2^2)\Delta x .$$

Thus, the volume of the conical sheet will be:

$$V = \int_0^{2.4} \pi(f(x)^2 - g(x)^2)dx .$$

Notice that because of the way we have defined the function g we do not
have to calculate the value of two integrals and add them together to find
the volume. That is, since $g(x) \equiv 0$ on $[0, .1]$, $f(x)^2 - g(x)^2 = f(x)^2$ on
that interval. When we compute $\mathbf{fnInt}(\pi(y1^2 - y2^2), x, 0, 2.4)$, we find that
the volume of our thin conical sheet is approximately 0.976053567562.

An Improper Integral as a Volume of Revolution

For the final exploration of Section 4.7, we find the volume of revolution generated by rotating the function $f(x) = x \ln x$ around the x-axis on the interval $[0, 3]$. A graph of f in the rectangle $[0, 3] \times [-3.5, 3.5]$ is shown in Figure 4.42 and a sketch of the surface of revolution is shown in Figure 4.43. Since f is not defined for $x \leq 0$, one way of solving this problem is to find the value (if it exists) of the improper integral:

$$\lim_{t \to 0+} \int_t^3 \pi (x \ln x)^2 dx.$$

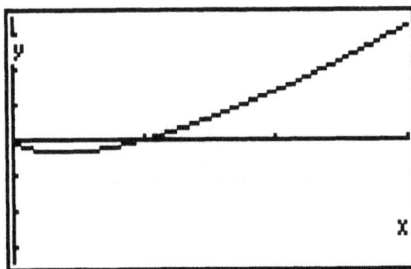

Figure 4.42. Graph of $f(x) = x \ln x$
in the rectangle $[0, 3] \times [-3.5, 3.5]$.

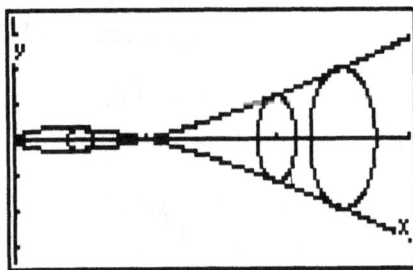

Figure 4.43. Sketch of the surface of revolution.

Since the integrand is defined and continuous for each value of $t > 0$, the definite integral exists and is finite for each of these values of t. We will use the TI-85 to help us explore the values of this integral for small values of t using Program 4.5. This program is designed to compute the value of the integral for $t = 0.1^k$ where $k = 1, 2, \cdots, 7$. In Program 4.5, the value of

```
 PROGRAM: TRIES
:7→NUM
:NUM→dimL ERR
:NUM→dimL TRIV
:NUM→TVAL
:.1→BEGIN
:1ε-13→tol
:For(K,1,NUM,1)
:fnInt(π(x ln x)²,x,BEGIN^K,3)→TRIV(K)
:fnIntErr→ERR(K)
:End
:ClLCD
:For(K,1,NUM,1)
:Disp "Lower bound: "
:BEGIN^K→TVAL(K)
:Disp TVAL(K)
:Disp "Integral ="
:Disp TRIV(K)
:Disp "fnIntErr ="
:Disp ERR(K)
:Pause
:ClLCD
:End
:
```

Program 4.5

NUM controls how many integrals will be evaluated; the largest and first
value of t used is stored in BEGIN. Thus, 0.1 has been stored in BEGIN and
7 in NUM. When you run this program on your TI-85, after several minutes
you should see the values displayed in Table 4.4 for t (i.e., Lower bound),
fnInt$(\pi(x \ln x)^2, x, t, 3)$ (i.e., Integral), and **fnIntErr**. These values are
stored in the **LIST** variables TVAL, TRIV, and ERR. By looking at Table
4.4, we see that the volume of revolution is approximately 19.7005; that is
that

$$\lim_{t \to 0+} \int_t^3 \pi(x \ln x)^2 dx \simeq 19.7005 .$$

Table 4.4

Values of $\int_t^3 \pi(x \ln x)^2 dx$
for $t = 1, \cdots, 0.1^7$.

t	Integral	fnIntErr
0.1	196931170798	-1.5324719E-11
0.01	19.7004837765	-1.592086E-11
0.001	19.7005093777	-6.37726843E-12
0.0001	19.7005094327	-8.274074044E-12
0.00001	19.7005094328	-1.181668419E-11
0.000001	19.7005094328	-1.2488527597E-11
0.0000001	19.7005094328	-1.273930043E-11

We can also approach this problem in another way by exploring

$$\lim_{x \to 0+} (x \ln x).$$

Since $x \ln x = \frac{\ln x}{\frac{1}{x}}$, $\ln x \to -\infty$ as $x \to 0^+$, and $\frac{1}{x} \to \infty$ as $x \to 0^+$, by l'Hôpital's Rule,

$$\lim_{x \to 0+} (x \ln x) = \lim_{x \to 0+} \frac{\ln x}{\frac{1}{x}} = \lim_{x \to 0+} \frac{\frac{1}{x}}{\frac{-1}{x^2}} = \lim_{x \to 0+} -x = 0.$$

Thus, if we define g as:

$$g(x) = \begin{cases} 0 & \text{for } x \leq 0 \\ x \ln x & \text{for } x > 0 \end{cases}$$

then g is defined and continuous on all of **R**, and so,

$$\int_0^3 \pi g(x)^2 dx$$

exists; $\mathbf{fnInt}(\pi(g(x)^2, x, 0, 3)$ is, once again, approximately 19.7005 with a possible error of $\mathbf{fnIntErr} = -1.2734500603E - 11$.

Explorations with Areas and Volumes

1. Find the area bounded by the function $f(x) = \sqrt{x} \ln x - 10 \sin x$ and the x-axis.

2. For some whole number values of a and b consider the functions $f(x) = \sin ax$ and $g(x) = \cos bx$. Explore the areas bounded by these two functions as we did in "A Different Area Problem."

3. Revolve one of the "left rabbit's ears" discussed in "A Different Area Problem" about the x-axis and find the volume of the resulting solid.

4. Find the volume of the solid of revolution that results when the curve $f(x) = x \ln x$ $(0 < x \leq 3)$ is revolved about the y-axis.

Reference

1. Edwards, C. H., Jr. and D. E. Penney, *Calculus and Analytic Geometry*, 2nd edition, Prentice-Hall, Englewood Cliffs, NJ, 1986.

5 Explorations in Calculus II

Wayne Roberts
Macalester College

The TI-85 can, to be sure, be used to make a triviality of many problems commonly found in calculus texts. This is not, however, the best use to make of it. Indeed, if you use your calculator to decimate an assignment, you should pause to ask what the writer of the problems – presumably someone not thinking about a student equipped with a TI-85 – hoped the student would learn from the problems.

There are, happily, better ways to use your calculator than in the destruction of well intentioned problem sets. It can be used to suggest to you an idea based on observations made possible by your calculator, to let you see what is going on by observing numerous intermediate steps as well as the culminating step of a solution, or to otherwise deepen your grasp of concepts. This is our goal here – to help you to employ your calculator to enhance your understanding.

Each section of this chapter is intended to accompany a similarly titled section in your calculus text. Since the order of topics will vary a bit from text to text, the sections here are independent of one another and can be taken up in any order.

We begin with one of the classic problems of mathematics, the finding of those values of x for which a function $f(x)$ assumes the value of zero. In this connection we explore the inverse of trigonometric functions, and we see how to implement Newton's Method. We turn in Section 2 to the graphing of conic sections, using the requirements of the TI-85 to underscore some important ideas about the concept of a function. We also see some advantages intrinsic to the parametric representation of curves.

Section 3 on polar graphing includes a program that lets you see, literally, "what's going on." Use of the TI-85 capability to graph two functions simultaneously enables us to give the best possible demonstration of the difference between the intersecting of paths and the colliding of bodies moving

Explorations with the Texas Instruments TI-85
John G. Harvey and John W. Kenelly (eds.), pp. 189-228
©1993 by Academic Press, Inc.
ISBN: 0-12-329070-8

on those paths. The same capabilities are utilized in Section 4 where we look more carefully at the parametric representations of curves.

With the TI-85 as an aid, functions defined by integrals can easily be evaluated for arbitrary x, and thus attain legitimacy in student's eyes as "real functions." In this context, we study $\ln x$ and Arctan x in Section 5. We discuss indeterminate forms in Section 6 and improper integrals in Section 7. Both are topics in which the abstraction of a limiting process can be illustrated by generating sequences of numbers.

Finally, in Section 8, we take an approach to Taylor series that is only practical with the aid of technology. We simply plot the sequence of approximating polynomials; arguments about the interval of convergence are replaced by pictures in which the convergence is plain for all to behold.

5.1. Zeroes of Functions

The values of x for which a function $f(x)$ is zero are called zeroes of the function, or roots of the equation $f(x) = 0$. To the numerous methods historically developed to solve such problems, modern technology now adds "the black box" with its $\boxed{\text{ROOT}}$ key.

In this section we first see that an initial idea of the location of the roots is essential, even when one is equipped with so sophisticated a black box as the TI-85. Then, because our object is not to get answers, but to understand calculus, we look rather closely at one of the historically developed methods that often underlies modern technology.

Using the inverse trigonometric functions

Suppose you are to find the zeroes of
$$f(x) = (2\sin x - 1)(\sin x + 2\cos x)$$
that lie in the interval $0 \leq x \leq 2\pi$. Drawing a graph of $y = f(x)$ is always a good way to begin, an activity for which your TI-85 offers quick help. Before beginning, press $\boxed{\text{MODE}}$ and make sure that the following settings are active: Normal, 4, Radian, RectC, Func, Dec, RectV, dxDer1. Go next to $\boxed{\text{GRAPH}}$ and use $\boxed{y(x) =}$ to set y1 equal to the function. For the interval of interest, in $\boxed{\text{RANGE}}$ set xMin0, xMax2, xScl1, yMin3, yMax6, and yScl1. Selecting $\boxed{\text{GRAPH}}$ should produce what you see in Figure 5.1. There are clearly four zeroes in the interval $[0.2\pi]$, and the trace feature enables you to get a quick approximation of them. They appear to be, from left to right,
$$x_1 = .54, \quad x_2 = 2.04, \quad x_3 = 2.64, \quad x_4 = 5.18.$$
Though it is not our purpose here to get more accurate values than we already have, this is a good place to point out that more accuracy is readily available by using the $\boxed{\text{ROOT}}$ key as follows. Having selected $\boxed{\text{GRAPH}}$, press

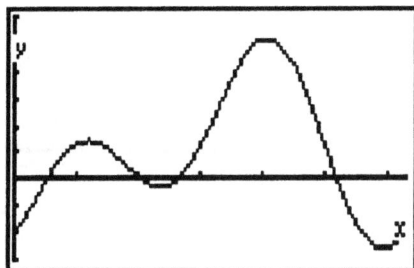

Figure 5.1. Graph of $f(x) = (2\sin x - 1)(\sin x + 2\cos x)$.

$\boxed{\text{MORE}}$ to display that part of the menu that enables you to press $\boxed{\text{MATH}}$, and then press $\boxed{\text{ROOT}}$. Use $\boxed{\blacktriangleright}$ to move the cursor to the leftmost root that we have called x_1. Press $\boxed{\text{ENTER}}$ and wait for the calculator to get at the root of the problem at hand. It will announce $x_1 = .5235987756$.

We return to our main purpose, that of how you can verify the roots we have found with the answers in the back of the book where they are commonly expressed as familiar multiples of π, or in terms of the inverse trigonometric functions.

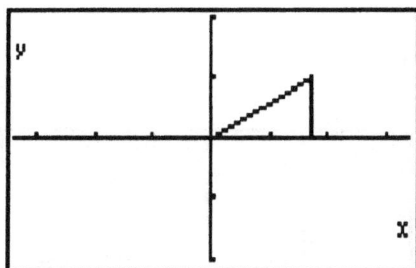

Figure 5.2(a). Angles for which $\sin x = \frac{1}{2}$.

An analytic solution is obtained by setting each of the factors defining $f(x)$ equal to zero. From the first, we get $2\sin x - 1 = 0$ or $\sin x = \frac{1}{2}$. The angles between 0 and 2π that have a sine of $\frac{1}{2}$ are familiar and are easily pictured as in Figure 5.2.

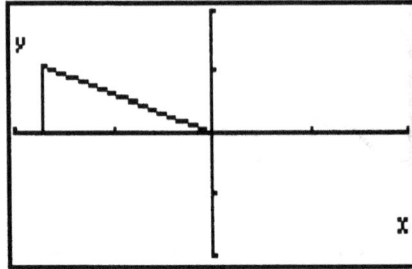

Figure 5.2(b). Angles for which $\sin x = \frac{1}{2}$.

Evidently we have found

$$x_1 = \frac{\pi}{6} = .5236 \quad \text{and} \quad x_3 = \frac{5\pi}{6} = 2.6180.$$

If we had not recognized the familiar $30° - 60° - 90°$ triangle, but had relied wholly on our calculator, we would have solved $\sin x = .5$ using $\sin^{-1}(.5)$ to obtain .5236 which is x_1. It would then have been necessary to draw upon what we know from trigonometry, observing that if the angle pictured in Figure 5.2(a) has a sine of .5, then so will the angle pictured in Figure 5.2(b), $\pi = \sin^{-1}(.5)$.

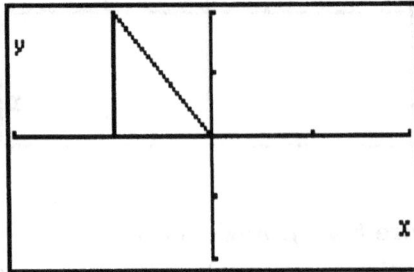

Figure 5.3(a). Angles for which $\tan x = -2$.

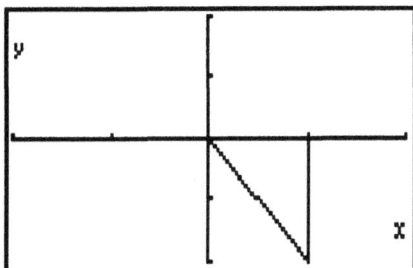

Figure 5.3(b). Angles for which $\tan x = -2$.

Setting the second factor equal to zero gives $\sin x + 2 \cos x = 0$; division by $\cos x$ gives $\tan x = -2$. We can again picture the angles, as in Figure 5.3. These are not familiar multiples of π, so this time we must turn to the $\tan^{-1}$ function. In doing so, we must anticipate that since $-\frac{\pi}{2} \leq \tan^{-1} y \leq \frac{\pi}{2}$. $\tan^{-1}(-2)$ will not be in the desired interval $[0, 2\pi]$. Indeed, we see from our calculator that $\tan^{-1}(-2) = -1.1071$. We did not get the positive angle x_4 of Figure 5.3(b), but the negative angle $\tan^{-1}(-2)$. Again we call upon our knowledge of trigonometry, this time to write

$$x_4 = \tan^{-1}(-2) + 2\pi \quad \text{and} \quad x_2 = \tan^{-1}(-2) + \pi$$

that is,

$$x_4 = 5.1760 \quad \text{and} \quad x_2 = 2.0344.$$

Newton's Method

We continue the example started above, using it to illustrate Newton's Method, a method of successive approximations based on the following idea. Having made an nth guess x_n as to where a real zero r of $f(x)$ might be (Figure 5.4), we write the equation of the line ℓ tangent to the graph of $y = f(x)$ at x_n. The place where ℓ crosses the x-axis is used as the next guess x_{n+1}.

$$x_{n+1} = x_n - \frac{f(x_n)}{f'(x_n)}$$

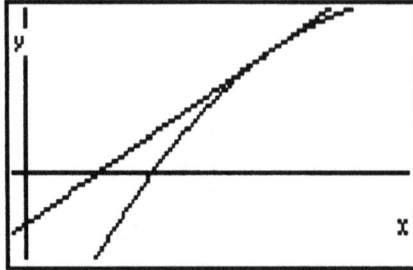

Figure 5.4. Line tangent at x_n.

An easy way to execute an iterative procedure is to proceed as follows. For the function f graphed in Figure 5.1 let us use a first guess of 3 to approximate the zero x_3. Using four decimal places, enter in one continuous line the instructions (listed here on separate lines for convenient reference)

$$3.0000 \to \text{GUESS:}$$
$$\text{der}1(y1, x, \text{GUESS}) \to \text{DER:}$$
$$\text{GUESS} \to x :$$
$$\text{GUESS} - y1/\text{DER}$$

The second instruction is obtained by pressing $\boxed{\text{CALC}}$, then $\boxed{\text{der1}}$; the $y1$ results from the sequence $\boxed{\text{2nd}}$ $\boxed{\text{ALPHA}}$ $\boxed{\text{Y}}$ $\boxed{\text{1}}$], and the x comes from $\boxed{\text{x-VAR}}$. Remember in entering the third instruction that the calculator is set for a capital following $\boxed{\blacktriangleright}$, so you must enter $\boxed{\text{ALPHA}}$ $\boxed{\text{X}}$ to get the lower case x.

Following these instructions, press $\boxed{\text{ENTER}}$. The calculator responds with 2.7102. The iteration may now be effected by pressing $\boxed{\text{2nd}}$ $\boxed{\text{ENTRY}}$ which recalls the last command. Use the $\boxed{\blacktriangle}$ and $\boxed{\blacktriangleleft}$ keys to position the cursor at the 3 in 3.0000 and type 2.7102. You do not need to be at the end of a command to execute it; press $\boxed{\text{ENTER}}$ to get 2.6292. Repeating this will give 2.6182. Once more (and forever after) will give 2.6180.

Explorations with Zeroes of Functions

1. Using the range settings specified at the beginning of this section, graph on the same axes $y1 = (2\sin x - 1)(\sin x + 2\cos x)$, $y2 = (2\sin x - 1)$,

and $y3 = (\sin x + 2\cos x)$. Watching the graphs drawn in sequence enables you to identify the specific factor that contributes each of the roots.

2. Use $\boxed{\text{MODE}}$ to reset your Float to 10. Beginning with our first estimate of $x_1 = .54$, use Newton's Method to get better estimates. Compare with the value obtained by using the $\boxed{\text{ROOT}}$ key on the TI-85.

3. How large a first guess can you use and still have Newton's Method converge to x_1? Explore the same question using the $\boxed{\text{ROOT}}$ key.

5.2. Graphing Conic Sections

There is a tendency for beginning students of mathematics to identify a function with an algebraic formula, an idea that develops from the fact that the first functions one meets are usually defined by a formula; $f(x) = x^2$ for instance.

The idea is wrong. Not every function can be described by a formula. We can, for example, define for each real number x the function that is 1 if x is an integer, 0 otherwise. This is a function; it assigns to every member of its domain a unique number. But it cannot be described by an algebraic formula.

The idea is wrong in another way. Not every algebraic formula defines a function. The formula $y^2 = x^2 - 9$ does not define a function. Most admissable values of x (that is, values of $x > 3$) do not determine a unique value of y.

Top half, bottom half

Persons inclined to think of $y^2 = x^2 - 9$ as defining a function, or to think that the distinction between a function and a formula is a technicality of interest only to mathematicians, should think about what they would need to enter into the TI-85 to get a graph representing this formula. It will quickly be seen that one must enter both

$$y1 = \sqrt{x^2 - 9} \quad \text{and} \quad y2 = -\sqrt{x^2 - 9}$$

in order to get the entire graph corresponding to $y^2 = x^2 - 9$. The calculator forces the conclusion that the graph of the given formula represents the graphs of two functions.

If you draw the graph of just one of these functions, say $y2 = -\sqrt{x^2 - 9}$ on a screen set for $-1 \leq x \leq 2$, another important idea about functions emerges. A function is not completely defined until you specify its domain, the set of values for which it makes sense. For the function under consideration, the domain is the set of all x such that $|x| \geq 3$. If the viewing screen is set for $-5 \leq x \leq 5$, the TI-85 indicates by the absence of any picture between -3 and 3 the fact that $|x| < 3$ is not in the domain (Figure 5.5).

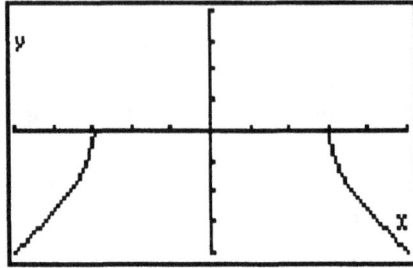

Figure 5.5. Graph of $y = -\sqrt{x^2 - 9}$.

An important class of curves studied in calculus, the conic sections, have equations of the form

$$Ax^2 + Bxy + Cy^2 + Dx + Ey + F = 0.$$

Except for the special case $C = 0$, these equations determine two functions of x. Consider, for example,

$$2x^2 + y^2 - 12x + 2y + 11 = 0.$$

You must begin by solving for y, either by completing the squares in x and in y, or by treating the given equation as a quadratic in y to be solved by the quadratic formula.

We illustrate use of the quadratic formula in our next example. In this example, we shall complete the squares, first by writing

$$2(x^2 - 6x \quad) + (y^2 + 2y \quad) = -11.$$

Inserting the required constant in each set of parentheses,

$$2(x^2 - 6x + 9) + (y^2 + 2y + 1) = -11 + 18 + 1$$
$$2(x - 3)^2 + (y + 1)^2 = 8$$
$$(y + 1)^2 = 8 - 2(x - 3)^2$$
$$(y + 1) = \pm\sqrt{8 - 2(x - 3)^2}.$$

We thus obtain the two functions

$$y1 = -1 + \sqrt{8 - 2(x - 3)^2} \quad \text{and} \quad y2 = -1 - \sqrt{8 - 2(x - 3)^2}.$$

The viewing window should be centered at $x = 3$, $y = -1$. Since there will be no picture if $2(x - 3)^2 > 8$, interest centers on $|x - 3| \leq 2$, and for such x, we will have

$$-1 \leq y1 \leq -1 + \sqrt{8} \quad \text{and} \quad -1 - \sqrt{8} \leq y2 \leq -1.$$

In light of this analysis, it seems reasonable to begin with a viewing screen described by xMin = 0, xMax = 6, yMin = 4, and yMax = 2 (Figure 5.6).

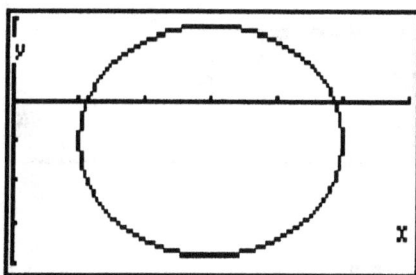

Figure 5.6. Somewhat distorted graph of
$$2x^2 + y^2 - 12x + 2y + 11 = 0.$$

You must keep in mind the distortion of scales on the two axes when viewing this graph. The major (long) axis of this ellipse is parallel to the y axis, something you will more easily see if you select $\boxed{\text{ZSQR}}$ from the $\boxed{\text{ZOOM}}$ menu.

For our final example, we shall graph

$$4y^2 - 5xy - x^2 + 9 = 0.$$

The presence of the xy term makes this a difficult curve to identify or graph by traditional methods. We shall treat it as a quadratic equation in y, using the quadratic formula to write

$$y = \frac{5x \pm \sqrt{25x^2 - 4(4)(9)}}{8}.$$

The two functions to be graphed are

$$y1 = \frac{5x + \sqrt{25x^2 - 144}}{8} \quad \text{and} \quad y2 = \frac{5x - \sqrt{25x^2 - 144}}{8}.$$

A suitable viewing screen may be found by analysis of the sort we did in the last example, or you may select the $\boxed{\text{ZSTD}}$ option from the $\boxed{\text{ZOOM}}$ menu as we did to obtain Figure 5.7.

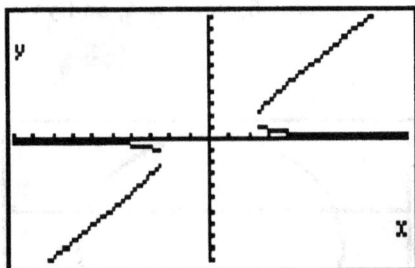

Figure 5.7. Graph of $4y^2 - 5xy - x^2 + 9 = 0$.

The true hyperbolic nature of the curve can then be examined more carefully using other options on the $\boxed{\text{ZOOM}}$ menu. The $\boxed{\text{ZIN}}$ option with $x \approx 3.97$, $y1 \approx .29$ produces Figure 5.8.

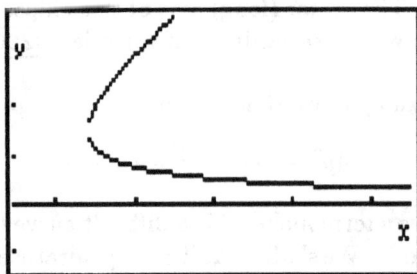

Figure 5.8. One branch of $4y^2 - 5xy - x^2 + 9 = 0$.

Parametric Representation

There is another way to represent the hyperbola $y^2 = x^2 - 9$ described above. Recalling that $\cosh^2 t - \sinh^2 t = 1$, we see that the parametric equations

$$x(t) = 3\cosh t \qquad \text{and} \qquad y(t) = 3\sinh t$$

will represent that part of the curve for which $x \geq 3$.

To obtain this graph on the TI-85, begin by pressing $\boxed{\text{MODE}}$ and activating Param. Then press in turn $\boxed{\text{GRAPH}}$, $\boxed{E(t)}$, $\boxed{3}$, $\boxed{2^{\text{nd}}}$, $\boxed{\text{MATH}}$, $\boxed{\text{HYP}}$, $\boxed{\text{COSH}}$, $\boxed{2^{\text{nd}}}$, $\boxed{t}$ to obtain $xt1 = 3\cosh t$. Similarly get $yt1 = 3\sinh t$. In the $\boxed{\text{RANGE}}$ menu, set the x and y bounds as before, and let tMin $= -3$, tMax $= 3$. Note (for example) that the choice of $t = \ell n2$ gives a point P with x and y coordinates given by $xt1 = 3.7500$, $yt1 = 2.2500$.

The slope of a tangent line at a point on a curve is obtained from the formula

$$\frac{dy}{dx} = \frac{dy}{dt} \bigg/ \frac{dx}{dt}.$$

At the point P on the curve under discussion, we may find the slope of the tangent line by using the derivative option as follows. Press $\boxed{\text{CALC}}$, $\boxed{\text{der1}}$, and then follow the steps above to access the cosh function. Continue in this way to obtain

$$\text{der1}(\sinh t, t, \ell n2)/\text{der1}(\cosh t, t, \ell n2) = 3.7500/2.2500 = 1.667\,.$$

Alternatively, find the slope from

$$\text{der1}(\sqrt{x^2 - 9}, x, 3.75) = 1.667\,.$$

"Seeing" Geometric Properties

The conic sections have many properties that make them important in applications and beautiful objects of study for mathematicians. Since we wish to illustrate some of these properties that can be "seen," all graphs drawn in this section use the $\boxed{\text{ZSQR}}$ option of $\boxed{\text{ZOOM}}$, enabling us to recognize equal lengths and equal angles.

Consider the ellipse graphed in Figure 5.6. Theory tells us that its foci are on the major axis, the line $x = 3$ in this case, at a distance $\sqrt{a^2 - b^2} = \sqrt{8 - 4} = 2$ from the center at $(3, -1)$. That is, the foci are $F_1(3, 1)$ and $F_2(3, -3)$.

You may now verify with your calculator the fact that any point P on the ellipse has the property that $F_1P + F_2P = 2a$. Use the trace feature to select a point such as $P(4.6997, 0.4906)$, and verify that

$$F_1 P + F_2 P = \sqrt{(4.6997 - 3)^2 + (0.4906 - 1)^2}$$
$$+ \sqrt{(4.6997 - 3)^2 + (0.4906 + 3)^2}$$
$$= 5.6569$$

is $2a$. The line segments $F_1 P$ and $F_2 P$ are shown in Figure 5.9.

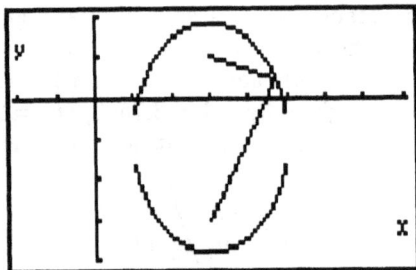

Figure 5.9. Line segments from foci to a point on an ellipse.

We next illustrate several properties of the parabola using the two functions

$$y1 = 2\sqrt{x} \qquad \text{and} \qquad y2 = -2\sqrt{x}$$

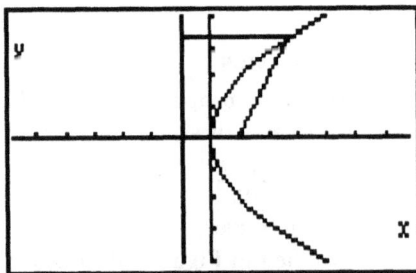

Figure 5.10. Line segments from foci and directrix to point on parabola.

to obtain the graph of $y^2 = 4x$ which has $F(1,0)$ as its focus and the line $x = -1$ as its directrix. Using the trace feature to determine a point such as $P(2.7267, 3.3025)$ on the parabola, you can easily show that $FP = \sqrt{(2.7267-1)^2 + (3.3025)^2} = 3.7267$, the same as the distance to the directrix that can be drawn by pressing $\boxed{\text{GRAPH}}$, $\boxed{\text{DRAW}}$, and $\boxed{\text{VERT}}$, then using $\boxed{\blacktriangleleft}$ to move the vertical line to $x = -1$ and pressing $\boxed{\text{ENTER}}$. (Figure 5.10).

Next, draw the tangent line at P by pressing $\boxed{\text{GRAPH}}$, $\boxed{\text{MATH}}$, and $\boxed{\text{TANLN}}$, then moving the cursor to a point P on the graph, and pressing $\boxed{\text{ENTER}}$. Then, draw the line from $F(1, 0)$ to P by pressing in sequence $\boxed{\text{GRAPH}}$, $\boxed{\text{MATH}}$, $\boxed{\text{DRAW}}$, and $\boxed{\text{LINE}}$; and with the same options draw from P the ray PR that is parallel to the x axis (Figure 5.11).

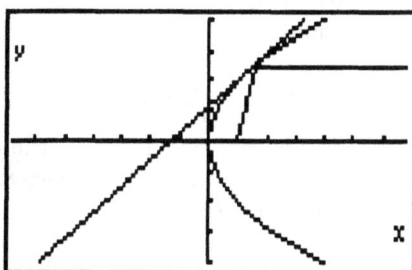

Figure 5.11. Ray from the focus reflected by a parabola.

The angle formed by FP with the tangent line should look equal to the angle formed by PR and the tangent line; the angle of incidence equals the angle of reflection. This illustrates the reflective property of a parabolic mirror. All rays from a source at P will be reflected in rays that are parallel to the axis of the parabola. Try this with other points.

Explorations with Conic Sections

1. Obtain the graph of the circle of radius 3 centered at the origin
   a) by using $y1 = \sqrt{9 - x^2}$ and $y2 = -\sqrt{9 - x^2}$.
   b) by using the parametric representation $xt1 = 3\cos t$, $yt1 = 3\sin t$.
   Make proper adjustments so that your graph looks like a circle.

2. Find the slope of the line tangent to the circle $x^2 + y^2 = 9$ at $(3\sqrt{3}/2, -3/2)$
   a) using the $\boxed{\text{der1}}$ key and the correct function of x.
   b) using the $\boxed{\text{der1}}$ key in conjunction with the parametric representation above.

    c) from elementary geometric considerations.

3. Graph $4xy - 3x^2 = 8$.

4. Using a point P of your own choosing on the graph of $2x^2 + y^2 - 12x + 2y + 11 = 0$, verify that $F_1P + F_2P = 2a$.

5. Referring to the parabola $y^2 = 4x$ graphed in Figure 5.11, use $\boxed{\texttt{der1}}$ to find the slope of the line tangent to the curve at $P(2.7267, 3.3025)$. Then write the equation of this line and determine the point Q where this tangent line intersects the x axis. Verify the reflective property of the parabola by showing that the $\triangle FPQ$ is isosceles.

6. Consider a ray that emanates from a focus of an ellipse. Can you "see" where it will be reflected? Start by drawing a line tangent to the ellipse at a point P of your choosing.

5.3. Polar Graphing

    Figure 5.12 shows a vector that extends from the origin to the point $P(3, 0)$. It shows the same vector, somewhat shorter, after it has been rotated to make a 30° angle with the positive x axis, and again when, still shorter, it makes a 60° angle with the x axis. Dots show other terminal points of this vector as it has rotated.

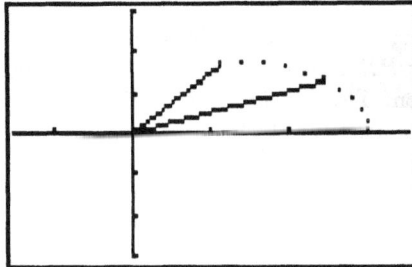

Figure 5.12. Terminal points of a rotating vector.

    Obviously the vector is getting shorter as the angle θ made with the x axis increases. The program given at the end of this section will enable you to see this picture develop as you watch, though the vectors are suppressed in order to focus attention on the terminal points. The vector pauses after each 30° interval; you must press $\boxed{\texttt{ENTER}}$ to get it to continue.

Examination of the program reveals the relationship between r and θ; $r = 1 + 2\cos\theta$. Graphs drawn in this way are called polar graphs. One thinks of a radius vector rotating around the origin (called the pole in this context), varying in length with θ. In the example we are considering, r decreases from 3 to 0 as θ goes from 0° to 120°, and then goes negative as θ goes from 120° to 240° (Figure 5.13).

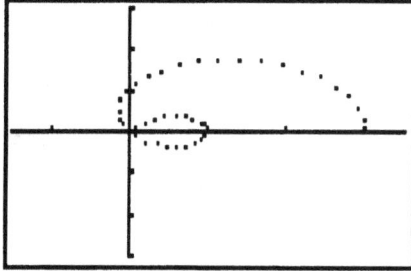

Figure 5.13. Graph of $r = 1 + 2\cos\theta$, $0 \le \theta \le 2\pi/3$.

Making Tracks

This same program, used with other polar equations, will enable you to visualize the relationship between r and θ; and of course the polar graphing capability (activate Pol from line 5 of the menu after you press $\boxed{\text{MODE}}$) of your TI-85 will enable you to draw these graphs without the pauses. The discrete dots in our figures result from our having chosen the DrawDot option in the graph format menu; the DrawLine option will give a more continuous appearing picture. If we think of the endpoint of the rotating vector as giving the location of moving body (say a child walking through a snow covered yard), then the graph can be thought of as the tracks left behind.

Intersecting Tracks and Colliding Bodies

Suppose two such graphs, $r1(\theta) = f(\theta)$ and $r2(\theta) = g(\theta)$ are drawn. Two questions can be asked:

1) Do the tracks made by these moving bodies intersect?

2) Do the bodies on these tracks collide?

If the answer to the first question is no, then the answer to the second question is also no, but it should be apparant that intersecting tracks do not necessarily imply colliding bodies.

Consider the specific example in which two moving bodies are described by

$$r1(\theta) = 1 + 2\cos\theta \quad \text{and} \quad r2(\theta) = \cos\theta - \frac{3}{2}.$$

Obtain their graphs by first setting the mode to Pol and the graph format to DrawLine as described above, and then pressing $\boxed{\text{GRAPH}}$ and $\boxed{r(\theta)}$ so as to be able to enter the two equations as $r1$ and $r2$, activating both. After pressing $\boxed{\text{RANGE}}$, set θMin = 0, θMax = 2π, xMin = -1.5, xMax = 3, yMin = -2, yMax = 2. The tracks of the two moving bodies we are considering obviously (Figure 5.14) intersect.

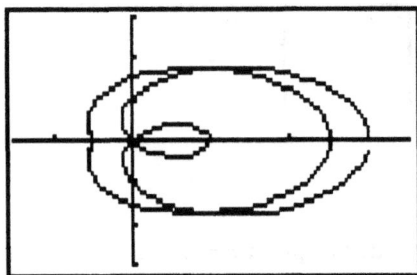

Figure 5.14. Graphs of $r1(\theta) = 1 + 2\cos$ and $r2(\theta) = \cos\theta - \frac{3}{2}$.

If we equate the two expressions for r, we get

$$1 + 2\cos\theta = \cos\theta - \frac{3}{2}$$

$$\cos\theta = -\frac{5}{2}.$$

There are no solutions. This means that there is no choice of θ that produces the same value of r in both equations. It is thus established that the bodies do not collide. This can be seen very nicely on your calculator if you press $\boxed{\text{GRAPH}}$, $\boxed{\text{FORMT}}$, and then activate the SimulG option. Then if you press $\boxed{\text{GRAPH}}$ again, the simultaneous drawing of the graphs will show you that as θ goes from 0 to $\pi/2$, the two graphs appear as in Figure 5.15, and as θ continues to increase, the leading edges of the curves are never in the same place at the same time.

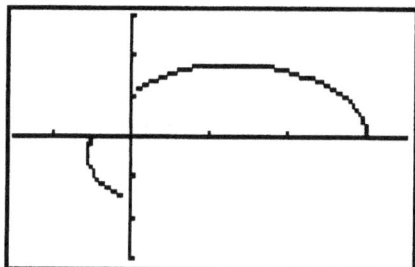

Figure 5.15. The two graphs of Figure 5.14 for $0 \leq \theta \leq \pi/2$.

To find the intersections apparent in Figure 5.14, it is necessary to use the fact that (r, θ) and $(-r, \theta + \pi)$ are polar coordinates of the same point. If you replace r by $-r$ and θ by $\theta + \pi$ in $r = \cos \theta - \frac{3}{2}$, you get

$$-r = \cos(\theta + \pi) - \frac{3}{2} = -\cos \theta - \frac{3}{2} \quad \text{or} \quad r = \cos \theta + \frac{3}{2}$$

A body whose motion is described by this equation makes the same track as does $r = \cos \theta - \frac{3}{2}$, but solving

$$1 + 2 \cos \theta = \cos \theta + \frac{3}{2}$$

gives the solutions of $\theta = \frac{\pi}{3}$ and $\theta = \frac{5\pi}{3}$. Once again the simultaneous graphing feature enables us to see what happens; in this case we have collisions.

Intervals of Integration for Finding Areas

To find the area A in the first quadrant between the outer and inner loops of the graph of $r = 1 + 2 \cos \theta$, refer to Figure 5.13. The outer loop is swept out as θ goes from 0 to $\frac{\pi}{2}$; the inner loop to be subtracted is traversed as θ goes from π to $\frac{4\pi}{3}$. Thus,

$$A = \frac{1}{2} \int_0^{\pi/2} r^2 d\theta - \frac{1}{2} \int_\pi^{4\pi/3} r^2 d\theta .$$

The first integral can be evaluated with your calculator, using

$$\text{fnInt}\left((1 + 2 \cos x)^2, x, 0, \frac{\pi}{2}\right)$$

to obtain 8.7124. Use $\boxed{2^{\text{nd}}}$ $\boxed{\text{Enter}}$ to recall this expression and edit to obtain fnInt $\left((1 + 2\cos x)^2, x, \pi, \frac{4\pi}{3}\right) = 0.5435$.

$$A = \frac{1}{2}[8.7124 - 0.5435] = \frac{1}{2}[8.1689] = 4.0844.$$

The program referred to in this section is as follows.

```
 PROGRAM: POLRGRAF
:ClLCD
:Radian
:Pol
:r1=1+2 cos θ
:FnOff
: -1→xMin
:3→xMax
:1→xScl
: -2→yMin
:2→yMax
:1→yScl
:0→J
:For(I,1,12,1)
:J+1→J
:For(θ,(π/6)(J-1),(π/6)J,π/24)
:(r1∠θ)→A
:PtOn(real A,imag A)
:End
:Pause
:End
```

Program 5.1.

Explorations with Polar Graphs

1. In Program 5.1, replace $r1 = 1 + 2\cos\theta$ by $r1 = 2\cos 3\theta$. Mark the points on the resulting graph that correspond to $\theta = \pi/6$, $\theta = 2\pi/6, \ldots, \theta = 12\pi/6$.

2. Repeat number 1 above with $r1 = \sqrt{1 - 2\cos\theta}$.

3. After pressing $\boxed{\text{GRAPH}}$, $\boxed{\text{FORMT}}$, and activating SeqG, draw graphs of both
$$r1 = 2 + 2\sin\theta \quad \text{and} \quad r2 = 2\sin\theta - 2.$$
How many graphs do you see? Why? What will happen if in the graphing format menu you choose SimulG?

4. Find the area enclosed by the graph of $r = \sqrt{1 - 2\cos\theta}$.

5.4. Parametric Equations

Even when the variable y depends on the single variable x, it often gives more information or makes the relationship easier to study if both x and y are expressed in terms of some third variable, say t. Such a dependency is expressed by writing $x(t) = f(t)$, $y(t) = g(t)$, and t is called a parameter.

A stone's throw

We propose to study the motion of a stone thrown from the edge of a ravine to a spot 150 feet below. We shall assume that it is thrown so that its initial horizontal and upward velocities are both 72 feet per second. If we impose axes on this scene, letting $(0, 150)$ be the point from which the stone is thrown, then its position at time t will be

$$x = 72t \quad \text{and} \quad y = 150 + 72t - \frac{1}{2}(32.2)t^2.$$

Both x and y depend on the variable t, the parameter. We could eliminate t from the equations, thus finding the equation of the path along which the stone travels. We would, however, lose some information in doing this, since the parametric equations tell us not only the coordinates of a point on the path, but also the time t at which the stone was at the point.

To get a picture of the path, press $\boxed{\text{MODE}}$ and activate Param on line five of the menu. Next enter the equations by pressing $\boxed{\text{GRAPH}}$ and $\boxed{E(t)}$. To set your range values, first observe that the stone will hit bottom, that is y will be 0, when $-16.1t^2 + 72t + 150 = 0$. Solving (press $\boxed{\text{POLY}}$, enter the coefficients for this order 2 polynomial, and press $\boxed{\text{SOLVE}}$) gives the positive root of $t = 6.0198$, meaning that a suitable range for x is 0 to something a bit larger than 72×6; we have used xMax = 450 in

Figure 5.16. The variable y is set between 0 and 250, and both scales are set at 50.

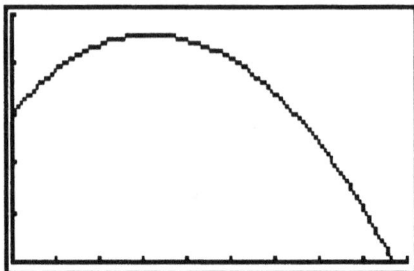

Figure 5.16. Parabolic path of a thrown stone.

Rolling Wheels

The parameter in parametric equations is not always thought of as time. A point on the rim of a wheel of radius 1, starting at the origin, will as the wheel rolls through an angle t trace out what is called a cycloid curve, described by the parametric equations

$$x = t - \sin t \quad \text{and} \quad y = 1 - \cos t.$$

For two revolutions of the wheel, t goes from 0 to 4π. In this case, we will need x to go from 0 to 4π, and of course y will vary from 0 to 2. This curve won't be in correct proportion unless you choose (as we have done in Figure 5.17) to press ZSQR found in the menu displayed after pressing GRAPH and ZOOM.

As a variation on rolling a circle of radius 1 along a straight line, consider rolling it along the rim of a larger circle, say the circle of radius 4 centered at the origin. If the small circle is exterior to the large circle, and if a point P on the smaller circle initially touches the large circle at $(4, 0)$, then with t measuring the angle through which the small circle is rotated, the point P traces out a path described by the equations

$$x = 5\cos t - \cos 5t \quad \text{and} \quad y = 5\sin t - \sin 5t.$$

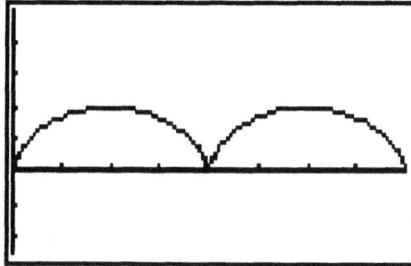

Figure 5.17. A cycloid.

The curve can be pictured on your TI-85 using xMin $= -6$, xMax $= 6$, yMin $= -6$, yMax $= 6$, and letting t go from 0 to 2. As described above, use ZSQR to make circles look like circles. You can then superimpose the larger circle of radius 4 on this picture by pressing GRAPH , DRAW , and CIRCL . Position the cursor at $(0, 0)$ and press ENTER , then move the cursor to as close as you can get to $(4, 0)$ and press ENTER again (Figure 5.18). The curve that lies exterior to the large circle is called an epicycloid.

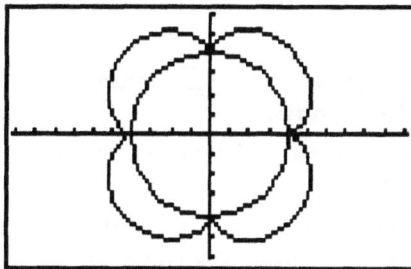

Figure 5.18. An epicycloid.

If the small circle is interior to the large circle, and if the point P again begins at $(4, 0)$, the curve traced out by P is called a hypocycloid.

Its equations are

$$x = \cos^3 t \quad \text{and} \quad y = \sin^3 t .$$

See Figure 5.19.

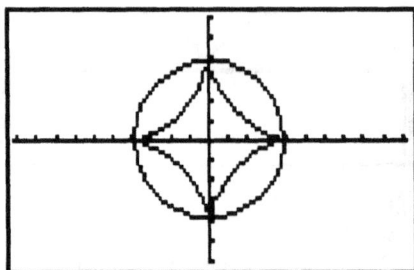

Figure 5.19. A hypocycloid.

Explorations with Parametric Equations

1. It is possible to explore in much greater detail than was done in this section the motion of the stone thrown into the ravine.

   a) If you followed with your calculator as the solution of $-16.1t^2 + 72t + 150 = 0$ was described, you obtained not only the announced root of 6.0198, but also -1.5477. What do you think is the significance of that number?

   b) Use $\boxed{\text{TRACE}}$, $\boxed{\text{ZOOM}}$ and $\boxed{\text{ZIN}}$ to estimate the maximum height to which the stone rises.

   c) Find the time when the stone reaches its maximum height; call this time tHI.

   d) Find the slope of the stone's trajectory at $t_1 = tHI - 2$ and also at $t_2 = tHI + 2$ by using the fact that

$$\text{slope} = \frac{dy}{dx} = \frac{dy}{dt} \bigg/ \frac{dx}{dt} .$$

Press $\boxed{\text{CALC}}$ and $\boxed{\text{der1}}$ in order to use

$$\text{der1}(-16.1t^2 + 72t + 150, t, t_1)/\text{der1}(72t, t, t_1)$$

and proceed similarly at $t2$.

2. We described in this section a stone that was thrown so that its initial horizontal and upward velocities were both 72 feet per second. This is equivalent to saying that it was thrown with an initial velocity of $72\sqrt{2}$ feet per second at an angle of 45° with the horizontal. If it had been thrown at an angle of θ with the horizontal, the parametric description of the path would have been

$$x = 72\sqrt{2}(\cos\theta)t \quad \text{and} \quad y = -16.1t^2 + 72\sqrt{2}(\sin\theta)t + 150.$$

Find the horizontal distance that the stone will have travelled when it hits the ground for $\theta = 20°$; $\theta = 30°$; $\theta = 40°$; $\theta = 50°$; $\theta = 60°$. What is the maximum height achieved for the listed values of θ ?

3. Using the procedure outlined in problem 1 above, explore the slope of the line tangent to the cycloid for various values of t between 0 and 2. Where is the slope most steep?

5.5. Functions Defined by Integrals

An integral of the form $\int_1^x f(t)dt$ may be interpreted to be the area under the graph of $y = f(t)$ from 1 to x. As such, it may be interpreted as a function of x. It is, in fact, a rather nice function of x; it is continuous; it is differentiable.

Still, teachers of calculus have trouble getting students to accept such expressions as an ordinary function, primarily because such functions are not easy to evaluate. We are conditioned, after all, to believe that a function g will be defined by a nice formula such as $g(x) = x + \frac{1}{x^2}$. We then know how to make a table of values, plot a graph of the function, and study its properties for $x > 0$ if asked to do so.

x	$g(x)$
$\frac{1}{2}$	$\frac{9}{2}$
1	2
2	$\frac{9}{4}$
3	$\frac{28}{9}$

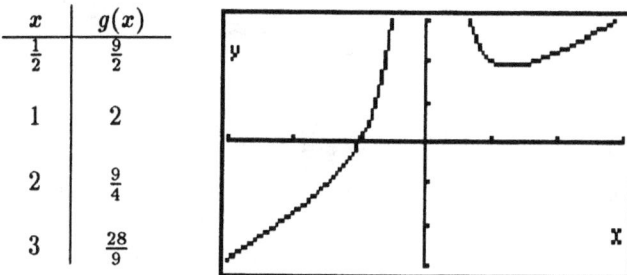

Figure 5.20. The graph of $g(x) = x + \frac{1}{x^2}$.

We would know how to enter such a function into our calculator and study it in that way.

The natural logarithm $\ln x$

What are we to do, however, if asked to study for $x > 0$,

$$L(x) = \int_1^x \frac{dt}{t}.$$

Armed with our calculator, we can proceed in much the same way as we did for g. We could begin by making a table of values. Type in the instruction

$$.5 \rightarrow x : \text{fnInt}(1/t, t, 1, x)$$

remembering that to obtain a lower case x after $\rightarrow$, you must enter $\boxed{\text{ALPHA}}$ $\boxed{\text{X}}$; the fnInt is obtained by pressing $\boxed{\text{CALC}}$, then $\boxed{\text{fnInt}}$. Following this expression by $\boxed{\text{ENTER}}$ will produce $-.6931$. We can in the usual way (press $\boxed{\text{2nd}}$ $\boxed{\text{ENTRY}}$ to recall the last command, use the $\boxed{\blacktriangle}$ and $\boxed{\blacktriangleleft}$ keys to position the cursor at the . in .5 and type 1) replace .5 with $1, 2, \ldots$ to obtain a table from which a graph can be drawn.

x	$L(x)$
.5	-0.6931
1	0.0000
2	0.6931
3	1.0986
4	1.3863
5	1.6094
6	1.7918
7	1.9459
8	2.0794

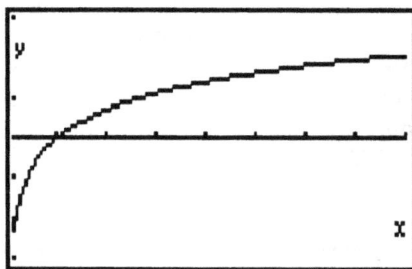

Figure 5.21. Graph of $y = L(x)$.

Moreover, the graph in Figure 5.21 can be obtained from your calculator just as surely as can the graph in Figure 5.20. Merely set $y(x) = \text{fnInt}(1/t, t, 1, x)$. You will notice that the graph takes longer to appear, reflecting the fact that there is a lot of summing to be done for each choice of x.

Drawing the graph of a function is the mechanical part of studying that function. The insight, the inspiration comes from the analysis that follows. If you note from the table above that

$$L(4) = 2L(2)$$
$$L(6) = L(2) + L(3)$$
$$L(8) = 3L(2)$$

then you are curious enough to read with anticipation what your calculus text has to say about this function. More importantly, you have taken the significant step of beginning to really believe that $L(x)$ is a legitimate function that can be graphed and studied like any other function.

The inverse tangent

Once in the frame of mind to accept functions defined in terms of integrals as legitimate creatures, recall from elementary anti-differentiation formulas that

$$\int_0^x \frac{dt}{1+t^2} = \tan^{-1}(x) - \tan^{-1}(0).$$

That is

$$\tan^{-1}(x) = \int_0^x \frac{dt}{1+t^2}.$$

Then since $\tan^{-1}(1) = \frac{\pi}{4}$, your calculator will give

$$\frac{\pi}{4} = \text{fnInt}(1/(1+t^2), t, 0, 1) = .7854$$

from which you obtain as the next step

$$\pi = 4(.7854) = 3.1416.$$

Explorations with Functions Defined by Integrals

1. Let $A(x) = \int_0^x \frac{dt}{\sqrt{1-t^2}}$. Find $A\left(\frac{1}{2}\right)$, $A\left(\frac{1}{\sqrt{2}}\right)$, $A\left(\frac{\sqrt{3}}{2}\right)$. How do your answers relate to π? Why?

2. Let $N(x) = \frac{1}{\sqrt{2\pi}} \int_0^x e^{-t^2/2} dt$. Find $N(0.5)$, $N(1.0)$, $N(1.5)$, $N(2.0)$. Compare your answers with a table in a statistics book showing areas under the normal distribution curve.

3. Let $S(x) = \int_0^x \left(1 - \frac{t^2}{2} + \frac{t^4}{24} - \frac{t^6}{720}\right) dt$. Find $S(\pi/6)$, $S(\pi/4)$, and $S(\pi/3)$. Compare these values with the values of $\sin(\pi/6)$, $\sin(\pi/4)$, and $\sin(\pi/3)$.

5.6. Indeterminate Forms

We come now to a topic that nicely illustrates the adage that fools rush in where angels fear to tread, and where a person equipped with a calculator can, with a little experimenting, avoid jumping to some erroneous conclusions.

Some questions

The derivative of $f(x) = \sqrt{x}$ at $x = 3$ is defined to be

$$\lim_{h \to 0} \frac{\sqrt{3+h} - \sqrt{3}}{h} . \tag{5.1}$$

An inexperienced person might reason that since the numerator will be very close to zero when h is chosen to be very small, the value of the fraction approaches zero. The problem is that as the numerator gets small, so also does the denominator. The experienced observer will be cautious, and will say that the result is not determined by the information in hand so far.

One way to motivate the formula for the derivative of the logarithm function is to suppose that for any base $b > 0$, $L(x) = \log_b x$ can be treated as a continuous function of x. Exploration of the derivative of L then leads to

$$\lim_{h \to 0} \frac{L(x+h) - L(x)}{h} = \frac{1}{x} \log_b \lim_{h \to 0} (1+h)^{1/h} . \tag{5.2}$$

In this case, an intelligent but inexperienced person will almost surely reason that for very small choices of h, $(1 + h)$ becomes 1 for all practical purposes, and 1 raised to any power will still be 1. It's an understandable conclusion. We shall see that it is also wrong.

If you wish to sort alphabetically 100 names, you may use almost any method you think of; it won't take too long. But if you had from a scrambled list to alphabetically sort the names in the Chicago phone book, you would be interested in finding an efficient method. One way that computer scientists evaluate algorithms for sorting, or for any other task, is to ask how fast the required time for the task will grow as the number n of items grows. Questions of this sort lead people to evaluate expressions such as

$$\lim_{n \to \infty} \frac{n \ell n n}{n^{3/2}} . \tag{5.3}$$

The limits labeled (5.1), (5.2) and (5.3) in this section are, respectively said to be indeterminate of the forms $\frac{0}{0}$, 1^∞, and $\frac{\infty}{\infty}$. Such forms come up frequently in the applications of mathematics. They are frequently counterintuitive, and they have given rise to numerous special methods designed to find answers to special problems.

Some suggested answers

Counter-intuitive answers, coupled with methods of analysis that appear to have been specifically designed for the single problem to which they are applied, leave the beginner wary of being asked to evaluate indeterminate forms. One may be excused for wishing that there was some general advice to help us to at least get started, to help us develop some sense of the size of the answers to be expected.

The calculator offers us a first approach of almost universal applicability. Begin by using the expression under investigation to define a function. Then create a table of values for this function, and observe what happens to the functional values as the variable changes in the prescribed way.

The TI-85 is set up in such a way that it is easiest (though not essential) to use x as the independent variable. Thus, for the three questions raised above, we define the three functions

$$y1 = \frac{\sqrt{3+x} - \sqrt{3}}{x}$$
$$y2 = (1+x)^{1/x}$$
$$y3 = \frac{x\ell nx}{x^{3/2}}.$$

Then press $\boxed{\text{GRAPH}}$ and $\boxed{\text{EVAL}}$ to create a table of values for each function.

x	$y1$	x	$y2$	x	$y3$
0.5	0.2775	0.5	2.2500	100	0.4605
0.1	0.2863	0.1	2.5937	1000	0.2184
0.01	0.2884	0.01	2.7048	5000	0.1205
0.001	0.2887	0.001	2.7169	10,000	0.0921

An algebraic trick (multiply numerator and denominator by $\sqrt{3+x} + \sqrt{3}$ and simplify) can be used on (5.1) to show that the limit is $1/2\sqrt{3} = 0.2887$. If encountered in a first investigation of logarithms and exponents, (5.2) calls for very sophisticated analysis; if encountered but not recognized after one knows about the natural logarithm, then analysis of the logarithm of the expression will show that the limit is $e = 2.7183$. With the aid of L'Hopital's rule, (5.3) is easily shown to be zero.

Each of the tables above, carried further, would give the desired limit to greater accuracy. Our purpose, however, is not to gain great accuracy or to suggest that the many techniques developed to study indeterminate forms should not be studied. Our purpose is to show that for any problem of this sort, your calculator can steer you away from the wrong conclusions to which your intuition might lead you, and toward a sense of what the right answer might be.

Two special cases

There are two indeterminate forms that deserve special attention because they come up in the derivation of the derivative of $\sin x$. In particular,

$$\lim_{h\to 0}\frac{\sin(x+h)-\sin x}{h}=\lim_{h\to 0}\frac{\sin x\cos h+\cos x\sin h-\sin x}{h}$$

$$=\lim_{h\to 0}\left[\sin x\frac{\cos h-1}{h}+\cos x\frac{\sin h}{h}\right].$$

In this way our attention is focused on

$$\lim_{h\to 0}\frac{\cos h-1}{h}\quad\text{and}\quad\lim_{h\to 0}\frac{\sin h}{h}.$$

By using tables as suggested above, you should have no trouble convincing yourself that the first limit is zero. We wish to focus attention on the second limit. The table that you construct this time will differ a great deal, depending on whether your calculator is set in degree or radian mode.

degrees		radians	
x	$\frac{\sin x}{x}$	x	$\frac{\sin x}{x}$
0.2	0.0175	0.2	0.9933
0.1	0.0175	0.1	0.9983
0.01	0.0175	0.01	1.0000

The simplicity of the limit when x is measured in radians gives a clear indication of what system of angle measurement will always be used in calculus; radians it shall be!

Explorations with Indeterminate Forms

1. Using the values of x used in the exploration of $\frac{\sin x}{x}$ as x, measured in radians, approached zero, create a similar table for $\frac{\cos x-1}{x}$.

2. We explored in this section the limit of $(1+x)^{\frac{1}{x}}$ as $x\to 0$, finding it to be the number $e=2.7183$. Similarly explore $(1+nx)^{\frac{1}{x}}$ for $n=2,3$, and 4. What relationship do these limits bear to e?

3. The techniques of this section can be used, of course, to find limits in cases that do not involve indeterminate forms. Consider, for example, the function

$$F(x)=\frac{4}{1+2^{\frac{1}{x-1}}}.$$

a) Find $F(2)$, $F(1.5)$, $F(1.1)$, and $F(1.01)$. Toward what limit does $F(x)$ tend as x approaches 1 from the right?

b) Toward what limit does $F(x)$ tend as x approaches 1 from the left?

c) Find $F(5)$, $F(10)$, $F(15)$, and $F(20)$. Toward what limit does $F(x)$ tend as $x \to \infty$?

d) Toward what limit does $F(x)$ tend as $x \to -\infty$?

5.7. Improper Integrals

There are many kinds of human behavior that are improper; we only need to worry about two kinds of improper integrals.

The first kind

A person who knows the fundamental theorem of calculus might, without giving thought to the potential problem, write

$$\int_0^1 x^{-1/2}dx = 2x^{1/2}\Big|_0^1 = 2 .$$

The TI-85 does the same thing; press $\boxed{\text{CALC}}$ and $\boxed{\text{fnInt}}$, allowing you to obtain $\text{fnInt}(x^{-1/2}, x, 0, 1) = 2.0000$.

Proceeding similarly, the fundamental theorem gives

$$\int_0^1 x^{-3/4}dx = 4x^{1/4}\Big|_0^1 = 4$$

but this time the TI-85 responds to $\text{fnInt}(x^{-3/4}, x, 0, 1)$ with an error message.

A potential problem was alluded to in our initial example. Specifically, the integrand in both examples is discontinuous at the left endpoint of the interval of integration. In both cases, because of the discontinuity, the integrals are properly called improper integrals of the first kind, and are defined in terms of limits.

$$\int_0^1 \frac{1}{\sqrt{x}}dx = \lim_{b \to 0+} \int_b^1 x^{-1/2}dx = \lim_{b \to 0+} [2(1) - 2b^{1/2}] = 2$$

$$\int_0^1 \frac{1}{x^{3/4}}dx = \lim_{b \to 0+} \int_b^1 x^{-3/4}dx = \lim_{b \to 0+} [4(1) - 4b^{1/4}] = 4 .$$

This idea of defining an improper integral in terms of a limit is of more than theoretic interest; it gives us a practical way to deal with the expression that produced the error message. Begin with $\text{fnInt}(x^{-3/4}, x, 0.2, 1)$ which returns 1.3250. Use $\boxed{\text{2nd}}$ $\boxed{\text{ENTER}}$ to return the last expression so that you can easily edit and find $\text{fnInt}(x^{-3/4}, x, 0.1, 1) = 1.7506$. Repeated use of this recall and edit technique enables us to see the effect of letting b approach zero.

b	$\text{fnInt}(x^{-3/4}, x, b, 1)$
0.2	1.3250
0.1	1.7506
0.05	2.1085
0.001	3.2887
0.0001	3.6000
0.00001	3.7751
0.000001	3.8735
0.0000001	3.9600

Our calculator thus demonstrates in a beautiful way the theory of improper integrals in which the integrand is discontinuous at one of the endpoints. While

$$\int_0^1 \frac{1}{x^{3/4}} dx = \text{fnInt}(x^{-3/4}, x, 0, 1)$$

cannot be evaluated, it is given meaning by

$$\lim_{b \to 0+} \int_0^1 x^{-3/4} dx = \lim_{b \to 0+} \text{fnInt}(x^{-3/4}, x, b, 1) = 4 \,.$$

One more example is instructive. The expression $\text{fnInt}(x^{-5/4}, x, 0, 1)$ again gives an error message, so once again we turn to the idea of limits.

b	$\text{fnInt}(x^{-5/4}, x, b, 1)$
0.2	1.9814
0.1	3.1131
0.05	4.4590
0.001	18.4937
0.0001	36.0000

Something different is happening this time. The values are rapidly increasing. This illustrates the result obtainable using the fundamental theorem:

$$\lim_{b \to 0+} \int_0^1 x^{-5/4} dx = \lim_{b \to 0+} (-4x^{-1/4}) \Big|_b^1$$

$$= \lim_{b \to 0+} \left(-4 + \frac{4}{b^{1/4}}\right) = \infty \,.$$

All of the examples we have looked at in this section have yielded to analysis using the fundamental theorem of calculus. We have used a calculator on two of the examples to illustrate numerically what it means to say that

$$\lim_{b \to 0+} \int_b^1 f(x)dx$$

does or does not exist as a finite number. It should be clear that these same numerical techniques could be used to analyze an integral such as $\int_b^1 \frac{dx}{\ell n x}$ for which an anti-derivative does not easily present itself.

The second kind

There is a second kind of integral that is called improper. Using the TI-85 calculator, we can once again illustrate the idea with a table of values using the function $f(x) = x^{-5/4}$.

b	$\text{fnInt}(x^{-5/4}, x, 1, b)$
10	1.7506
50	2.4958
100	2.7351
500	3.1541
1000	3.2887
10000	3.6000

We are clearly investigating $\int_1^b x^{-5/4}dx$ as b gets big, but the observation that the last calculation takes the TI-85 over 20 seconds raises an obvious question. How big can we choose b before our calculator takes a prohibitive amount of time or balks altogether? Will we be able within the limits of our calculator to see a pattern?

The answer to these last questions might surprise you. Let's continue the table we started above.

b	$\text{fnInt}(x^{-5/4}, x, 1, b)$
50,000	3.7325
100,000	3.7751
250,000	3.8211
500,000	3.8496
1,000,000	3.8735
5,000,000	3.9154
100,000,000	3.9600
1,000,000,000,000	3.9960

This last calculation takes about one minute; it also prepares us to understand and believe the statement

$$\lim_{b \to \infty} \int_1^b x^{-5/4}dx = \lim_{b \to \infty} \left(-4x^{-1/4} \right) \Big|_1^b$$

$$= \lim_{b \to \infty} \left(\frac{-4}{\sqrt[4]{b}} + \frac{4}{1} \right) = 4 .$$

This situation is commonly summarized by writing

$$\int_1^\infty x^{-5/4} dx = 4.$$

Again, besides giving us a feel for whether a limit exists, what its value might be, and how large a choice of b we need to approximate the limit, the numeric calculations provide a way to analyze functions not having an easily determined anti-derivative. Consider

$$f(x) = \frac{1}{\sqrt[3]{x} + \sqrt{x}}$$

b	fnInt$(1/(\sqrt[3]{x} + \sqrt{x}), x, 1, b)$
10	2.4070
50	7.3370
100	11.2679
1,000,000	1744.7715

The evidence suggests a fact that is not easily established using the fundamental theorem alone, namely that

$$\int_1^\infty \frac{dx}{\sqrt[3]{x} + \sqrt{x}} = \infty.$$

Comparison of this integral with

$$\int_1^\infty \frac{dx}{2\sqrt{x}}$$

can be used to establish the limit, but it will not give to us the sense of how fast the integral grows that we get from the table.

Explorations with Improper Integrals

1. We suggested in this section that the methods illustrated could be used to analyze

$$\lim_{b \to 0+} \int_b^1 \frac{dx}{\ell n x}.$$

Do so.

2. We noted that fnInt($x^{-1/2}, x, 0, 1$) = 2.0000, the correct value of $\int_0^1 x^{-1/2}dx$, but that fnInt($x^{-3/4}, x, 0, 1$) gives an error message rather than the correct value of 4 for $\int_0^1 \frac{1}{x^{3/4}}dx$. For what values of r, $-\frac{3}{4} \leq r \leq -\frac{1}{2}$, does fnInt($x^r, x, 0, 1$) give the correct values?

3. Investigate the behavior of $\int_2^\infty \frac{dx}{x\ell nx}$.

5.8. Polynomials that Approximate Functions

If one picture is worth a thousand words, consider what can be learned from a series of pictures, or more precisely in the present section, from pictures of series.

The interval of convergence

On a screen set to show $-3 \leq x \leq 3$, $-5 \leq y \leq 5$, look at the graphs of $y = f(x) = \frac{1}{1+x}$ and the graphs, in turn, of each of the following four Taylor polynomials that correspond to f at $x_0 = 0$.

$$P_1(x) = 1 - x$$
$$P_2(x) = 1 - x + x^2$$
$$P_3(x) = 1 - x + x^2 - x^3$$
$$P_4(x) = 1 - x + x^2 - x^3 + x^4$$

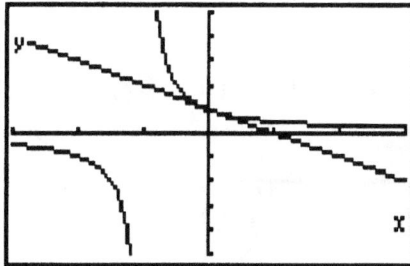

Figure 5.22. Graph of $f(x)$ and $P_1(x)$.

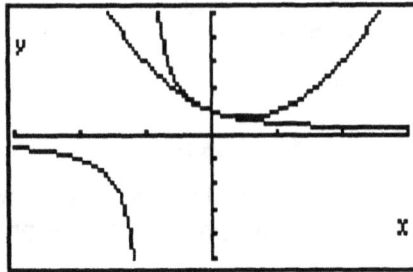

Figure 5.23. Graph of $f(x)$ and $P_2(x)$.

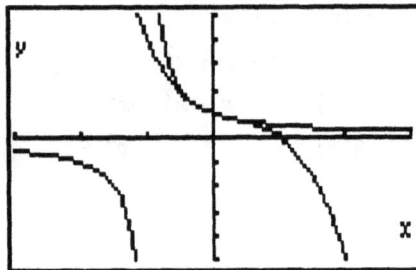

Figure 5.24. Graph of $f(x)$ and $P_3(x)$.

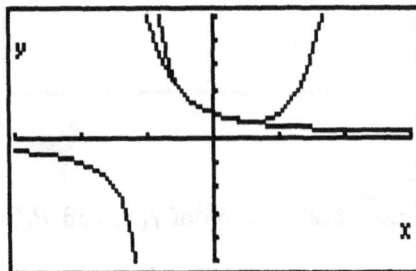

Figure 5.25. Graph of $f(x)$ and $P_4(x)$.

It is clear from the graphs that the polynomials provide good approximations to $f(x) = \frac{1}{1+x}$ so long as $-1 \leq x \leq 1$, and that the approximations
improve as the degree of the polynomials go up.

The Taylor polynomials corresponding to $S(x) = \sin x$ at $x_0 = 0$ behave differently, as can be seen by graphs that show $S(x)$ and each of the
following polynomials for $-2\pi \leq x \leq 2\pi$, $-1.5 \leq y \leq 1.5$.

$$Q_1(x) = x$$

$$Q_2(x) = x - \frac{x^3}{3!}$$

$$Q_3(x) = x - \frac{x^3}{3!} + \frac{x^5}{5!}$$

$$Q_4(x) = x - \frac{x^3}{3!} + \frac{x^5}{5!} - \frac{x^7}{7!}$$

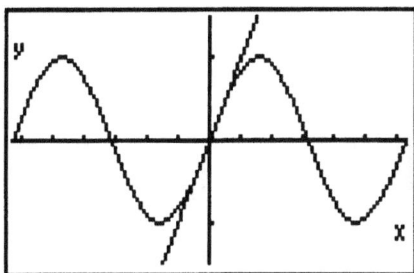

Figure 5.26. Graph of $S(x)$ and $Q_1(x)$.

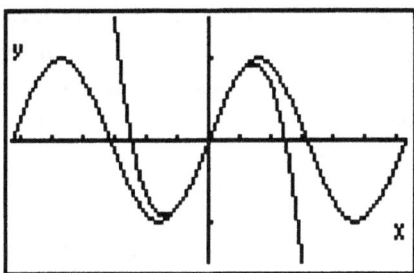

Figure 5.27. Graph of $S(x)$ and $Q_2(x)$.

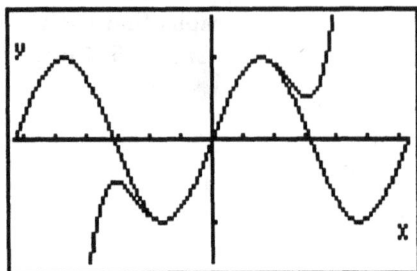

Figure 5.28. Graph of $S(x)$ and $Q_3(x)$.

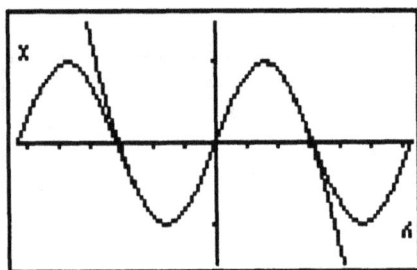

Figure 5.29. Graph of $S(x)$ and $Q_4(x)$.

In this case the polynomials seem to be identical with the given function for progressively wider intervals centered at the origin. They really are not identical; tracing the two functions S and Q_4 will show you, for instance, that for $x_1 = 0.6981317008$, $S(x_1) = 0.64278760969$ while $Q_4(x_1) = 0.64278750161$. It will also show that $Q_3(x_1) = 0.6428035389$; $Q_4(x_1)$ is a better approximation than $Q_3(x_1)$.

The broad conclusion is still correct; the most noticeable change as i increases is the increase in the size of the interval centered at the origin on which Q_i approximates S.

In the first case examined, there was a fixed, finite interval about the origin on which the given function was approximated ever more closely as the degree of the Taylor polynomials increased. In the second case, it was clear that we not only got better approximations, but that the approximations could be used on ever widening intervals about the origin.

Methods taught in calculus enable you to determine the interval on which a given function can be approximated by its Taylor polynomials. These methods will show you that $f(x) = \frac{1}{1+x}$ is approximated for $-1 \leq x \leq 1$ while $S(x) = \sin x$ is approximated on the entire real axis as the degree of the Taylor polynomial increases.

How good are the approximations? That is another question worthy of attention.

The error

We have observed that for $1 \leq x \leq 1$, the polynomials P_i, $i = 1, \ldots, 4$ defined above provide increasingly good approximations to $f(x) = \frac{1}{1+x}$. To better appreciate the meaning of "increasingly good," set the ranges of your calculator to $-1 \leq x \leq 1$, $0 \leq y \leq 2$, and draw graphs of f and all four polynomials on the same set of axes (Figure 5.30).

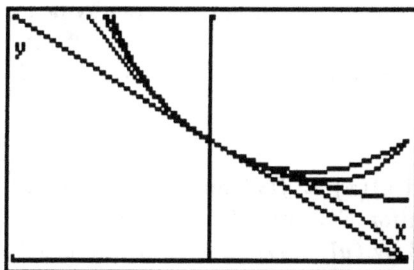

Figure 5.30. Graph of $f(x)$ and four polynomial approximations.

In Figure 5.30, we have numbered the graphs of the polynomials, a numbering you can follow by watching the order in which these graphs appear on your screen. It is evident that the graph of each polynomial, in its turn, follows more closely the graph of f.

The actual error between f and a particular polynomial approximation can be estimated for a particular choice of x by using the ZOOM and TRACE keys of your calculator. If you are interested in how well P_4 approximates

f across the entire interval, look at the graph of just these two functions (Figure 5.31).

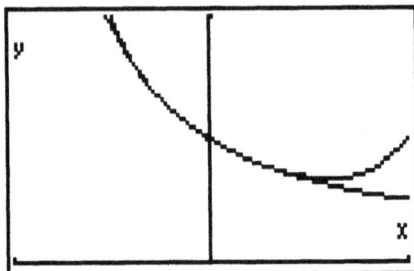

Figure 5.31. Graph of $f(x)$ and $P_4(x)$.

You might also explore the maximum error for $-0.5 \leq x \leq 0.5$, or ask yourself for the largest interval in which the error will be less than 0.05. A wide variety of such questions can be answered by using the $\boxed{\text{ZIN}}$, $\boxed{\text{TRACE}}$, and $\boxed{\text{EVAL}}$ keys of your calculator.

You might also explore these questions by first defining the Error function

$$E(x) = |f(x) - P_4(x)|.$$

If you have entered $y1(x) = f(x)$ and $y5(x) = P_4(x)$ in your calculator, then the expression for $E(x)$ may be entered by pressing $\boxed{\text{MATRX}}$, $\boxed{\text{CPLX}}$, and $\boxed{\text{abs}}$, followed by $(y1 - y5)$ entered in the usual way, giving $E(x) = \text{abs}(y1 - y5)$. Its graph is shown in Figure 5.32.

For a given function and its Taylor polynomial P_n, numerous methods exist for estimating the maximum value of the Error function on an interval $[a, b]$ since the maximum of E bounds the error that will be made if f is replaced by P_n on that interval. Find such a method in your calculus text, and use it to bound E on $\left[-\frac{3}{4}, \frac{3}{4}\right]$ for the functions f and P_4 of this section. Compare your answer with Figure 5.32.

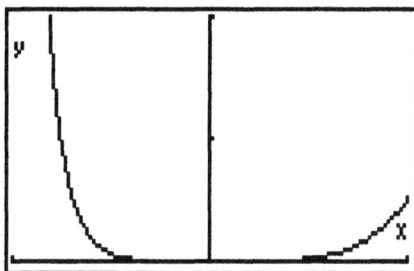

Figure 5.32. Graph of $E(x) = |f(x) - P_4(x)|$.

Explorations with Polynomials that Approximate

The following questions all refer to the function $L(x) = 1 + x$ and the approximating Taylor polynomials

$$R_1(x) = x$$

$$R_2(x) = x - \frac{x^2}{2}$$

$$R_3(x) = x - \frac{x^2}{2} + \frac{x^3}{3}$$

$$R_4(x) = x - \frac{x^2}{2} + \frac{x^3}{3} - \frac{x^4}{4}.$$

1. With the range set for $-3 \le x \le 3$ and $-3 \le y \le 3$, plot on the same axes the function $L(x)$ and the approximating Taylor polynomials. For what values of x do the polynomials seem to be approximating $L(x)$?

2. What is the lowest degree polynomial that can be used as an approximation for $L(x)$ on the interval $-\frac{1}{3} \le x \le \frac{1}{3}$ if you desire to keep the error less than .005?

3. What is the maximum error that might be made if $R_4(x)$ is used to approximate $L(x)$ on the interval $-\frac{1}{2} \le x \le \frac{1}{2}$?

6 Explorations in Linear Algebra

Don LaTorre
Clemson University

This chapter is devoted to exploring linear algebra with the TI-85 graphics, programmable calculator. Our aim here is not to teach introductory linear algebra, but rather to show how the TI-85 may be effectively used in a study of linear algebra and its applications.

The material is divided into four major sections. Section 6.1, Matrices on the TI-85, is a summary of elementary procedures for entering and using matrices on the calculator, along with a brief discussion of determinants, inverses and the built-in random matrix generator. The single calculator program in this section builds a random symmetric matrix over Z_{10}.

Section 6.2 explores Systems of Linear Equations, especially the various elimination methods for solving these systems: Gaussian elimination with back substitution, LU-factorizations, Gauss-Jordan pivoting and reduced row-echelon forms. We include several calculator programs to assist with the basic pivoting process as well as back and forward substitution.

Section 6.3 considers Orthogonality, concentrating on the construction of orthonormal bases and QR-factorizations, with applications to least squares solutions of linear systems. Utility programs for converting vectors into the rows of a matrix and for calculating projections are provided.

Section 6.4, Eigenvalues and Eigenvectors, begins with a program for calculating the characteristic polynomial of a matrix, then discusses the TI-85's built-in facility for finding eigenvalues and eigenvectors. It concludes with diagonalization, a favorite application of elementary linear algebra.

Each section concludes with a short set of "Explorations," a sampling of calculator-based activities, exercises and projects which complement or extend the basic material. These explorations are designed to show the power of high-level calculators like the TI-85 to significantly contribute to experiencing, understanding and applying the concepts of linear algebra.

Explorations with the Texas Instruments TI-85
John G. Harvey and John W. Kenelly (eds.), pp. 229–264
©1993 by Academic Press, Inc.
All rights of reproduction in any form reserved.
ISBN: 0-12-329070-8

6.1. Matrices on the TI-85

We begin with a brief overview of the basic procedures for entering and using matrices on the TI-85. Before proceeding further, check to make sure your calculator has the following mode settings highlighted: Normal, Float, RectC, Dec and RectV.

Entering Matrices and Vectors

Rectangular arrays of real or complex numbers are called *arrays*. A vector is a one-dimensional array represented in the calculator by a sequence of numbers in square brackets, as in [0 2 4 6] or [(0, 2) (4, 6)], where (4, 6) is the usual representation of the complex number $4 + 6i$. A matrix is a two-dimensional array characterized by an initial opening bracket [, followed by the row vectors in turn, and ending with a final closing bracket]. Thus, for example, the real matrix

$$\begin{bmatrix} 1 & 8 & 5 \\ 3 & 6 & 4 \end{bmatrix}$$

will appear on the calculator's screen as

$$\begin{matrix} [[& 1 & 8 & 5 &] \\ [& 3 & 6 & 4 &]] \end{matrix}$$

and the complex matrix

$$\begin{bmatrix} 6 & 8 + i & 5 - 2i \\ 3 - 4i & 1 & 7i \end{bmatrix}$$

will appear as

$$\begin{matrix} [[& (6,0) & (8,1) & (5,-2) &] \\ [& (3,-4) & (1,0) & (0,7) &]] \end{matrix}.$$

The quickest way to enter a vector, say [0 2 4 6], is to use direct keyboard entry with the keystrokes $\boxed{[}$ 0, 2, 4, 6 $\boxed{\textbf{ENTER}}$. Notice that pressing the **ENTER** key automatically produces the closing bracket for you. To store this vector in the calculator's memory under the variable name **V**, simply press $\boxed{\text{STO>}}$ **V** $\boxed{\text{ENTER}}$. To evaluate **V** quickly, press $\boxed{\text{ALPHA}}$ **V** $\boxed{\text{ENTER}}$. You may wish to verify that [0 2 4 6] is stored as the vector variable **V** by accessing the VECTOR NAMES menu with $\boxed{\text{VCTR}}$ $\boxed{\text{M1}}$ followed by the menu key (one of the M1 – M4 keys) appearing below the boxed **V** (press $\boxed{\text{MORE}}$) to go to the next page of the menu if necessary), then $\boxed{\text{ENTER}}$.

In a similar way matrices can be entered directly from the keyboard and stored as matrix variables, but entering the opening and closing brackets for each row vector and the commas which are needed to separate entries within the rows slows the process considerably. Thus for matrices, it is more efficient to use the Matrix Editor feature of the TI-85. To use the Matrix Editor to enter and store

$$[[\ \ 1 \ \ 8 \ \ 5 \ \ ]$$
$$[\ \ 3 \ \ 6 \ \ 4 \ \ ]]$$

as matrix variable A, begin by pressing $\boxed{\textbf{MATRX}}$ $\boxed{\textbf{M2}}$. The blinking cursor prompts you for the variable name, and the "A" within the cursor signifies that alpha entry mode is active, so there is no need to press the ALPHA key to generate alphabetical characters. Simply press A $\boxed{\textbf{ENTER}}$ to create the matrix variable name A, then 2 $\boxed{\textbf{ENTER}}$ 3 $\boxed{\textbf{ENTER}}$ to specify the dimensions. Now enter the individual entries in row order, being sure to press $\boxed{\textbf{ENTER}}$ after each entry except the last one, when you should press $\boxed{\textbf{QUIT}}$. Press $\boxed{\textbf{ALPHA}}$ A $\boxed{\textbf{ENTER}}$ to evaluate matrix variable A. To see this variable in the MTRX menu, simply press $\boxed{\textbf{MATRX}}$ $\boxed{\textbf{M1}}$; another $\boxed{\textbf{M1}}$ $\boxed{\textbf{ENTER}}$ will evaluate it from the MTRX menu.

Matrices having complex number entries are entered in the same way, except that you must begin each complex number with a left parenthesis and use a comma to separate the real and imaginary parts. You need not insert the right parenthesis since each press of the ENTER key does that for you. If any entry of an array is a complex number then all entries in that array will be complex numbers. Knowing this, you may simply enter any real number b into a complex array as a real number; it will automatically be recorded as $(b, 0)$.

You cannot overwrite a complex matrix variable with a real matrix; you must delete it from the MTRX menu and start over. To delete, go to the Memory menu with $\boxed{\textbf{MEM}}$, press $\boxed{\textbf{M2}}$ $\boxed{\textbf{MORE}}$ $\boxed{\textbf{M1}}$, move the cursor to the variable you want to delete and press $\boxed{\textbf{ENTER}}$. Return to the Home screen with $\boxed{\textbf{QUIT}}$.

Using Matrices

Chapter 13 of the TI-85 Owner's Manual shows how to edit matrices and vectors using the Matrix and Vector Editors, and describes the various features found on the submenus of the MATRX and VECTR menus. We will not elaborate upon this material here except to make a few points of general interest.

(i) Although vectors and matrices are recognized as two distinct object types by the TI-85, there is one occasion where vectors are treated like matrices: any n-vector x can be premultiplied by any $m \times n$ matrix A to obtain the m-vector Ax. But, this is the only case where vectors are treated like matrices. As an example, store

$$
\begin{bmatrix}
[& 1 & 2 & 3 &] \\
[& 4 & 5 & 6 &] \\
[& 7 & 8 & 9 &]
\end{bmatrix}
$$

as matrix A and $[1\ 1\ 1]$ as vector X. Then press $\boxed{\text{ALPHA}}$ A $\boxed{\times}$ $\boxed{\text{ALPHA}}$ X $\boxed{\text{ENTER}}$ to see $A * X = [6\ 15\ 24]$.

Notice that we were careful to include the multiplication symbol $*$ in the expression $A * X$ above. To have omitted it would be asking the calculator to evaluate a variable AX, which may or may not exist. The TI-85 will return the correct value for $A*X$ if you omit the $*$ symbol but replace it with a space. (Try it for yourself; the space can be obtained by using the key sequence $\boxed{\text{ALPHA}}$ $\boxed{\sqcup}$). Our best advice, however, is to avoid the implied multiplications which appear in such common mathematical notations as AB, AX, AA^{-1}, etc. and always use $A * B$, $A * X$, $A * A^{-1}$, etc.

(ii) Given a square matrix A, we are often interested in its integer powers A^n ($n > 0$) or, more generally, evaluating a polynomial $f(t) = a_n t^n + a_{n-1} t^{n-1} + \cdots a_1 t + a_0$ at A: $f(A) = a_n A^n + a_{n-1} A^{n-1} + \cdots + a_1 A + a_0 I$. Unlike its predecessor, the TI-81, the TI-85 will calculate powers A^n using the syntax $A^\wedge n$. For example, using the matrix A in (i) above, you may quickly check that

$$
A^\wedge 5 =
\begin{bmatrix}
[& 121824 & 149688 & 177552 &] \\
[& 275886 & 338985 & 402084 &] \\
[& 429948 & 528282 & 626616 &]
\end{bmatrix} .
$$

(Use the right arrow key to scroll right to see all of the entries.) Similarly, you may check that

$$
A^\wedge 3 - 5^* A^\wedge 2 + 2^* A - 6I =
\begin{bmatrix}
[& 314 & 400 & 480 &] \\
[& 740 & 404 & 1080 &] \\
[& 1160 & 1420 & 1674 &]
\end{bmatrix} .
$$

(To get the 3×3 identity matrix I, go to the MATRX OPS menu and press $\boxed{\text{M3}}$ 3.)

(iii) Students of linear algebra quickly encounter the result that a square matrix A is invertible if and only if the determinant of A, det A, is non-zero. This is a powerful theoretical result, but one which presents computational difficulties in non-exact arithmetic. Consider, for example, the now familiar matrix

$$
A =
\begin{bmatrix}
[& 1 & 2 & 3 &] \\
[& 4 & 5 & 6 &] \\
[& 7 & 8 & 9 &]
\end{bmatrix} .
$$

If you go to the MATRX MATH menu (press $\boxed{\texttt{MATRX}}$ $\boxed{\texttt{M3}}$) and ask for det A (press $\boxed{\texttt{M1}}$ $\boxed{\texttt{ALPHA}}$ A $\boxed{\texttt{ENTER}}$), you will see det A evaluated as $-2.4E - 12$. Reassured that A is invertible, ask for A^{-1} with $\boxed{\texttt{ALPHA}}$ A $\boxed{x^{-1}}$ $\boxed{\texttt{ENTER}}$. Are the entries in A^{-1} a little larger than you might have expected? As a quick check, calculate $A^{-1} * A$ with $\boxed{\texttt{ANS}}$ $\boxed{\times}$ $\boxed{\texttt{ALPHA}}$ A $\boxed{\texttt{ENTER}}$. The answer,

$$[[\quad .8 \quad -.4 \quad -.6 \quad]$$
$$[\quad .5 \quad 1 \quad 1.5 \quad] \quad ,$$
$$[\quad -.3 \quad 0 \quad .2 \quad]]$$

doesn't seem correct ... or does it? Try $A * A^{-1}$ as a double check. If you are now ready to trash your TI-85, DON'T! Remember, "behind every successful calculation lurks a thinking mind." Just what is the problem here?

The truth of the matter is that det $A = 0$ (see Exploration 2 at the end of this section), and A is not invertible. Has your calculator failed you? Not really ... it did the best that it could, given its built-in routine (the Crout reduction) for finding determinants and inverses and its floating point environment. We will say more about the Crout reduction in the next section. But a little thought about cofactor expansions tells us that the determinant of a matrix with integer entries will also be an integer, so we should have recognized $-2.4E - 12$ as being 0, except for some round-off error. Of course your calculator cannot make that sort of mental judgement, so with the calculated value of det A being non-zero, it responded by trying to find a matrix to qualify for A^{-1}. High-level, professional microcomputer software for matrices would most certainly include enough checks so that the determinant of this simple matrix would have been calculated correctly. But the lesson to be learned here is readily apparent. *Calculating determinants in a floating point environment is tricky business and should be avoided if at all possible. Likewise, the calculation of matrix inverses should not be done as a practical matter.* Although determinants and inverses are of paramount theoretical importance in linear algebra, they are not computationally significant.

(iv) In addition to the ident command for building an identity matrix, the MATRX OPS menu contains a number of other useful commands, several of which will be mentioned in due course. But the one appearing on the third page of this menu (press $\boxed{\texttt{MORE}}$ twice to get to the third page) deserves mention now. We are referring to the randM function, which is extremely useful for creating matrices having low-order, random integer entries, each entry being $0, \pm 1, \pm 2, \ldots,$ or ± 9.

For convenience, we shall refer to such matrices as random matrices over Z_{10}. Here's how the randM function works. To build, say a random 5×5 matrix over Z_{10}, go to the third page of the MATRX OPS menu and select the randM function by pressing the $\boxed{\text{M1}}$ key; now press 5, 5 $\boxed{\text{ENTER}}$. In the event that you wish to have everyone in a group of TI-85 users (say, in an introductory linear algebra class) generate the same random matrix, have them seed their calculator's random number generator with the same integer-valued seed. The random number generator is found on the MATH PROB menu. To seed the generator with 2, go to the MATH PROB menu and store 2 to rand with 2 $\boxed{\text{STO>}}$ $\boxed{\text{M4}}$ $\boxed{\text{ENTER}}$. If you now generate a random 4×5 matrix over Z_{10}, it is guaranteed to be

$$\begin{bmatrix} -6 & 2 & -7 & -3 & 9 \\ -9 & 4 & -5 & 8 & 2 \\ 1 & -5 & -5 & -6 & -6 \\ -2 & 8 & -1 & 4 & 0 \end{bmatrix}.$$

You may be interested in building your own version of a random matrix generator to generate a special class of matrices. For example, the following TI-85 program, SYMMETRC, can be used to build random symmetric matrices over Z_{10}. When executed, it prompts you for the size n of a square matrix and then returns a random $n \times n$ symmetric matrix over Z_{10}.

```
PROGRAM: SYMMETRC
:Disp "ENTER size"
:Input N
:randM(N,N)→ztem
:For(I,1,N-1)
:For(J,I+1,N)
:ztem(I,J)→ztem(J,I)
:End
:End
:ztem→ztem
:Disp ztem
:
```

Program 6.1.

By way of example, if you seed your random number generator with 1 and then ask program SYMMETRC to build a random 6×6 symmetric matrix over Z_{10} you will get

$$
\begin{bmatrix}
1 & -6 & 9 & 7 & 9 & -8 \\
-6 & 6 & 5 & 4 & 3 & 9 \\
9 & 5 & 5 & 8 & -8 & 6 \\
7 & 4 & 8 & -9 & -3 & -7 \\
9 & 3 & -8 & -3 & 2 & -8 \\
-8 & 9 & 6 & -7 & -8 & 5
\end{bmatrix}
$$

Explorations with Matrices

1. Begin by accessing the MODE menu and setting the numeric display format to Normal with 3 decimal digits. Let $x = [.2\ .1\ .4\ .3]$ and let

$$
A =
\begin{bmatrix}
.2 & .1 & .6 & .5 \\
.4 & .3 & 0 & .1 \\
.1 & .5 & .2 & 0 \\
.3 & .1 & .2 & .4
\end{bmatrix}
$$

(a) Examine the sequence $A, A^2, A^3, \ldots$ and find $\lim_{n \to \infty} \{A^n\}$ (to 3 decimal places).

(b) Now examine the sequence $Ax, A^2x, A^3x, \ldots$ and find $y = \lim_{n \to \infty} \{A^n x\}$ (to 3 decimal places). [Suggestion: After calculating Ax, calculate $A * \text{Ans}$ repeatedly.]

(c) How is y related to the limit in (a)?

(d) Calculate $A * y$.

(e) Repeat (a) – (d) using $x = [.3\ .2\ .2\ .3]$.

2. (a) Begin by resetting the numeric display mode to Normal, Float. Use row operations to convert

$$
A =
\begin{bmatrix}
1 & 2 & 3 \\
4 & 5 & 6 \\
7 & 8 & 9
\end{bmatrix}
$$

to an upper triangular matrix U. Do this as follows: go to the second page of the MATRX OPS menu and use the mRAdd function. For example, to add -4 times row 1 of A to row 2 of A, the syntax will be $\text{mRAdd}(-4, A, 1, 2)$. After performing this first row operation, perform the others on your last Ans each time.

(b) What is U? What is det U? What is det A?

3. (a) Build a random 4×4 complex matrix C over $Z_{10} \times Z_{10}$ by building
       random 4×4 matrices A and B over Z_{10} and calculating $C =$
       $A + (0, 1) * B$.

   (b) Is the product of two $n \times n$ symmetric matrices symmetric? Exper-
       iment with random symmetric matrices over Z_{10} with $n = 3, 4, 5$
       to formulate your conclusion.

4. Build a random 5×6 matrix over Z_{10} and store it as matrix variable A.

   (a) Get row 1 of matrix A by entering the expression $A(1)$; also get
       row 3 and column 5.

   (b) To extract a submatrix of A, use the syntax $A(I, J, K, L)$ where I
       and J indicate the beginning row and column, respectively, and K
       and L indicate the ending row and column. Extract the submatrix
       of A consisting of the last three rows; also extract the submatrix
       C of A consisting of columns 2 thru 4. Now calculate $C^T C$ by
       using $\text{Ans}^T * \text{Ans}$ (the transpose instruction T is on the MTRX
       MATH menu).

6.2. Systems of Linear Equations

Gaussian Elimination

The most straightforward approach to solving a system of n linear
equations in n unknowns is simple Gaussian elimination followed by back
substitution. Gaussian elimination systematically adds multiples of certain
equations to other equations in order to obtain an equivalent upper triangular system. Back substitution then solves the upper triangular system
by solving the last equation for its single unknown, substituting its value
back into the next-to-last equation and solving for its single unknown, and
so on. If the matrix formulation of the original system is $Ax = b$ then the
elimination process begins by applying row operations to the augmented
matrix $[A \mid b]$ and ends with the matrix $[U \mid b']$, where U is upper triangular. An introductory study of linear algebra typically begins with a
careful look at this process because linear algebra, as we know it today, is
a highly developed outgrowth of the very basic need to solve systems of
linear equations.

As we have described it, Gaussian elimination requires only two types
of row operations: add a scalar multiple of one row to another and, if
necessary, interchange two rows. The MATRX OPS menu on the TI-85
includes commands that will perform these operations for you, the mRAdd

and rSwap functions. Let's apply them to solve the system

$$
\begin{array}{rcrcrcrcr}
x_1 & + & 2x_2 & - & x_3 & + & 3x_4 & = & 4 \\
-2x_1 & - & 4x_2 & + & 5x_3 & - & 7x_4 & = & -8 \\
3x_1 & + & 8x_2 & - & 2x_3 & + & 10x_4 & = & 13 \\
x_1 & + & 4x_2 & + & 6x_3 & + & 3x_4 & = & 1
\end{array}
$$

Begin by entering and storing the augmented matrix as matrix variable A:

$$
\begin{bmatrix}
1 & 2 & -1 & 3 & 4 \\
-2 & -4 & 5 & -7 & -8 \\
3 & 8 & -2 & 10 & 13 \\
1 & 4 & 6 & 3 & 1
\end{bmatrix}
$$

Then, with A visible on the home screen, go to the second page of the MATRX OPS menu and build mRAdd $(2, A, 1, 2)$. Press $\boxed{\text{ENTER}}$ to see

$$
\begin{bmatrix}
1 & 2 & -1 & 3 & 4 \\
0 & 0 & 3 & -1 & 0 \\
3 & 8 & -2 & 10 & 13 \\
1 & 4 & 6 & 3 & 1
\end{bmatrix}
$$

Now eliminate the last two entries in column 1 with mRAdd $(-3,\ \text{Ans}, 1, 3)$ followed by mRAdd $(-1,\ \text{Ans}, 1, 4)$. At this point you should see

$$
\begin{bmatrix}
1 & 2 & -1 & 3 & 4 \\
0 & 0 & 3 & -1 & 0 \\
0 & 2 & 1 & 1 & 1 \\
0 & 2 & 7 & 0 & -3
\end{bmatrix}
$$

To continue, you must interchange row 2 and a lower row, say row 3. Do this with rSwap $(\text{Ans}, 2, 3)$ $\boxed{\text{ENTER}}$. Now execute mRAdd $(-1,\ \text{Ans}, 2, 4)$ to obtain

$$
\begin{bmatrix}
1 & 2 & -1 & 3 & 4 \\
0 & 2 & 1 & 1 & 1 \\
0 & 0 & 3 & -1 & 0 \\
0 & 0 & 6 & -1 & -4
\end{bmatrix}
$$

Finally, mRAdd $(-2,\ \text{Ans}, 3, 4)$ $\boxed{\text{ENTER}}$ will return the upper triangular system

$$
\begin{bmatrix}
1 & 2 & -1 & 3 & 4 \\
0 & 2 & 1 & 1 & 1 \\
0 & 0 & 3 & -1 & 0 \\
0 & 0 & 0 & 1 & -4
\end{bmatrix}
$$

You may now use back substitution by hand to obtain the solution
$[8.\bar{3}, 3.1\bar{6}, -1.\bar{3}, -4]$. In fraction form the solution is:

$$[25/3, 19/6, -4/3, -4].$$

Whether you are a teacher of linear algebra whose goal may be to teach
Gaussian elimination or a student of linear algebra whose goal may be to
obtain a better understanding of the process, by now you may see the need
for two calculator programs:

(1) One to assist with the elimination phase, wherein all entries in a column below a certain element-the *pivot* – are converted to zeros by row
operations; and

(2) another to assist with the back substitution process.

We have included two such programs: PIVOT and BkSUB. PIVOT will
perform the row operations necessary to put zeros below the pivot, while
BkSUB will perform the back substitution process, step-by-step. Since back
substitution uses an upper triangular matrix and a vector b (vector b is the
"right-hand side" of the linear system), we have also included a simple
program SPLITb used to split off the rightmost column of a matrix.

Program PIVOT takes as input an $m \times n$ matrix from the reserved
variable Ans. It prompts the user to enter the location (K, L) of the pivot
element, and pivots on that element (i.e., does the necessary row operations)
to produce zeros below the pivot. The result is stored in matrix variable
zmat. In the event the pivot element is zero, the program pauses and
displays the message "0 PIVOT, press ENTER." When the ENTER key
is pressed, the last zmat is displayed along with the query "WHERE TO
PIVOT?"

Program SPLITb takes as input an $m \times n$ matrix stored in Ans and
splits off the rightmost column, which is then stored as vector variable bb.
The remaining $m \times (n-1)$ matrix is stored as Ans and displayed on the
Home screen for convenience. (Note: We store the rightmost column to
vector variable bb since b is a "reserved name" variable.)

Program BkSUB takes as inputs an $n \times n$ invertible upper triangular
matrix from Ans and an n vector from vector variable bb. It backsolves the
linear system $(Ans)x = bb$ for the last variable x_n, then pauses with the
message "press ENTER". When the ENTER key is pressed, the program
backsolves for the next-to-last variable x_{n-1} and pauses as before. When all
the components $x_n, x_{n-1}, \ldots, x_1$ of the solution vector have been obtained,

```
 PROGRAM: PIVOT
:Ans→zmat
:dim zmat→zlst
:zlst(1)→M
:Lbl PV1
:Disp "WHERE TO PIVOT?"
:Prompt K,L
:If real zmat(K,L)==0
:Then
:If imag zmat(K,L)==0
:Then
:Disp "0 PIVOT,press ENTER"
:Pause
:Disp zmat
:Goto PV1
:Else
:Goto PV2
:End
:Else
:Lbl PV2
:For(I,K+1,M,1)
:mRAdd(-zmat(I,L)/zmat(K,L),zmat,K,I)→zmat
:round(zmat,11)
:Ans→zmat
:End
:Disp zmat
:
```

Program 6.2.

```
PROGRAM: SPLITb
:Ans→zmat
:dim zmat→zlst
:zlst(1)→M
:zlst(2)→N
:zmatᵀ
:Ans(N)→bb
:zmat(1,1,M,N-1)
:Disp Ans
:
```

Program 6.3.

```
PROGRAM: BkSUB
:Ans→zmat
:dim zmat→zlst
:zlst(1)→N
:N→dim zvec
:Fill(0,zvec)
:N→J
:Lbl Bk1
:(bb(J)-dot(zmat(J),zvec))/zmat(J,J)
:round(Ans,11)
:Disp Ans
:Disp "press ENTER"
:Pause
:Ans→zvec(J)
:DS<(J,1)
:Goto Bk1
:zvec→zvec
:Disp zvec
:
```

Program 6.4.

a final ENTER returns the solution vector $[x_1, x_2, \ldots, x_n]$. (Note: The command zvec $\rightarrow$ zvec insures that the solution vector zvec is stored in Ans.)

Assuming that you have entered these programs into your TI-85, let's now return to our earlier linear system and use them. As you work, notice that the programs carry all the computational burden, enabling you to focus more clearly on the elimination process itself.

Begin by pressing [ALPHA] A [ENTER]. This will insure that the augmented matrix A we begin with is in Ans. Go to the PRGM NAMES menu and press the menu key beneath program PIVOT, then [ENTER]. Respond to the calculator prompts with 1 [ENTER] ... 1 [ENTER]. You should see

$$\begin{bmatrix} [& 1 & 2 & -1 & 3 & 4 &] \\ [& 0 & 0 & 3 & -1 & 0 &] \\ [& 0 & 2 & 1 & 1 & 1 &] \\ [& 0 & 2 & 7 & 0 & -3 &] \end{bmatrix}.$$

To continue you must swap rows 2 and 3 of this Ans as before, to obtain

$$\begin{bmatrix} [& 1 & 2 & -1 & 3 & 4 &] \\ [& 0 & 2 & 1 & 1 & 1 &] \\ [& 0 & 0 & 3 & -1 & 0 &] \\ [& 0 & 2 & 7 & 0 & -3 &] \end{bmatrix}.$$

Now run program PIVOT again, this time pivoting on the $(2,2)$-entry. You will get

$$\begin{bmatrix} [& 1 & 2 & -1 & 3 & 4 &] \\ [& 0 & 2 & 1 & 1 & 1 &] \\ [& 0 & 0 & 3 & -1 & 0 &] \\ [& 0 & 0 & 6 & -1 & -4 &] \end{bmatrix}.$$

A final application of program PIVOT with $(3,3)$ as the pivot location will return the desired triangular system

$$\begin{bmatrix} [& 1 & 2 & -1 & 3 & 4 &] \\ [& 0 & 2 & 1 & 1 & 1 &] \\ [& 0 & 0 & 3 & -1 & 0 &] \\ [& 0 & 0 & 0 & 1 & -4 &] \end{bmatrix}.$$

To do the back substitution, begin by returning to the PRGM NAMES menu and running SPLITb. This will store $[4\ 1\ 0\ -4]$ as vector variable bb, and the remaining upper triangular submatrix as the reserved variable Ans. Now you are ready to run BkSUB. When you do, the routine returns -4 as the value for x_4, pauses and prompts you to backsolve again by pressing ENTER. Two more ENTER's will produce the remaining components of the solution, and a final ENTER produces the solution vector. In order to scroll right to see all the components, first hit [Ans] [ENTER], then use the right arrow key. If you are interested in seeing this answer in fractional form, go to the MATH MISC menu with $\boxed{\text{MATH}}$ $\boxed{\text{M5}}$ $\boxed{\text{MORE}}$ followed by $\boxed{\text{Ans}}$ $\boxed{\text{M1}}$ $\boxed{\text{ENTER}}$ to see $[25/3 \quad 19/6 \quad -4/3 \quad -4/1]$.

In the above example, after pivoting on the $(1,1)$-entry we had to interchange rows 2 and 3 in order to have a non-0 pivot in the $(2,2)$-entry. Actually, it is just as important to avoid extremely small pivots because division by such numbers may eventually lead to considerable round-off error. Thus, it is common practice to adopt a strategy known as *partial pivoting*, whereby the actual pivot is chosen to be any element in the pivot column on or below the pivot position whose absolute value is maximum. Although partial pivoting during hand computations can lead to cumbersome, tedious arithmetic, it is easy to affect on the calculators and is routinely done in professional elimination codes.

A variant of Gaussian elimination which is popular with students and teachers who still rely upon hand computations calls for making sure each pivot is 1. This can be done by a simple rescaling of the pivot row. In our example, the first pivot was already 1. But the second pivot was 2, so we could have rescaled by multiplying row 2 of

$$\begin{bmatrix} [& 1 & 2 & -1 & 3 & 4 &] \\ [& 0 & 2 & 1 & 1 & 1 &] \\ [& 0 & 0 & 3 & -1 & 0 &] \\ [& 0 & 2 & 7 & 0 & -3 &] \end{bmatrix}$$

by .5 to obtain

$$\begin{bmatrix} [& 1 & 2 & -1 & 3 & 4 &] \\ [& 0 & 1 & .5 & .5 & .5 &] \\ [& 0 & 0 & 3 & -1 & 0 &] \\ [& 0 & 2 & 7 & 0 & -3 &] \end{bmatrix} .$$

The next row operation would then have been mRAdd$(-2, \text{Ans}, 2, 4)$ to produce

$$\begin{bmatrix} [& 1 & 2 & -1 & 3 & 4 &] \\ [& 0 & 1 & .5 & .5 & .5 &] \\ [& 0 & 0 & 3 & -1 & 0 &] \\ [& 0 & 0 & 6 & -1 & -4 &] \end{bmatrix}$$

as before. Finally, rescaling row 3 and then applying mRAdd$(-6, \text{Ans}, 3, 4)$ would have produced

$$\begin{bmatrix} [& 1 & 2 & -1 & 3 & 4 &] \\ [& 0 & 1 & .5 & .5 & .5 &] \\ [& 0 & 0 & 1 & -1/3 & 0 &] \\ [& 0 & 0 & 0 & 1 & -4 &] \end{bmatrix} .$$

ready for back substitution. For hand computations, such rescalings of the pivot rows make it easy to see what multiples of the pivot row should be added to the lower rows, and it simplifies the arithmetic of the back substitution process a bit. But rescalings are certainly not necessary with calculators and we recommend avoiding them.

LU-factorizations

When Gaussian elimination is systematically applied without row interchanges to convert an invertible matrix A to upper triangular from U, there is an associated factorization of $A : A = LU$. Here, L is a lower triangular matrix with 1's the main diagonal and with the negatives of the multipliers used in the reduction process filling the lower triangle. If row interchanges are used to avoid 0 pivots or to implement partial pivoting, the factorization $A = LU$ is replaced by $PA = LU$ where P is a permutation matrix which accounts for the row interchanges. Factorizations into lower triangular and upper triangular matrices are important because they are widely used in professional computer codes to handle linear systems.

There is an LU-factorization which differs slightly from the one we have just described. In the above reduction of A to an upper triangular matrix U, the pivots appear on the main diagonal of U. Let D be the diagonal matrix $D = \text{diag}[u_{11}, u_{22}, \ldots, u_{nn}]$, so $D^{-1} = [u_{11}^{-1}, u_{22}^{-1}, \ldots, u_{nn}^{-1}]$. Then we have the factorization $PA = LU = LDD^{-1}U = L_1 U_1$, where $L_1 = LD$ and $U_1 = D^{-1}U$ are lower and upper triangular factors. This factorization differs from the original one in that row i of $U(i = 1, \ldots, n)$ has been rescaled by u_{ii}^{-1} (in particular, U_1 has 1's on its main diagonal) and column j of L ($j = 1, \ldots, n$) has been rescaled by u_{jj} (so, in particular, L_1 has the u_{jj}'s on its main diagonal). This new LU-factorization is known as the *Crout reduction* and is especially well-suited to calculator and microcomputer implementation.

With the TI-85, you can obtain an LU-factorization using the Crout reduction by executing the LU instruction on the MATRX MATH menu. The instruction requires 4 arguments, LU(arg1, arg2, arg3, arg4). The first argument, arg1, is the matrix variable containing the matrix to be factored. Arguments 2, 3 and 4 are, respectively, the matrix variables to which you want to store the lower triangular, upper triangular and permutation matrices. I suggest L, U and P for these last 3 variables. We shall illustrate with the coefficient matrix A of the linear system we studied earlier,

$$\begin{bmatrix} [& 1 & 2 & -1 & 3 &] \\ [& -2 & -4 & 5 & -7 &] \\ [& 3 & 8 & -2 & 10 &] \\ [& 1 & 4 & 6 & 3 &] \end{bmatrix}$$

Assuming that you have this matrix stored as matrix variable A, go to the second page of the MATRX MATH menu and use the $\boxed{\text{M3}}$ key to create the instruction $LU(A, L, U, P)$. Press $\boxed{\text{ENTER}}$ to execute the instruction. When your screen reads "Done", press $\boxed{\text{ALPHA}}$ L $\boxed{\text{ENTER}}$ to see

$$
\begin{bmatrix}
3 & 0 & 0 & 0 \\
-2 & 1.33333333333 & 0 & 0 \\
1 & 1.33333333333 & 3 & 0 \\
1 & -.666666666667 & 1.5 & -.5
\end{bmatrix}'
$$

$\boxed{\text{ALPHA}}$ U $\boxed{\text{ENTER}}$ to see

$$
\begin{bmatrix}
1 & 2.66666666667 & -.66666666667 & 3.33333333333 \\
0 & 1 & 2.75 & -.25 \\
0 & 0 & 1 & 3.33333333333E-14 \\
0 & 0 & 0 & 1
\end{bmatrix}
$$

(note that, except for a little round-off error, $3.33333333333E - 14$ is 0) and $\boxed{\text{ALPHA}}$ P $\boxed{\text{ENTER}}$ to see

$$
\begin{bmatrix}
0 & 0 & 1 & 0 \\
0 & 1 & 0 & 0 \\
0 & 0 & 0 & 1 \\
1 & 0 & 0 & 0
\end{bmatrix}.
$$

Because P is a permutation matrix $P^{-1} = P^T$, so from $PA = LU$ we obtain $A = P^T LU$. Clear your Home screen and then calculate $P^T * L * U$ to see the original matrix A.

We should point out that the Crout algorithm programmed into the TI-85 uses partial pivoting throughout. In fact, you can obtain the calculator's U by using partial pivoting to determine the pivot, then rescaling the pivot row to make the pivot element 1 before pivoting.

To use the factors P, L and U in the factorization $PA = LU$ to solve a linear system, first solve $Ly = Pb$ by forward substitution, then solve $Ux = y$ by back substitution. The importance of this approach can be seen by considering different b's for the same A: you need only reduce A to U once, then apply the results of that reduction (i.e., the P, L and U) to the different b's. Forward substitution for lower triangular systems is similar to back substitution for upper triangular systems: solve the first equation for its single unknown x_1, then substitute it forward into the second equation and solve for x_2, etc.

Here is a TI-85 program, program FwSUB, to do forward substitution. It requires as its inputs an $n \times n$ invertible lower triangular matrix from

Ans and an n vector from vector variable bb, and solves $(Ans)x = bb$. As with program BkSUB, it pauses after each step and prompts the user to continue.

```
 PROGRAM: FwSUB
:Ans→zmat
:dim zmat→zlst
:zlst(1)→N
:N→dim zvec
:Fill(0,zvec)
:1→J
:Lbl Fw1
:(bb(J)-dot(zmat(J),zvec))/zmat(J,J)
:round(Ans,11)
:Disp Ans
:Disp "Press ENTER"
:Pause
:Ans→zvec(J)
:IS>(J,N)
:Goto Fw1
:zvec→zvec
:Disp zvec
:
```

Program 6.5.

To solve a linear system $AX = b$ using an LU-factorization on the TI-85, follow these steps:

Step 1. Store the coefficient matrix as variable A and then execute the instruction $LU(A, L, U, P)$.

Step 2. Store $P*b$ into variable bb and then store L into Ans with [ALPHA] L [ENTER]; now run program FwSUB to solve the linear system $Ly = Pb$ for y.

Step 3. Store y into bb and then store U into Ans with [ALPHA] U [ENTER]; now run program BkSUB to solve the linear system $Ux = y$ for x.

Note: At the end of step 3, you should examine your answer to see if there is any round-off error that needs clean-up. If so, go to the MATH NUM menu and execute the instruction round (Ans, d), where d is the number (≤ 11) of decimal places to which you wish to round the answer.

We illustrate by solving the linear system with which we began this section.

Step 1. Goto the MATRX EDIT menu and create the matrix

$$A = \begin{bmatrix} [& 1 & 2 & -1 & 3 &] \\ [& -2 & -4 & 5 & -7 &] \\ [& 3 & 8 & -2 & 10 &] \\ [& 1 & 4 & 6 & 3 &] \end{bmatrix}.$$

Go to the MATRX MATH menu and execute the instruction $LU(A, L, U, P)$.

Step 2. Enter $P * [4, -8, 13, 1]$, then press $\boxed{\text{STO>}}$ $\boxed{\text{ALPHA}}$ $\boxed{\text{2nd}}$ $\boxed{\text{ALPHA}}$ $\boxed{\text{ALPHA}}$ bb $\boxed{\text{ENTER}}$. Now press $\boxed{\text{ALPHA}}$ L $\boxed{\text{ENTER}}$ and run program FwSUB to see [4.33333333333   .49999999999 -1.33333333333 -3.99999999998].

Step 3. Press $\boxed{\text{STO>}}$ $\boxed{\text{ALPHA}}$ $\boxed{\text{2nd}}$ $\boxed{\text{ALPHA}}$ $\boxed{\text{ALPHA}}$ bb $\boxed{\text{ENTER}}$, then $\boxed{\text{ALPHA}}$ U $\boxed{\text{ENTER}}$. Now run program BkSUB to see the solution [8.33333333331 3.166666666665 -1.33333333333 -3.9999999998]. You may round this answer to 10 decimal places to obtain [8.3333333333 3.16666666666 -1.3333333333 -4].

Our discussion up to this point has focused on using the TI-85 to gain an understanding of Gaussian elimination and back substitution, partial pivoting and LU-factorizations, and how these concepts may be applied to solve a system of n linear equations in n unknowns. With this background, you may now appreciate the calculator's built-in, automatic linear system solver, the SIMULT application. This routine automatically solves $n \times n$ linear systems by applying an LU-factorization obtained from an implementation of the Crout reduction using partial pivoting. To use SIMULT, begin by pressing $\boxed{\text{SIMULT}}$. When prompted by the message "Number =" enter the number of equations in the system, e.g., 4 $\boxed{\text{ENTER}}$. You will then be prompted to enter the numbers $a_{11}, a_{12}, \ldots, a_{1n}, b_1$ in the first equation $a_{11}x_1 + a_{12}x_2 + \cdots + a_{1n}x_n = b_1$ (press $\boxed{\text{ENTER}}$ after each number). After entering all the numbers in all of the equations, press $\boxed{\text{M5}}$, the SOLVE menu key, to activate the routine and view the components of the solution vector. You should try out SIMULT now on the linear system we have been using. The $\boxed{\text{STOx}}$ menu key enables you to store the components into a vector of your choice. Likewise, the $\boxed{\text{STOa}}$ and $\boxed{\text{STOb}}$ menu keys enable you to store the coefficients to an $n \times n$ matrix and the right-hand components to an n vector.

Two final comments about the use of LU-factorizations. If $PA = LU$ then $A = P^T LU$, so det A is given as the product of det P^T, det L

and det U. Since P is a permutation matrix, det $P^T = \pm 1$, where the $-$ sign appears if P came from an odd number of row interchanges. In the Crout reduction, U is upper triangular with 1's on its diagonal, so det $U = 1$. Thus, det $A = \pm$ det L, where det L is the product of the diagonal entries of $L \ldots$ the pivots before rescaling. This is, in fact, how the TI-85 calculates determinants. If, for example, you obtain an LU-factorization for the matrix

$$\begin{bmatrix} [\ 1 & 2 & 3\ ] \\ [\ 4 & 5 & 6\ ] \\ [\ 7 & 8 & 9\ ] \end{bmatrix}$$

of Section 6.1, you will note that

$$L = \begin{bmatrix} [\ 7 & 0 & 0 &] \\ [\ 1 & .857142857143 & 0 &] \\ [\ 4 & .428571428572 & -4E-13 &] \end{bmatrix}.$$

The product of the diagonal entries is $-2.4\ E - 12$, precisely the value for det A returned by the calculator. If A is invertible (understood by the TI-85 to mean that L has no 0's on its main diagonal), then from $PA = LU$ comes $A^{-1}P^{-1} = U^{-1}L^{-1}$ so that $A^{-1} = U^{-1}L^{-1}P$. Thus A^{-1} is calculated as the product of U^{-1}, L^{-1} and P. Because U and L are triangular, their inverses are easily found by back and forward substitutions.

General Systems and Row Echelon Matrices

Gaussian elimination may also be applied to obtain all solutions to an arbitrary system $Ax = b$ of m equations in n unknowns. Typically, the two basic row operations (add a multiple of one row to another; interchange rows) are systematically applied to the augmented matrix $[A \mid b]$ to convert it to a *row echelon* matrix:

(i) Any zero rows lie at the bottom.

(ii) The pivot entry in any non-zero row lies to the right of the pivot entry in any preceding row.

If the last non-zero row looks like $[0 \quad 0 \quad \cdots \quad 0 \quad *]$, where $*$ is non-zero, then $Ax = b$ has no solution; but in every other case $Ax = b$ has at least one solution. Any variable associated with a pivot is called a *pivot variable* and any other variables are called *free variables*. There is exactly one solution if every variable is a pivot variable, obtained by simple back substitution. But, if there is at least one free variable then there are infinitely many solutions; in this case, back substitution is generalized to express each pivot variable in terms of the free variables whose values may be chosen arbitrarily.

For example, consider the linear system

$$
\begin{aligned}
x_1 + 2x_2 + 3x_3 &= -2 \\
2x_1 + x_2 + x_3 &= 2
\end{aligned}.
$$

The augmented matrix is

$$
\begin{bmatrix}
[& 1 & 2 & 3 & -2 &] \\
[& 2 & 1 & 1 & 2 &]
\end{bmatrix}
$$

and adding -2 times row 1 to row 2 will produce the row-echelon matrix

$$
\begin{bmatrix}
[& 1 & 2 & 3 & -2 &] \\
[& 0 & -3 & -5 & 6 &]
\end{bmatrix}.
$$

Thus x_1 and x_2 are pivot variables while x_3 is a free variable. Using the second row to express x_2 in terms of x_3 we have $x_2 = -2 - 1.\bar{6}x_3$. Substitution of this expression for x_2 into the equation implied by the first row and solving for x_1 gives $x_1 = 2 + .\bar{3}x_3$.

Unless further restrictions are imposed, the two conditions given above do not specify a unique row echelon form for a given matrix. But, following the lead of several elementary texts, the TI-85 will produce a row-echelon matrix which meets in addition to the above two conditions, also these:

(iii) Partial pivoting is used throughout (ties for the pivot element are broken by using the first available candidate).

(iv) Each pivot element is rescaled to 1 before applying the pivoting process.

For a given matrix A, the row-echelon matrix produced by conditions (i) – (iv) is uniquely associated with A. On the TI-85, the required instruction is ref (for row-echelon form), found on the MATR OPS menu. If, for example, we let A be the augmented matrix of the above 2×3 linear system, then ref A returns

$$
\begin{bmatrix}
[& 1 & .5 & .5 & 1 &] \\
[& 0 & 1 & 1.\bar{6} & -2 &]
\end{bmatrix}.
$$

An even more refined row-echelon form, widely known as the *reduced row-echelon form* (RREF) requires, in addition to (i) – (iv), also

(v) Each pivot is the only non-zero entry in the pivot. column.

The RREF is available on the TI-85 through the rref command on the MATR OPS menu. For the matrix A above, rref A returns

$$
\begin{bmatrix}
[& 1 & 0 & -.\bar{3} & 2 &] \\
[& 0 & 1 & 1.\bar{6} & -2 &]
\end{bmatrix}.
$$

Although neither of the two row-echelon forms provided by the ref and rref commands is in general use by professional codes for handling linear systems, the RREF is especially popular with beginning students of linear algebra who rely upon hand computation because it makes the solutions of a linear system obvious. Just compare our earlier solutions to the 2×3 linear system with the above RREF.

A strong word of caution is in order regarding the use of the ref and rref instructions. In much the same way that the built-in LU-factorization may fail to recognize round-off error and thus lead to serious errors in, say, the calculation of matrix inverses, so also may the built-in ref and rref routines fail to recognize round-off error and produce results which deviate substantially from the correct ones. As a case in point, consider the matrix

$$A = \begin{bmatrix} [& 1 & 2 & 3 &] \\ [& 4 & 5 & 6 &] \\ [& 7 & 8 & 9 &] \end{bmatrix}$$

we encountered earlier. We have already seen that although A is not invertible, the LU-routine fails to recognize the round-off error appearing on the diagonal of L and returns a ridiculous candidate for A^{-1}. When the ref and rref instructions are applied to matrix A, we obtain

$$\text{ref } A = \begin{bmatrix} [& 1 & 1.14285714286 & 1.28571428571 &] \\ [& 0 & 1 & 2 &] \\ [& 0 & 0 & 1 &] \end{bmatrix}$$

and

$$\text{rref } A = \begin{bmatrix} [& 1 & 0 & 0 &] \\ [& 0 & 1 & 0 &] \\ [& 0 & 0 & 1 &] \end{bmatrix}$$

which deviate considerably from the correct forms,

$$\begin{bmatrix} [& 1 & 8/7 & 9/7 &] \\ [& 0 & 1 & 2 &] \\ [& 0 & 0 & 0 &] \end{bmatrix} \quad \text{and} \quad \begin{bmatrix} [& 1 & 0 & -1 &] \\ [& 0 & 1 & 2 &] \\ [& 0 & 0 & 0 &] \end{bmatrix}.$$

If, for example, A were the coefficient matrix in a homogeneous system $Ax = 0$, then each of the two calculated forms would imply that $x = [0\ 0\ 0]$ is the only solution whereas, in fact, there are infinitely many solutions given by $x = x_3[1\ -2\ 1]$. To see where the problem occurs, use the rSwap and multR instructions and program PIVOT to convert A to its row-echelon form. After pivoting on the $(2, 2)$-entry you will have

$$\begin{bmatrix} [& 1 & 1.14285714286 & 1.28571428571 &] \\ [& 0 & 1 & 2.00000000001 &] \\ [& 0 & 0 & -1E - 11 &] \end{bmatrix}.$$

We clearly recognize the round-off error in the $(3,3)$-entry. But, if we fail to make it 0 and then rescale row 3 to make the $(3,3)$-entry 1, you can see how the calculated ref and rref appeared. Because of such considerations, *we suggest that you use the ref and rref commands sparingly*, initially relying upon program PIVOT to help step through the reduction process and enabling you to clean-up round-off error. The ref command can be invoked, appropriately, later in the process.

Explorations with Linear Systems

1. Consider the linear system

$$
\begin{array}{rcrcrcr}
2x_1 & + & 2x_2 & - & 4x_3 & = & -1 \\
-3x_1 & - & 6.5x_2 & - & 10x_3 & = & 2 \\
4x_1 & + & 8x_2 & + & 12x_3 & = & -3
\end{array}
$$

(a) Solve this system using Gaussian elimination with partial pivoting and back substitution; use the rSwap and mRAdd instructions to perform the elimination phase and program BkSUB for the back substitution.

(b) Repeat part (a), but this time use rSwap and program PIVOT to perform the elimination phase.

(c) Solve this system by finding and applying an LU-factorization.

(d) Solve this system by using the ref instruction and program BkSUB.

(e) Solve this system by using the rref instruction.

2. Repeat Exploration 1 for the linear system

$$
\begin{array}{rcrcrcrcr}
-x_1 & + & 8x_2 & - & 6x_3 & - & 7x_4 & = & -10 \\
-2x_1 & + & 19x_2 & - & 7x_3 & - & 12x_4 & = & 20 \\
3x_1 & - & 24x_2 & + & 24x_3 & - & 27x_4 & = & -30 \\
-2x_1 & + & 20x_2 & - & 8x_3 & + & 6x_4 & = & 40
\end{array}
$$

3. (a) Find all solutions to the linear system consisting of the first three equations from the system in Exploration 2.

(b) Find all solutions to the linear system consisting of equations $1, 3$, and 4 from the system in Exploration 2.

4. Find all solutions to the homogeneous system $AX = 0$, where

$$
A = \begin{bmatrix}
[& -9 & 6 & 6 & -14 &] \\
[& 6 & -17 & -14 & 16 &] \\
[& -5 & 12 & 9 & -13 &] \\
[& 8 & -1 & -1 & 11 &]
\end{bmatrix}.
$$

First try the rref command, then adopt another strategy.

6.3. Orthogonality

The analytic formulation of perpendicularity in R^2 and R^3 is straightforward: vectors u, v are perpendicular iff their dot product is zero: $u \bullet v = 0$. More generally, vectors u, v in R^n are called *orthogonal* if $u \bullet v = 0$. The simplicity of this concept belies its importance, for orthogonality is a powerful tool in linear algebra applicable in a variety of settings. The instruction to calculate dot products on the TI-85 is found on the VECTR MATH menu and uses the syntax dot(u, v). You may have noticed this instruction earlier in programs BkSUB and FwSUB.

We are often concerned with orthogonal sets of vectors, that is, sets $\{v_1, v_2, \ldots, v_k\}$ where $v_j \bullet v_j = 0$ for all $i \neq j$. If we let matrix A have $v_1, v_2, \ldots v_k$ as its columns then the condition that $\{v_1, v_2, \ldots v_k\}$ be orthogonal is simply that $A^T * A$ be a diagonal matrix. The main diagonal of $A^T * A$ contains the dot products $v_j \bullet v_j$, $j = 1, \ldots, k$. You can glimpse the importance of orthogonality once you realize that orthogonal sets are linearly independent, the vectors thus forming a basis for span $[v_1, v_2, \ldots, v_k]$.

The VECTR MATH menu contains two other instructions which are important to us in linear algebra: norm w calculates the Euclidean vector norm of vector $w = [x_1 \ x_2 \cdots x_n]$:

$$\|w\|_2 = \left[\sum_{j=1}^{n} |x_j|^2 \right]^{\frac{1}{2}} \; ;$$

and unitV w returns the unit vector $w/\|w\|^2$ obtained by normalizing w. Orthogonal sets of unit vectors are called *orthonormal*, and are of the greatest importance. In terms of matrices, the set $\{v_1, v_2, \ldots v_k\}$ is orthonormal iff the matrix Q having $v_1, v_2, \ldots, v_k$ as its columns satisfies $Q^T * Q = I$.

We shall soon encounter a situation where vectors $v_1, v_2, \ldots, v_k$ are constructed on the TI-85 and stored as vector variables. It will thus be helpful to have a utility program that will make them into the rows of a matrix. Program VtoM ("Vectors to Matrix"), given below, will assemble vectors as the rows of a matrix. The program begins by prompting the user to enter the number of rows in the matrix, then the row vectors in order. The vectors may be entered directly from the keyboard or as named vector variables.

```
PROGRAM: VtoM
:Disp "ENTER no. of rows:"
:Prompt M
:Disp "ENTER first row"
:Input zvec
:dim zvec→N
:(M,N)→dim zmat
:zvec→zmat(1)
:2→J
:Lbl VM1
:Disp "ENTER next row"
:Input zvec
:zvec→zmat(J)
:IS>(J,M)
:Goto VM1
:zmat
:Disp zmat
:
```

Program 6.6.

To illustrate the use of program VtoM, enter and store the following vectors: $X1 = [1\ 1\ 3\ -1]$, $X2 = [-2\ 1\ 0\ -1]$ and $X3 = [-1\ -4\ 1\ -2]$. Now run program VtoM; when prompted with "Enter no. of rows: $M =$?", press 3 [ENTER]. When prompted with "ENTER first row?", press [ALPHA] X1 [ENTER]; respond to the next two prompts by entering $X2$ and $X3$, respectively. When done, the Home screen will display

$$\begin{bmatrix} [& 1 & 1 & 3 & -1 &] \\ [& -2 & 1 & 0 & -1 &] \\ [& -1 & -4 & 1 & -2 &] \end{bmatrix}$$

and this matrix will be stored in the reserved variable Ans. Calculate Ans $*$ Ans^T to see

$$\begin{bmatrix} [& 12 & 0 & 0 &] \\ [& 0 & 6 & 0 &] \\ [& 0 & 0 & 22 &] \end{bmatrix},$$

showing that $\{X1\ X2\ X3\}$ is an orthogonal set. Now go to the VECTR MATH menu and use the unitV instruction to normalize $X1$, $X2$ and $X3$, storing the results as $Q1$, $Q2$, and $Q3$, respectively. You may now use program VtoM to check that $\{Q1\ Q2\ Q3\}$ is an orthonormal set. Note that the dot product of $Q1$ and $Q3$ shows a little round-off error $(-8E - 15)$, which is typical.

Gram-Schmidt process

Any orthonormal set of vectors in R^n, say $\{q_1, q_2, \ldots, q_k\}$, is a highly specialized basis for the subspace W that it spans. Geometrically, it generalizes the familiar $\{\epsilon_1, \epsilon_2, \epsilon_3\}$-coordinate system for R^3: each q_j has length 1 and distinct q_j's are orthogonal. Not only can any vector w in W be written in exactly one way in terms of this basis, $w = c_1 q_1 + c_2 q_2 + \cdots + c_k q_k$, but the coefficients c_j in this linear combination are especially easy to calculate: $c_j = w \cdot q_j$. Because orthonormal bases are so desirable, it is standard practice in introductory linear algebra to study the Gram-Schmidt process for converting an existing basis to an orthonormal basis.

The key idea involved in the Gram-Schmidt process is that of the orthogonal projection, $\text{proj}_w v$, of one vector v onto another vector w:

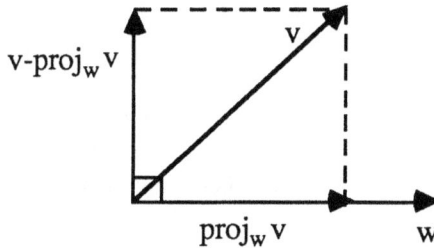

Figure 6.1. Orthogonal projection of one vector v onto another vector w.

Analytically, $\text{proj}_w v = \frac{(v \cdot w)}{\|w\|_2} w$, and a short calculation will verify that $v - \text{proj}_w v$ is orthogonal to w. Because of the repeated need to calculate projections, a calculator program will be helpful. Program PROJ prompts the user for two input vectors v and w, then returns the projection of v onto w.

```
PROGRAM: PROJ
:Disp "ENTER vectors v,w:"
:Prompt v,w
:dot(v,w)/dot(w,w)*w
:Disp "PROJECTION:",Ans
```

Program 6.7.

Use program PROJ with $v = [1\ 2\ 3]$ and $w = [2\ 0\ -1]$ to see $\text{proj}_w v = [-.4\ 0\ .2]$. Now go to the VECTR MATH menu and use the dot instruction to verify that $v - \text{proj}_w v$ is orthogonal to w.

Given a basis $\{x_1, x_2, \ldots, x_k\}$ for a subspace W of R^n, the Gram-Schmidt process constructs an orthonormal basis $\{q_1, q_2, \ldots, q_k\}$ for W as follows:

(i) Let $q_1 = \frac{x_1}{\|x_1\|}$

(ii) then, having constructed $q_1, \ldots, q_j$ construct q_{j+1} by normalizing the vector $x_{j+1} -$ (the sum of the projections of x_{j+1} onto $q_1, \ldots, q_j$).

The first several steps of the process are:

Step 1. x_1, normalized, is q_1

Step 2. $x_2 - (x_2 \bullet q_1)q_1$, normalized, is q_2

Step 3. $x_3 - (x_3 \bullet q_1)q_1 - (x_3 \bullet q_2)q_2$, normalized, is q_3

$$\vdots$$

etc.

Because each q_j has length 1, no denominators appear in the construction of q_{j+1}.

To accomplish all this on the TI-85, we recommend that you begin with the original basis vectors stored as variables $X1, X2, \ldots$.

Step 1. Go to the VECTR MATH menu and press $\boxed{\text{unitV}}$ X1 $\boxed{\text{ENTER}}$ to normalize $X1$; store the result as vector variable $Q1$.

Step 2.   (i) Use program PROJ to calculate the projection of $X2$ onto $Q1$;

    (ii) Go to the VECTR MATH menu and press $\boxed{\text{unitV}}$ (X2-Ans) $\boxed{\text{ENTER}}$; store the result as $Q2$.

Step 3.   (i) Use program PROJ to calculate the projection of $X3$ onto $Q1$, storing the result as $P1$; then calculate the projection of $X3$ onto $Q2$, storing the result as $P2$.

    (ii) Go to the VECTR MATH menu and press $\boxed{\text{unitV}}$ (X3-P1-P2) $\boxed{\text{ENTER}}$; store the result as $Q3$.

$$\vdots$$

etc.

As an example, apply the Gram-Schmidt process to the vectors $X1 = [1\ 0\ -2\ 1]$, $X2 = [2\ 1\ -1\ 0]$ and $X3 = [1\ 2\ 2\ -1]$. When done, your TI-85 should show

$Q1 = [.408248290464\quad 0\quad -.816496580928\quad .408248290464]$,

$Q2 = [.73029674334\quad .547722557505\quad .182574185835\quad -.36514837167]$,

$Q3 = [-.22360679775\quad .67082039325\quad .22360679775\quad .67082039325]$.

Now use program VtoM to see how close the calculated $Q1$, $Q2$ and $Q3$ are to orthonormality. If you put $Q1$, $Q2$ and $Q3$ as rows 1, 2, and 3 of a matrix with VtoM, Ans $*$ Ans^T will show

$$\begin{bmatrix} [& 1 & 4E-14 & -1.62E-13 &] \\ [& 4E-14 & 1 & -5E-14 &] \\ [& -1.62E-13 & -5E-14 & 1 &] \end{bmatrix},$$

Thus, the effects of round-off are apparent. You can clean-up this round-ff error with the instruction round (Ans,11).

As this example shows, the Gram-Schmidt process is numerically unstable, in the sense that round-off errors may produce vectors that are not, numerically, orthonormal. There is a modified version of the Gram-Schmidt process which is more stable numerically, but it is not usually included in an introductory study of linear algebra because it is somewhat awkward to implement except in a pre-packaged program.

QR-factorizations

If matrix A has independent column vectors $x_1, x_2, \ldots, x_k$, then these vectors form a basis for the column space $CS(A)$ of A. Thus, if the Gram-Schmidt process is applied to this basis we obtain an orthonormal basis $q_1, q_2, \ldots, q_k$. It is interesting to note that such an application of the Gram-Schmidt process also produces a factorization of A. Indeed, $A = QR$ where Q is the matrix having $q_1, q_2, \ldots, q_k$ as its column vectors, and R is an invertible upper triangular matrix with entries $r_{ij} = q_i \bullet x_j$ for $i \leq j$.

To see this take place on the TI-85, consider the orthonormal vectors $Q1$, $Q2$ and $Q3$ we constructed from the $X1$, $X2$ and $X3$ above. Use program VtoM to put $Q1$, $Q2$ and $Q3$ as rows 1, 2 and 3 of a matrix, transpose the result and store as matrix Q. Now go to the MATRX EDIT menu and build a 3×3 upper triangular matrix R, where the upper triangular entries are given as follows: for $I \leq J$, the (I, J)-entry is dot(QI, XJ). If you now calculate $Q * R$, you will see

$$A = \begin{bmatrix} [& 1 & 2 & 1 &] \\ [& 0 & 1 & 2 &] \\ [& -2 & -1 & 2 &] \\ [& 1 & 0 & -1 &] \end{bmatrix},$$

after cleaning-up the round-off error in the (3,2)-entry.

For an $m \times n$ matrix A with $m \geq n$, it can be proved that the QR-factorization of A obtained by the Gram-Schmidt process is unique; any process which produces a factorization $A = QR$, where Q has orthonormal columns and R is upper triangular with positive diagonal entries, yields the same Q and R, theoretically. However, because of the numerical instability of the traditional Gram-Schmidt process, professional computer codes use more sophisticated methods to produce such factorizations.

A classic application of QR-factorizations is to overdetermined linear systems, i.e. to systems $Ax = b$ where A is $m \times n$, $m \geq n$. Such systems often fail to have solutions in the usual sense. But we may settle for a bit less and seek a vector x^* such that $Ax^* - b$ is close to O_v in the sense that $\|Ax^* - b\|$ is as small as possible. Minimizing $\|Ax^* - b\|$ amounts to minimizing $\|Ax^* - b\|^2$, a sum of squares, so such "solutions" are called *least squares solutions*. We shall not go into the mathematical theory (which, incidentally, uses projections and orthonormal bases); but it can be shown that least squares solutions always exist and are precisely the solutions to the linear system $A^T Ax = A^T b$, widely known as the system of *normal equations*. In fact, in the case that A has maximal rank n, there is a unique least squares solution. However, in many applications the coefficient matrix $A^T A$ is notoriously ill-conditioned (in the sense that small errors in $A^T A$ or in $A^T b$ may generate large errors in the solutions), so good computational practice suggests that the normal equations $A^T Ax = A^T b$ not be solved by the techniques of Gaussian elimination. Armed with $A = QR$, $Q^T Q = I$ and R is invertible, a little algebra reduces the normal equations to $Rx = Q^T b$, which can be readily solved by back substitution.

We illustrate with the overdetermined linear system

$$
\begin{array}{rrrrrr}
x_1 & + & 2x_2 & + & x_3 & = & 1 \\
 & & x_2 & + & 2x_3 & = & 1 \\
-2x_1 & - & x_2 & + & 2x_3 & = & 1 \\
x_1 & & & - & x_3 & = & 1
\end{array}
$$

The augmented matrix is

$$
\begin{bmatrix}
[& 1 & 2 & 1 & 1 &] \\
[& 0 & 1 & 2 & 1 &] \\
[& -2 & -1 & 2 & 1 &] \\
[& 1 & 0 & -1 & 1 &]
\end{bmatrix}
$$

and its RREF is

$$
\begin{bmatrix}
[& 1 & 0 & 0 & 0 &] \\
[& 0 & 1 & 0 & 0 &] \\
[& 0 & 1 & 1 & 0 &] \\
[& 0 & 0 & 0 & 1 &]
\end{bmatrix}
$$

so the system has no solution. The coefficient matrix A is the one whose QR-factorization was obtained earlier, and a quick calculation shows that $Q^{T*}[1\ 1\ 1\ 1] = [0\ 1.09544511501\ 1.3416407865]$. Store this as variable *bb*, recall matrix R into Ans with [ALPHA] R [ENTER] and run program BkSUB. You will find the unique least squares solution to be $x^* = [2 \ -1.5 \ 1.5]$.

Explorations with Orthogonality

1. (a) Generate a random 5×3 matrix over Z_{10} whose column vectors (from left-to-right) will be called u, v and w.

   (b) Find $\text{proj}_v u$ and verify that $u - \text{proj}_v u$ is orthogonal to v.

   (c) Find $\text{proj}_w u$ and verify that $u - \text{proj}_w u$ is orthogonal to w.

2. (a) Generate a random 5×4 matrix over Z_{10} whose column vectors (from left-to-right) will be called x_1, x_2, x_3, x_4.

   (b) Construct an orthonormal basis $\{q_1, q_2, q_3\}$ for $W = \text{span}[x_1, x_2, x_3]$.

   (c) Calculate the projection of x_4 onto W: $\text{proj}_w x_4 = (x_4 \bullet q_1)q_1 + (x_4 \bullet q_2)q_2 + (x_4 \bullet q_3)q_3$.

   (d) Verify that $x_4 - \text{proj}_w x_4$ is orthogonal to W by checking that it is orthogonal to each of the original basis vectors x_1, x_2 and x_3.

3. (a) Begin by seeding your random number generator with 4, then generate a random 4×4 matrix over Z_{10} which you may regard as the augmented matrix $[A \mid b]$ of a linear system $Ax = b$.

   (b) Use the rref instruction to verify that $AX = b$ has no solution and A has independent column vectors.

   (c) Obtain a QR-factorization of A; check your results.

   (d) Apply your QR-factorization to obtain a least squares solution of $Ax = b$.

6.4. Eigenvalues and Eigenvectors

Given an $n \times n$ matrix A, a non-zero vector x for which Ax is a scalar multiple of x, say $Ax = \lambda x$, is called an *eigenvector* of A. The scalar λ, a real or complex number, is called an *eigenvalue* of A associated with x. Eigenvalues and eigenvectors are of central importance, arising in many applications of linear algebra to science and engineering.

The defining equation $Ax = \lambda x$ is equivalent to $(A - \lambda I)x = O_v$, from which we see that the pair (λ, x) is an eigenvalue-eigenvector pair iff x is a non-zero solution to the homogeneous linear system having coefficient matrix $A - \lambda I$. This system has a non-zero solution iff $A - \lambda I$ is not invertible, i.e., iff $\det(A - \lambda I) = 0$. The determinant $\det(A - \lambda I)$ is a polynomial of degree n in λ, known as the *characteristic polynomial*. Some authors prefer to use $\det(\lambda I - A)$ instead, because it is a monic polynomial in λ. But the difference is trivial since $\det(\lambda I - A) = (-1)^n \det(A - \lambda I)$.

Both of these polynomials have the same roots, the eigenvalues of A, and for any eigenvalue λ the associated eigenvectors are the non-zero solutions to $(A - \lambda I)x = 0$.

```
PROGRAM: Cpoly
:Ans→zmat
:dim zmat→zlst
:zlst(1)→N
:zmat→ztem
:1→dimL zlst
:1→zlst(1)
:For(J,1,N)
:0→tr
:For(K,1,N)
:tr+ztem(K,K)→tr
:End
:J⁻¹*(-tr)→zlst(J+1)
:(J⁻¹*(-tr)*zmat+zmat*ztem)→ztem
:End
:zlst→zlst
:Disp zlst
:
```

Program 6.8.

For an $n \times n$ matrix A stored in Ans, the above program Cpoly returns a list $\{1\ a_{n-1} \cdots a_1\ a_0\}$ of the coefficients of the characteristic polynomial

$$\det(\lambda I - A) = \lambda^n + a_{n-1}\lambda^{n-1} + \cdots + a_1\lambda + a_0.$$

Consider, for example, the matrix

$$\begin{bmatrix} \begin{bmatrix} 5 & -2 & -3 & -1 \end{bmatrix} \\ \begin{bmatrix} -4 & 4 & 4 & 2 \end{bmatrix} \\ \begin{bmatrix} 8 & -5 & -7 & -4 \end{bmatrix} \\ \begin{bmatrix} -7 & 5 & 8 & 5 \end{bmatrix} \end{bmatrix}$$

Store this as variable A, then press [ALPHA] **A** [ENTER] to store A into Ans. Now run program Cpoly to see $\{1\ -7\ 18\ -22\ 12\}$ as the list of coefficients of the characteristic polynomial of A. Thus the characteristic polynomial is $\det(\lambda I - A) = \lambda^4 - 7\lambda^3 + 18\lambda^2 - 22\lambda + 12$. You may now use the TI-85's built-in polynomial root finder to find the eigenvalues as the roots of this polynomial. Press $\boxed{\textbf{POLY}}$ and enter 4 when prompted by "order=", then enter the coefficients from the list $\{1 \ -7 \ 18 \ -22 \ 12\}$ in turn, left-to-right, and press the $\boxed{\textbf{M5}}$ menu key to activate the solver. You will find the roots to be $3, 2, 1+i$ and $1-i$.

Using a polynomial root finder to obtain eigenvalues as the roots of the characteristic polynomial is permissible for examples like the above because the characteristic polynomial of a matrix having integer entries will always have integer coefficients. But this procedure is not recommended in general for the matrices that occur in most applications in science and engineering. The preferred technique for finding eigenvalues numerically uses the QR-algorithm, a sophisticated iterative technique based upon QR-factorizations and well beyond the scope of an introductory study of linear algebra. Nevertheless, the TI-85 uses a built-in QR-algorithm to obtain eigenvalues (and eigenvectors). If, for example, you go to the MATRX MATH menu and execute the instruction eigVl A using the $\boxed{\textbf{M4}}$ key, the calculator will return a list of the eigenvalues, $\{(3,0) \ (1,1) \ (1,-1) \ (2,0)\}$.

As an interesting exercise, find the eigenvalues of

$$B = \begin{bmatrix} -9 & 2 & -6 & 0 \\ 5 & 1 & 5 & 7 \\ 6 & 0 & 3 & 1 \\ 3 & -2 & 3 & -1 \end{bmatrix}$$

as the roots of the characteristic polynomial via Cpoly and also by using the eigVl instruction.

Returning to matrix A above, for any eigenvalue λ the associated eigenvectors are precisely the non-zero solutions to the system $(A - \lambda I)x = 0$. Using $\lambda = 2$, calculate $A - 2I$ (use, e.g., $\boxed{\textbf{ALPHA}}$ $\textbf{A}$ $\boxed{-}$ 2 $\boxed{\times}$ $\boxed{\textbf{ident}}$ 4 $\boxed{\textbf{ENTER}}$) and then use the rref instruction ($\boxed{\textbf{rref}}$ $\boxed{\textbf{Ans}}$) to obtain its reduced row echelon form

$$\begin{bmatrix} 1 & 0 & 0 & 0 \\ 0 & 1 & 0 & -1 \\ 0 & 0 & 1 & 1 \\ 0 & 0 & 0 & 0 \end{bmatrix}.$$

Thus the linear system $(A - 2I)x = 0$ has x_4 as a free variable, and all solutions are given in terms of x_4 by $x = x_4[0 \ 1 \ -1 \ 1]$. Any non-zero value for x_4 will generate an eigenvector associated with $\lambda = 2$, and certainly $x_4 = 1$ is the most convenient choice.

When you obtain the reduced row echelon form of $A - 3I$, you will see

$$\begin{bmatrix} 1 & 0 & 0 & .5714285712429 \\ 0 & 1 & 0 & -.857142857143 \\ 0 & 0 & 1 & 1.28571428571 \\ 0 & 0 & 0 & 0 \end{bmatrix}.$$

To better recognize this matrix, go to the 2nd page of the MATH MISC menu and view the matrix in fraction form by pressing ⟨ Ans ⟩ ⟨ M1 ⟩ ⟨ ENTER ⟩. The result,

$$[[\ 1/1 \quad 0 \quad 0 \quad 4/7 \quad]$$
$$[\ \ 0 \quad 1/1 \quad 0 \quad -6/7 \quad]$$
$$[\ \ 0 \quad 0 \quad 1/1 \quad 9/7 \quad]\ ,$$
$$[\ \ 0 \quad 0 \quad 0 \quad 0 \quad]]$$

shows that all solutions to $(A - 3I)x = 0$ are given by $x = x_4[-4/7,\ 6/7,\ -9/7,\ 1]$. Thus, taking $x_4 = 1$ produces the eigenvector

$$[-4/7,\ 6/7,\ -9/7,\ 1]\,.$$

You can readily verify that all solutions to $[A - (1 + i)I]x = 0$ appear as $x = x_4[(-1/3, 0)\ (1/3, -1/3)\ (-1, 1/3)\ (1, 0)]$ and that all solutions to $[A - (1 - i)I]x = 0$ appear as $x = x_4[(-1/3, 0), (1/3, 1/3), (-1, -1/3), (1, 0)]$. (But, be careful in your use of the rref command!)

In addition to the eigVl instruction, the MATRX MATH menu also includes the eigVc instruction, which uses the QR-algorithm to return a matrix whose columns are eigenvectors associated with the eigenvalues appearing in the list returned by eigVl. But, because these eigenvectors are calculated by an iterative process, you will not usually recognize them in the simple form presented above. When you execute the eigVc A instruction for the matrix A above, you will obtain a matrix whose first column is $[-.400720693076, .601081039614, -.90162155942, .701261212883]$.

Divide this vector by its fourth component .701261212883 and then convert the result to fractional form. You will recognize it as the eigenvector $[-4/7,\ 6/7,\ -9/7,\ 1]$ associated with $\lambda = 3$ that we found earlier. Likewise, the last column in the matrix returned by eigVc A is an eigenvector associated with $\lambda = 2$, in fact, $(1.40308748884)\ [0\ 1\ -1\ 1]$; and the two middle columns are eigenvectors associated with $\lambda = 1 + i$ and $\lambda = 1 - i$, scalar multiples of the ones we noted above.

Diagonalization

When an $n \times n$ matrix A can be converted to a diagonal matrix D by means of an invertible matrix P, as in $P^{-1}AP = D$, we call A *diagonalizable*. Being diagonalizable is equivalent to having n linearly independent eigenvectors (the columns of P... the corresponding eigenvalues form the diagonal of D), so we naturally direct our attention to eigenvectors. An elementary result asserts that eigenvectors associated with distinct eigenvalues are linearly independent, so if A has n distinct eigenvalues then A is certainly diagonalizable. But A can be diagonalizable even when there are fewer than n distinct roots to the characteristic polynomial.

The eigenvectors associated with an eigenvalue λ are just the non-zero solutions to the linear system $(A - \lambda I)x = 0$. Together with the zero vector, they form a subspace of R^n, the nullspace $NS(A - \lambda I)$ of matrix $A - \lambda I$. The key to determining whether or not A has n independent eigenvectors, and thus whether or not A is diagonalizable, rests with the nullspaces $NS(A - \lambda_j I)$ for the *distinct* eigenvalues $\lambda_1, \lambda_2, \ldots, \lambda_k$. Indeed, $\dim NS(A - \lambda_j I) \leq m_j$, where m_j is the multiplicity of λ_j as a root of the characteristic polynomial, and A has n independent eigenvectors iff $\dim NS(A - \lambda_j I) = m_j$ for every $j = 1, \ldots, k$. In this event we choose, for each λ_j, a basis for the nullspace $NS(A - \lambda_j I)$; the basis vectors so chosen can then be used as the columns of a diagonalizing matrix P, and $P^{-1}AP = D$ where the λ_j's on the diagonal of D match up with the columns of P. On the other hand, if for some j we have $\dim NS(A - \lambda_j I) < m_j$ then A is not diagonalizable; there simply aren't enough independent eigenvectors.

We illustrate with three matrices.

(i) Consider the matrix

$$A = \begin{bmatrix} [& 1 & 2 & 3 & 4 & 5 &] \\ [& 2 & 3 & 4 & 5 & 1 &] \\ [& 3 & 4 & 5 & 1 & 2 &] \\ [& 4 & 5 & 1 & 2 & 3 &] \\ [& 5 & 1 & 2 & 3 & 4 &] \end{bmatrix}.$$

The instruction eigVl A returns the list $\{15 \; - 4.25325404176$ $4.25325404176 \; - 2.6286555606 \; 2.6286555606\}$; thus A has five distinct eigenvalues λ_j, $j = 1, \ldots, 5$. For each j, $\dim NS(A - \lambda_j I) = 1$, so to diagonalize A we choose an eigenvector for each λ_j, and let these vectors form the columns of P. The instruction eigVc A returns such a matrix, and the calculation $\text{Ans}^{-1} * A * \text{Ans}$ shows a diagonal matrix, which after clean-up with round(Ans, 11) is the diagonal matrix containing the eigenvalues from the previous list.

(ii) Now consider the matrix

$$B = \begin{bmatrix} [& 1 & 0 & -1 & 0 &] \\ [& 0 & -1 & 0 & 1 &] \\ [& -1 & 0 & 1 & 0 &] \\ [& 0 & 1 & 0 & -1 &] \end{bmatrix}.$$

The instruction eigVl B returns the list $\{2 \; 0 \; 0 \; - 2\}$ of eigenvalues, so to determine whether or not B is diagonalizable we must find the dimension of $NS(B - 0I) = NS(B)$. The instruction rref B returns

$$\begin{bmatrix} [& 1 & 0 & -1 & 0 &] \\ [& 0 & 1 & 0 & -1 &] \\ [& 0 & 0 & 0 & 0 &] \\ [& 0 & 0 & 0 & 0 &]] \end{bmatrix},$$

which shows that the system $(B - 0I)x = 0$ has two free variables, x_3 and x_4. Equivalently, $NS(B - 0I)$ has dimension 2, the multiplicity of $\lambda = 0$ as a root of the characteristic polynomial. Thus B is diagonalizable. To build a diagonalizing matrix P, we first choose a basis for $NS(B - 0I)$. All eigenvectors associated with $\lambda = 0$ are of the form $x = [x_3, x_4, x_3, x_4] = x_3[1, 0, 1, 0] + x_4[0, 1, 0, 1]$, so $[1\ 0\ 1\ 0]$ and $[0\ 1\ 0\ 1]$ form a basis. For $\lambda = 2$, the reduced row-echelon form of $B - 2I$ is

$$\begin{bmatrix} [& 1 & 0 & 1 & 0 &] \\ [& 0 & 1 & 0 & 0 &] \\ [& 0 & 0 & 0 & 1 &] \\ [& 0 & 0 & 0 & 0 &]] \end{bmatrix},$$

showing that all eigenvectors associated with $\lambda = 2$ are scalar multiples of $[-1\ 0\ 1\ 0]$. For $\lambda = -2$, $B + 2I$ has reduced echelon form

$$\begin{bmatrix} [& 1 & 0 & 0 & 0 &] \\ [& 0 & 1 & 0 & 1 &] \\ [& 0 & 0 & 1 & 0 &] \\ [& 0 & 0 & 0 & 0 &]] \end{bmatrix},$$

so vector $[0\ -1\ 0\ 1]$ spans the eigenspace. Putting these basis vectors as the columns of a matrix

$$P = \begin{bmatrix} [& -1 & 1 & 0 & 0 &] \\ [& 0 & 0 & 1 & -1 &] \\ [& 1 & 1 & 0 & 0 &] \\ [& 0 & 0 & 1 & 1 &]] \end{bmatrix},$$

you can quickly check that $P^{-1} * B * P = \text{diag}[2\ 0\ 0\ -2]$.

(iii) Finally, we let

$$C = \begin{bmatrix} [& -9 & 6 & 6 & -14 &] \\ [& 6 & -17 & -14 & 16 &] \\ [& -5 & 12 & 9 & -13 &] \\ [& 8 & -1 & -1 & 11 &]] \end{bmatrix}.$$

Program Cpoly returns the list $\{1\ 6\ 9\ 0\ 0\}$ of coefficients of the characteristic polynomial and the polynomial root finder shows that C has eigenvalues $\lambda = 0, 0, -3$ and -3. The reduced row echelon form of $[C - (-3)I]$ is

$$\begin{bmatrix} 1 & 0 & 0 & 0 \\ 0 & 1 & 1 & 0 \\ 0 & 0 & 0 & 1 \\ 0 & 0 & 0 & 0 \end{bmatrix},$$

showing that $\dim NS[C + 3I] = 1$, so that C is not diagonalizable.

Explorations with Eigenvalues

1. Do the following, for $n = 3, 4, 5$ and 6.

   (a) Generate a random $n \times n$ matrix A over Z_{10}, and calculate $\operatorname{tr}A$ and det A (the *trace* of A, $\operatorname{tr}A$, is defined as the sum of the diagonal entries; you may use program TR at the end of this activity to calculate $\operatorname{tr}A$). Record your results.

   (b) Find the characteristic polynomial $p(\lambda) = \det(\lambda I - A)$ of A; how are $\operatorname{tr}A$ and det A related to this polynomial?

   (c) Evaluate $p(A)$; what do you observe?

2. (a) What can you observe about the eigenvalues of a real symmetric matrix? Experiment with random $n \times n$ symmetric matrices over Z_{10} with $n = 3, 4, 5$ and 6.

   (b) What can you observe about the eigenvalues of a skew-symmetric matrix? Since every skew-symmetric matrix has the form $A - A^T$ for some square matrix A, experiment with such matrices of orders $3, 4, 5$ and 6 over Z_{10}.

3. Which of the following matrices A are diagonalizable? For any one that is, find matrices P and D such that $P^{-1}AP = D$.

   (a) $\begin{bmatrix} 6 & -2 & -4 & -2 \\ -2 & 3 & 2 & 1 \\ 2 & 1 & 0 & -3 \\ -2 & -1 & 2 & 5 \end{bmatrix},$

   (b) $\begin{bmatrix} -3 & 0 & 6 & 5 \\ 1 & -2 & -6 & -5 \\ -6 & 0 & 9 & 5 \\ 5 & 0 & -5 & -2 \end{bmatrix},$

$$\text{(c)} \quad \begin{bmatrix} 7 & 0 & -1 & 9 \\ 0 & -3 & -4 & -4 \\ 6 & 0 & 0 & 9 \\ -6 & 0 & 1 & -8 \end{bmatrix}.$$

The following program returns the trace of a matrix in Ans.

```
PROGRAM: TR
:Ans→zmat
:dim zmat→zlst
:zlst(1)→N
:0→tr
:For(K,1,N)
:tr+zmat(K,K)→tr
:End
:Disp "Trace:"
:Disp tr
:
```

Program 6.9.

7 Exploring Differential Equations

T. G. Proctor
Clemson University

The mathematical models for many applied problems contain differential equations. Examples of these problems include the position and velocity of a falling body, the size of a population, and the amount of a particular contaminant within an aqueous solution. Many of these quantities depend on time, and the differential equation describes how changes in these quantities occur over time. A distinctive feature of the TI-85 is a built-in program for calculating and displaying approximate solutions to one or more differential equations with initial conditions. The calculator has reserved name variables under the GRAPH key for the solutions to differential equations – t and Q – when in the DifEq MODE; these variables are analogous to the named variables for functions – x and y – when in Func MODE.

To approximate a solution to a differential equation with an associated initial condition requires that the following information be entered into your calculator:

MODE	DifEq
GRAPH	$Q'(t) =$	$Q1' = \_\_$
RANGE	values for tMin, tMax, tStep, tPlot, xMin, xMax, xScl, yMin, yMax, yScl, difTol
INITC	$QI1 = \_\_$
AXES	$x = t, y = Q1$

Then to obtain a graph, press GRAPH if only one menu row is showing on the screen, press 2nd GRAPH if two rows are showing.

Example 7.1

Press GRAPH $Q(t) =$ and enter $Q1' = Q1$, press EXIT, RANGE and set these values: tMin $= 0$, tMax $= 1$, tStep $= .1$, tPlot $= 0$, xMin $= -.5$,

Explorations with the Texas Instruments TI-85
John G. Harvey and John W. Kenelly (eds.), pp. 265-306
©1993 by Academic Press, Inc.

xMax = 1.5, xScl = 1, yMin = −.5, yMax = 3, yScl = 1, difTol = .001. Press INITC and set $Q/1 = 1$; finally press GRAPH. If you want to see the solution value at $t = 1$, press the EVAL key (made visible by pressing MORE), press 1 for the value of t and press the ENTER key.

This chapter contains many exercises in which information must be entered into the TI-85 after pressing GRAPH, $Q(t) =$, RANGE, INITC, AXES or other keys obtained by pressing MORE. After making the required entry, it may be necessary to press EXIT or 2nd. We will not repeat the instructions to press one of these latter two keys in the exercises.

The study of differential equations is greatly enhanced when students can graph the concepts being presented. The first course in differential equations usually contains many problems that can be solved explicitly. Thus, students may use the calculator in Func MODE to construct graphs of exact solutions (e.g. $y(x) = 2 * \text{Exp}(-x) - 3 * \text{Exp}(-3x)$) or in DifEqn MODE to construct graphs of approximate solutions. The generic names t and Q (T and q in the programs given later for elementary algorithms) may present an additional problem step to beginning students. Since many textbook problems use variables such as x and y or t and y or t and x, the first step in a problem must be a translation to the new variables, and the last step must be a translation back to the original set of variables.

The student in a first course in differential equations should already know some methods for approximating the solution of a differential equation with an appropriate initial condition. In this chapter we will present programs for two elementary algorithms that demonstrate the programming capabilities for the TI-85 and to provide a comparison with the solution values obtained by the built-in program. For most problems, the student can use the TI-85 program to display solution graphs and values to draw conclusions. There is also a section on the solution of first-order initial value problems in which we explore how the graphs of several solutions can be collected in one picture, allowing the reader to study the sensitivity of the solution to the initial condition and the mathematical model. Some of the functions defining the differential equation are piecewise defined – an easy task for this calculator.

It is easy to exhibit on one screen both the graph of a solution to a first order differential equation and the graph of an input forcing function allowing users to study the dependence of the solution on the input. Material relating to this idea is presented in Section 7.3. The next two sections contain examples and explorations that lead to graphs of the solution of an initial value problem with two differential equations, first for $Q1$ or $Q2$ versus time, then in the form $(Q1(t), Q2(t))$. Section 7.6 explores analytical solutions of linear vector problems created by finding eigenvalues and eigenvectors of an appropriate matrix with calculator key strokes. Finally, programs to construct the trail of an iterative planar map and the direction field for two differential equations are given in the last section.

7.1. Euler and Improved Euler Algorithms

The Euler algorithm for the solution of an initial value problem results from assuming that the slope of the solution of a differential equation $dq/dT = f(T,q)$ is well approximated by the constant $f(T_k, q_k)$ in the interval $T_k \leq T \leq T_k + H$. If q_k is the approximation of $q(T_k)$ and values T_k and q_k and the step size H are known, the algorithm is given by $T_{k+1} = T_k + H, q_{k+1} = q_k + Hf(T_k, q_k)$. Our TI-85 program is called EULER and is given by:

```
PROGRAM: EULER
:Input "STEPSIZE=",H
:Input "INITIAL T=",T
:Input "INITIAL q=",q
:Lbl A
:FSD
:T+H→T
:q+H*W→q
:Disp T
:Disp q:Pause
:Goto A
:
```

Program 7.1.

As written, Program 7.1 does not terminate, that is, at the pause statement, the current value of T and q are on display. When the $\boxed{\text{ENTER}}$ key is pressed, another step is taken, etc. To terminate the program, press the $\boxed{\text{ON}}$ key. To execute this program, we need a subprogram producing $f(T,q) = W$ that we will store with name FSD (abbreviation of f side). Our first example (for solving $q' = q$) is

```
PROGRAM: FSD
:q→W
:
```

Program 7.2.

To record Program 7.1, press $\boxed{\text{PRGM}}$, and $\boxed{\text{EDIT}}$, enter the program name, press $\boxed{\text{ENTER}}$ and insert the program as written. To exit press $\boxed{\text{QUIT}}$. To execute the program, press $\boxed{\text{PRGM}}$, press $\boxed{\text{NAME}}$, and $\boxed{\text{EULER}}$.

If we use the FSD program as given above in Program 7.1 and if we enter stepsize $H = .1$ an initial T value of 0 and an initial q value of 1, the results that we get are shown in Table 7.1. EULER gives a crude approximation 2.59 of q at $T = 0$; the correct solution value for q is 2.71828...

Table 7.1. Results from EULER for $q' = q, q(0) = 1$.

T	=	0.1	0.2	0.3	0.4	0.5	0.6	0.7	0.8	0.9
q	$\approx$	1.10	1.21	1.33	1.46	1.61	1.77	1.95	2.14	2.36

The Improved Euler algorithm is another method for approximating the solution of an initial value problem. This method results from assuming the slope of the solution is well approximated by the average of $f(T_k, q_k)$ and a guess for $f(T_{k+1}, q_{k+1})$ in the interval $T_k \leq T \leq T_k + H$. The algorithm is given by

$$T_{k+1} = T_k + H, q_{k+1} = q_k + \frac{H}{2}(f(T_k, q_k) + f(T_{k+1}, q_k + Hf(T_k, q_k))).$$

Again, q_k is the approximation of $q(T_k)$ and $T_0, q_0,$ and H are given so the algorithm may be initiated. Our TI-85 program is named IULER and is given by:

```
PROGRAM: IULER
:Input "STEPSIZE=",H
:Input "INITIAL T=",T
:Input "INITIAL q=",q
:Lbl A
:FSD
:W→W1
:T+H→T:q→qT
:q+H*W→q
:FSD
:qT+H*(W+W1)/2→q
:Disp T
:Disp q:Pause
:Goto A
:
```

Program 7.3.

To enter this program into the calculator quickly, press $\boxed{\text{PRGM}}$ $\boxed{\text{EDIT}}$ and name the program IULER. Then press $\boxed{\text{RCL}}$, type EULER, and press $\boxed{\text{ENTER}}$. Make the necessary additions and alterations to the program listing that results.

Use IULER and Program 7.2 with a stepsize $H = .1$ an initial T value of 0 and an initial q value to get $q \approx 2.71408$ at $T = 1$ and the values shown in Table 7.2. The values given by the TI-85 are shown in Table 7.3:

Table 7.2. Results from IULER for $q' = q, q(0) = 1$.

T =	0.1	0.2	0.3	0.4	0.5	0.6	0.7	0.8	0.9
q ≈	1.105	1.221	1.349	1.491	1.647	1.820	2.012	2.223	2.456

Table 7.3. Results from the TI-85 for $q' = q, q(0) = 1$.

t =	0.1	0.2	0.3	0.4	0.5	0.6	0.7	0.8	0.9
Q ≈	1.106	1.221	1.350	1.492	1.649	1.822	2.013	2.225	2.459

and Q at $t = 1$ given by 2.7176 as compared with correct value of 2.71828... To get these values, graph the solution, then press TRACE and press the right arrow key as needed.

Suppose we want to execute N steps of the EULER algorithm and observe the output only at $T = T_0 + NH$. The following program, called RPTE (for repeat Euler) requires that values of H, initial values of T and q and the value of N be supplied on input, the output shows the final values of T and q.

```
PROGRAM: RPTE
:Input  "STEPSIZE=",H
:Input  "INITIAL T=",T
:Input  "INITIAL q=",q
:Input  "N STEPS=",N
:0→K
:Lbl  A
:FSD
:T+H→T
:q+H*W→q
:K+1→K
:If  K<N
:Goto  A
:Disp  T
:Disp  q
:
:
```

Program 7.4.

To enter this program quickly, press PRGM , EDIT ,and name the program RPTE. Then press RCL , type EULER, press ENTER , then modify the program listing that appears.

Explorations for Euler and Improved Euler Programs

1. Execute the program EULER to obtain approximate values for the solution of the initial value problem $q' = q^2, q(0) = 1$, using stepsize $H = .1$. Press $\boxed{\text{ENTER}}$ 10 times and observe the T, q values. Repeat the problem for $H = .05$, except press $\boxed{\text{ENTER}}$ twenty times. Now use the technique of separating the variables to obtain an analytical form of the solution for comparison. Repeat the problem using the improved Euler algorithm.

2. Execute the program RPTE for the differential equation $q' = q$ with stepsize .05, $N = 20$ and initial conditions $T = 0, q = 1$. Compare with the results given above for $H = .1$ (Generally accuracy improves when H is reduced, but execution time is increased: check your textbook for an explanation.)

3. Construct a RPT program using the improved Euler algorithm, and repeat the previous exercise.

7.2. Solution Graphs of First Order Initial Value Problems

Graphs of several different solutions to a differential equation obtained by changing the problem's initial conditions can be collected in one picture. This composite graph can reveal a solution structure that may surprise the beginning student. For example, there may be *special solutions* that portray the long range behavior of particular initial value solutions. Consider the equation

$$\frac{dQ}{dt} = \sin^2(t - Q).$$

A change of variables $w = t - Q$ transforms the differential equation into the new system $dw/dt = 1 - \sin^2 w$. From this new equation we see that $Q = t - (2n+1)\pi/2, n = 0, \pm 1, \pm 2, \ldots$ is a family of solutions of the original problem. Even though we can obtain an analytical form of other solutions, we want to investigate their graphical behavior here. For this we use the TI-85 differential equations program to *present the graphs of several solutions in one picture*.

Example 7.2

Enter the following settings into your TI-85:
$\boxed{\text{GRAPH}}$ (DifEqn MODE):
$\boxed{Q'(t) =}$ $Q'1 = (\sin(t - Q1))\wedge 2$, $\boxed{\text{INITC}}$ $QI1 = -1.5708$, $\boxed{\text{AXES}}$ $x = t$, $y = Q$

$\boxed{\text{RANGE}}$ tMin= 0, tMax= 6, tStep= .05, tPlot= 0, xMin= −.3, xMax = 6, xScl= 1, yMin= −5, yMax= 5, yScl= 1, difTol= .001

Press $\boxed{\text{GRAPH}}$ to obtain a graph of the special solution $Q = t - \pi/2$. Press $\boxed{\text{MORE}}$ (twice), press $\boxed{\text{STPIC}}$, name the picture $P1$ and press $\boxed{\text{ENTER}}$. Return to $\boxed{\text{INITC}}$, enter $QI1 = -3\pi/2$, and press $\boxed{\text{GRAPH}}$. After the graph is drawn, press $\boxed{\text{MORE}}$ twice, press $\boxed{\text{RCPIC}}$, and then press $\boxed{P1}$ to obtain a composite picture of two solutions. Store this graph as $P1$. Continue in this way to accumulate the solution trajectories with initial conditions $QI1$ from the set $\{-3\pi/2, -3, -\pi/2, 0, 1.3, \pi/2, 3\}$ on one graph. An easy conjecture is the solutions with $Q(0)$ between $-\pi/2$ and $\pi/2$ approach $Q = t - \pi/2$ as t increases. What about solutions with $Q(0)$ between $-\pi/2$ and $-3\pi/2$?

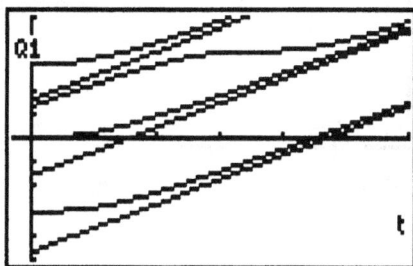

Figure 7.1. Solutions of $Q' = \sin^2(t - Q)$.

It is possible to collect several solutions on one graph by another device. The user could enter duplicates of the differential equation in the $Q'(t)$ menu. By this we mean that the first copy of the differential equation is written for the variable $Q1$, the second for the variable $Q2$, and so forth. Then all equations are selected for graphing, and appropiate initial conditions are entered for each equation. Then $\boxed{\text{GRAPH}}$ will show the solution of each initial value problem. However, in many cases appropriate values of the initial conditions are chosen by inspection of the first graph. Moreover, if solutions of two differential equations and the same initial condition are to be graphed, it may be difficult to determine the graph arising from each differential equation.

Using methods suggested by Example 7.2 we can study quickly the graphical solution structures of $dQ/dt = |\sin(t - Q)|^r$ for different values of r or other systems with or without means of finding the analytical form of the solutions. We should point out that the methods of obtaining the graph are approximate and errors may accumulate for large t.

Sometimes the graphs of solutions to a differential equation show interesting structure and scientists can investigate the problem theoretically even though no analytical form of the solutions can be found. A modification of the example studied by Mills, Weisfeiler and Krall "Discovering

Theorems with a Computer," in *The American Mathematical Monthly*, volume 86 (1979) pages 733-739, is the differential equation

$$dQ/dt = \sin(t * (Q - t/2)).$$

The study gives a way of predicting the number of peaks on a trajectory and the rate at which the solution approaches its limiting form.

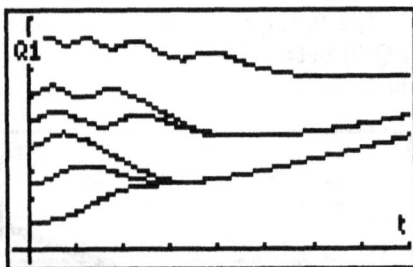

Figure 7.2. Solutions of $Q' = \sin(t * (Q - t/2))$.

Example 7.3

Use the following setting to accumulate six trajectories in one graph:

GRAPH (DifEqn MODE):

$Q'(t) =$ $Q'1 = \sin(t * (Q1 - t/2))$, INITC $QI1 =$ see below, AXES $x = t, y = Q$

RANGE tMin= 0, tMax= 8, tStep= .05, tPlot= 0, xMin= −.2, xMax= 8, xScl= 1, yMin= −.5, yMax= 9, yScl= 1, difTol= .001

Take as initial conditions $QI1$ the elements in $\{1, 2.5, 4, 5, 6, 8\}$.

Mathematical models to simulate the behavior of physical situations contain simplifications of complicated processes. In many cases the form of these simplifications are determined by laboratory measurements. It is important that the solutions of the mathematical models not be extremely sensitive to the form of the simplification. We can illustrate this idea by comparing the graphs of solutions for different models of a problem.

Example 7.4

A mathematical model for the velocity $Q(t)$ of a *particle falling from rest* under the force of gravity in a resisting medium has the form

$$dQ/dt = g - f(Q), Q(0) = 0,$$

where the force exerted on the particle by the resisting medium, $f(Q)$, is determined by experimental means. We take physical units so that $g = 2$ and compare the trajectories resulting from two different functions $f(Q)$.

GRAPH (DifEqn MODE):

$Q'(t) =$ $Q'1 =$ see below, INITC $QI1 = 0$, AXES $x = t, y = Q$

RANGE   tMin= 0, tMax= 4, tStep= .05, tPlot= 0, xMin= −.5, xMax = 4, xScl= 1, yMin= −.5, yMax= 3, yScl= 1, difTol= .001

The first model is the commonly studied linear model $Q'1 = 2 - Q1$. Obtain a graph of the solution and save the graph in $P1$. Then change to a quadratic model $Q'1 = 2 - .5 * Q1 \wedge 2$ and save the graph showing the solution trajectories from both models in the $P1$ variable. Notice that each of these initial value problems can be solved analytically and the terminal velocity in each case is $Q1 = 2$.

Graphs from different mathematical models of a falling body problem show that a change in the form of the resistance force may give different values for the velocity at intermediate times even though the limiting velocities are the same. The displacement of the body is given by the integral of the velocities and such differences could give significant differences in the body position.

Suppose that the $f(Q)$ function exhibits linear behavior for small values of Q and nonlinear behavior for larger values of Q. For example, suppose

$$f(Q) = Q \quad \text{for} \quad 0 \le Q \le 1 \quad \text{and} \quad f(Q) = (2*Q - 1)^{.5} \quad \text{for larger} \quad Q.$$

This resistance force is continuous. The initial value problem may be solved analytically for small t giving: $Q(t) = 2 * (1 - \text{Exp}(-t))$ for $0 < t < \text{Ln } 2$. The problem from this point may be integrated to get an implicitly defined function Q given by $\sqrt{(2 * Q - 1)} + \text{Ln}(2 - \sqrt{(2Q - 1)})^2 = 1 + \text{Ln } 2 - t$. We will use the calculator to show a graph of the resulting trajectory!

The calculator test function $(x < x_0) = 1$ for $x < x_0$ and $= 0$ for $x \ge x_0$ (found under the calculator menu TEST) may be used to construct a **switch off** function and a **switch on** function $(x > x_0) = 0$ for $x < x_0, = 1$ for $x \ge x_0$ can be used to represent functions that are given by piecewise definitions.

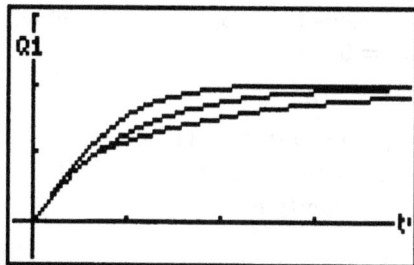

Figure 7.3. Solutions of $Q' = 2 - f(Q)$ for
(a) $f(Q) = Q$, (b) $f(Q) = .5Q^2$, (c) $f(Q) = Q * (Q \le 1) + (2Q)^{.5} * (Q > 1)$.

Example 7.5

Consider the corresponding initial value problem is given by:

$$Q1' = (2 - Q1) * (Q1 \leq 1) + (2 - \sqrt{abs(2 * Q1)}) * (Q1 > 1), Q1(0) = 0,$$

graph the solution for $0 \leq t \leq 4$, using the settings given above for falling particles. Then combine the graph with $P1$ (showing the graphs for the Example 7.4 models).

Consider a population of animals or bacteria, etc. of size $Q(t)$ that varies in time. Births and deaths will be larger when the population is big; however births and deaths may also be affected by overpopulation (competition for space and food). If the population is large, it may be approximated by a continuous function $Q(t)$: we suppose that the growth may be approximated by $dQ/dt = f(Q)$, where $f(0) = 0, f(K) = 0$ and $f(Q) > 0$ for $0 < Q < K$ and $Q = K$ is the largest population the environment will support indefinitely. Two such models are: (a) $f(Q) = Q * (K - Q)$ which is called the logistic model and (b) $f(Q) = Q * \text{Ln}(K/Q)$ which is called the Gompertz model.

Example 7.6

Use the following settings to accumulate the graphs of the solutions of these models for $Q(0) = .2$ in one picture. Save the picture for use in the next example.

GRAPH	(DifEqn MODE):
$Q'(t) =$	$Q'1 = $ see below, $\boxed{\text{INITC}}$ $QI1 = .2$, $\boxed{\text{AXES}}$ $x = t, y = Q$
RANGE	tMin= 0, tMax= 6, tStep= .05, tPlot= 0, xMin= $-.5$, xMax
	$= 6$, xScl= 1, yMin= $-.3$, yMax= 1.2, yScl= 1, difTol= .001

First case: $Q'1 = Q1 * (1 - Q1)$, second case: $Q'1 = -Q1 * \text{Ln}Q1$.

Inspection of the solution graphs of these models shows that the inflection points differ as well as the approach to the limiting population. Sometimes it is possible to specify a model that will produce solutions that have particular properties. The following example is an illustration. First we notice that the solutions to a population model will have an inflection point whenever $f'(Q) = 0$.

Example 7.7

Suppose that experimental data for another similar population suggests that the inflection point should occur at $Q = 2K/3$. A model to achieve this particular property is

$$Q' = f_1(Q) * (Q < 2 * K/3) + f_2(Q) * (Q \geq 2 * K/3)$$

with

$$f_1(Q) = (2 * K/3) \wedge 2 - (Q - 2 * K/3) \wedge 2,$$

$$f_2(Q) = (2 * K/3) \wedge 2 - 4(Q - 2 * K/3) \wedge 2.$$

Graph the solution for this model with limiting population parameter $K = 1$ and with $Q(0) = .2$ using the settings given for the logistic and Gompertz models in Example 7.6. Then compare the graphs of three models of constrained population growth.

Models from various applications may lead to conditions that cannot be solved analytically (at least not by elementary means). In some cases, the TI-85 calculator may be used to numerically determine a solution and we may thus get on to other features of the problem. We will consider such an example.

Example 7.8

An object of mass m in kilograms is released from rest at a height of x meters above the ground. We assume the force on the object exerted by air resistance is kv where v is the velocity and k has units kilograms/sec (so $mdv/dt = mg - kv$, $v(0) = 0$). When does the object strike the ground? We solve the initial value problem for $v = dx/dt$ and integrate for $x(t)$. Then the time t when the object lands satisfies $x_1 = mgt/k - (m/k)^2 g(1 - e^{-kt/m})$. Using $\tau = mt/k$ as a new time scale we obtain the following transcendental equation for τ

$$\left\{ 1 + (\frac{k}{m})^2 \frac{x_1}{g} \right\} - \tau = e^{-\tau}.$$

The left side of this equation gives a straight line graph that intersects the horizontal axis at $\tau_g = 1 + (k/m)^2 x_1/g$. Whenever the number τ_g is less than, say 2.5, we find that the value of τ satisfying the τ equation is significantly different than τ_g. Consequently to find the time when the object strikes the ground we must solve a transcendental equation for τ. To do this we can graph both sides of the equation and find where the graphs intersect or we can use the TI-85 SOLVER program.

To be specific we take $m/k = 10$ (seconds), $x_1 = 500$ (meters), $g = 9.81$ meters/sec^2. Press $\boxed{\text{SOLVER}}$ and enter the equation $1.51 - x = e^{-x}$. Set $x = .5$ and bound $= .1, 1.5$, return the cursor to the x level, and press $\boxed{\text{SOLVE}}$ to obtain $\tau = 1.2126$ as the *scaled time* of intersection.

Explorations of the Solution Graphs of First Order Initial Value Problems

1. Consider the differential equation $dQ/dt = \sin(t - Q)$ and choose the RANGE parameters as before except yMin $= -8$, yMax $= 8$. Then show the solutions of the differential equation with initial conditions

in the set $-5\pi/2, -\pi/2, 3\pi/2, -3, 4, 6.5$. Since solutions don't intersect (this would violate the uniqueness theorem), much information about the solution structure can be deduced from the composite graph that may not be apparent from the analytical form of these solutions.

2. Consider the differential equation $dQ/dt = |\sin(t - Q)|$ and choose the RANGE parameters as before except yMin $= -5$, yMax $= 5$. Then show the solutions of the differential equation with initial conditions in the set $-3\pi/2, -3, -\pi/2, 0, \pi/2, 3$. The solution structure resembles that of $dQ/dt = \sin 2(t - Q)$. For comparison accumulate the graphs of the solution of each of these systems with initial condition $Q(0) = 0$ in a single picture with settings as before except yMin $= -.5$, yMax $= 2$. Note: to find the analytical form of the solutions of $dQ/dt = |\sin(t - Q)|$ takes considerable time.

3. Accumulate the graphs of the solutions of the differential equation dQ/dt
$= \sin(t * (Q + t/2))$ for $0 \le t \le 8$ and for which $Q(0)$ is in the set, $-1, 0, 1, 2, 3, 4$ in one picture. Compare the composite graph with the graph obtained in Example 7.3.

4. Initially a tank holds V_0 gallons of pure water. An inlet stream of r_i gallons per minute with salt content of c_i # /gallon is input into the tank and r_o gallons per minute of the well mixed solution is drained from the tank. Assume $r_i \ne r_o$ so that the volume of solution in the tank is thus $V_0 + (r_i - r_o)t$ gallons where t is time in minutes. The question we want to be able to answer is when is there Q_T # of salt in the tank? The model $dQ/dt = $ input rate $-$ output rate gives

$$1 - N + \frac{r_i - r_o}{V_0}t = \left\{ \frac{V_0}{V_0 + (r_i - r_o)t} \right\}^{\sigma}$$

where

$$N = \frac{Q_T(\sigma + 1)(r_i - r)_o)}{V_0 c_i r_i}, \quad \text{and} \quad \sigma = \frac{r_o}{r_i - r_o}.$$

Case 1.

$r_i > r_o$. The left side of the equation is a straight line with positive slope and the right side is a decreasing positive with value at $t = 0$ above that of the left side: there will be an intersection if the tank doesn't overflow before that time.

Case 2.

$r_i < r_o$. Here the tank will become empty at time $t_e = V_0/(r_o - r_i)$. Since $\sigma < -1, N > 0$, the left side of the equation is a straight line with height below 1 at $t = 0$ and intersecting the horizontal axis at a time smaller than t_e. Since $\sigma < -1$, the right side of the equation has value 1 at $t = 0$ and vanishes at t_e; consequently there may be no intersection, one intersection or two intersections.

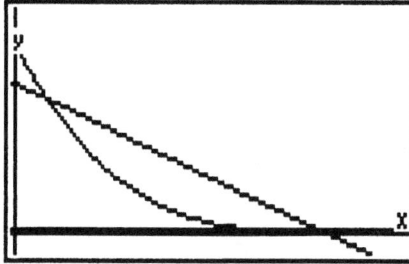

Figure 7.4. Solutions of $.8 - x/500 = (1 - x/500)^4$.

For $V_0 = 500, r_i = 3, r_o = 4$, and $c_i = 1$, use the SOLVER to determine when $Q = 100$.

7.3. First Order Input/Output Problems

Several different physical problems have a common mathematical model that consists of a linear first order nonhomogeneous differential equation $dQ/dt + kQ = f(t)$ and an appropriate initial condition. Here k is a positive constant and $f(t)$ may be controlled by external means and is called the input function. Examples include electrical circuits containing only resistance and inductive elements (Q is current flow, $f(t)$ is proportional to the voltage supplied to the circuit), heat flow to an object from an exterior source (Q is the temperature, $f(t)$ represents the exterior temperature), and flow of a contaminant in a reservoir (Q is proportional to the amount of the contaminant, $f(t)$ the contaminant input rate to the reservoir).

Here we want to study the graphical response of the model $Q(t)$ corresponding to various input functions $f(t)$. After some work we see that the solution is given by

$$Q(t) = Q(0)e^{-kt} + \int_0^t e^{-k(t-s)} f(s) ds.$$

The first term on the right side vanishes as t increases, that is, the system forgets the effect $Q(0)$ over time. The second term may (with appropriate $f(t)$) provide significant long range output. Moreover, if $f(t)$ can be broken into parts $f(t) = f_1(t) + f_2(t)$, the second term splits into output due to $f_1(t)$ added to output due to $f_2(t)$. We denote the second term in the expression given above as $Tf(t)$. Sometimes $Tf(t)$ can be obtained using the method of undetermined coefficients. For example, if $f(t)$ is the constant K, then $Tf(t)$ is the constant K/k, or if $f(t)$ is given by $K\cos\omega t$, then $Tf(t)$ is given by $K\sin(\omega t + \phi)/\sqrt{(\omega^2 + k^2)}$ for the phase angle ϕ defined by $\sin\phi = k/\sqrt{(\omega^2 + k^2)}$, and $\cos\phi = \omega/\sqrt{(\omega^2 + k^2)}$. These special cases are combined in the following example.

Example 7.9

Use the calculator to graph the input $f(t) = 1 - \cos 6.283t$ and the output for $k = 1, Tf(t) = 1 - \sin(6.283t + .1578)/6.362$. Warning: the reserved name variables in the Func MODE are x and y. In our case $t \to x$ and $f(t)$ and $Tf(t) \to y(x)$.

$\boxed{\text{GRAPH}}$ (Func MODE):

$\boxed{y(x) =}$ $y1 = 1 - \cos(6.283 * x), y2 = 1 - \sin(6.283 * x + .1578)/6.362$

$\boxed{\text{RANGE}}$ xMin= 0, xMax= 1, xScl= 1, yMin= −.3, yMax= 2.1, yScl= 1

Notice that the differential equation system changed the input by shifting the cosine function in time and changing the amplitude.

What can be done for more complicated input functions $f(t)$? In many cases, the input $f(t)$ may be periodic, say with period T and we will concentrate on this case. Then the output will undergo a transient phase and settle into a periodic response. If we wish to avoid waiting on the system to settle into this steady state response, we can choose the initial condition $Q(0)$ so that $Q(T) = Q(0)$. (Since the system has period T, this condition guarantees the output will have period T.) This leads to the condition

$$Q(0) = \frac{1}{1 - e^{-kT}} \int_0^T e^{-k(T-s)} f(s)\,ds.$$

For the case when $f(t)$ is proportional to sine or cosine functions, we can use the method of undetermined coefficients as we did above to find the steady state solution. For other periodic input functions we may or may not be able to evaluate the integrals in $Q(0)$ and $Tf(t)$ using integral tables and thus obtain the periodic response. In either case it may be convenient to proceed as follows: obtain the value of $Q(0)$ above by using the **calculator's numerical integration program**, then use that initial condition in the difEqn MODE to obtain the graph of the periodic response.

Example 7.10

Consider the input with period 1 and given on $0 \le t < 1$ by $f(t) = 5t$ for $0 \le t \le .2, f(t) = 5 * (1 - 2t)/3$ for $.2 \le t \le .8, f(t) = 5 * (t - 1)$ for $.8 \le t < 1$ in a system with $k = 1$. The steps given below allow us to integrate the function

$$e^{-(1-t)}\{5t(t \le .2) + 5(1 - 2t)/3(t > .2)(t \le .8) + 5(t - 1)(t > .8)\}$$

in t from 0 to 1 using the TI-85. Press $\boxed{\text{CALC}}$, then press $\boxed{\texttt{fnInt}}$ enter the function listed above, followed by $, t, 0, 1)$ and press $\boxed{\text{ENTER}}$. We divide the result by $1 - 1/e$ to get $Q(0) = -.131$. Then we use the calculator setting given by:

$\boxed{\text{GRAPH}}$ (DifEqn MODE):

$\boxed{Q'(t) =}$ $Q'1 = (5*t*(t.2)+5/3*(1-2*t)*(t > .2)*(t \le .8)+5*(1-t)*(t > .8)) - Q1,$

$\boxed{\text{INITC}}$ $QI1 = -.131,$ $\boxed{\text{AXES}}$ $x = t, y = Q$

$\boxed{\text{RANGE}}$ tMin= 0, tMax= 1, tStep= .025, tPlot= 0, xMin= 0, xMax= 1, xScl= 1, yMin= $-.3$, yMax= .3, yScl= 1, difTol= .001

Press $\boxed{\text{EVAL}}$, enter 1 and press $\boxed{\text{ENTER}}$ to see that the solution returns approximately to its initial $Q1$ value. Save the graph as $P1$. Then overlay the graph of the input function by using the setting:

GRAPH) (Func MODE):

$\boxed{y(x) =}$ $y1 = 5x(x < .2) + 5(1 - 2x)/3(x.2)(x.8) + 5(x - 1)(x > .8)$

$\boxed{\text{RANGE}}$ xMin= 0, xMax= 1, xScl= 1, yMin= -1.2, yMax= 1.2, yScl= 1

Press $\boxed{\text{RCPIC}}$, press $\boxed{\text{P1}}$ then $\boxed{\text{STPIC}}$ (if desired) as $P1$. Notice the output rises while the input is positive, then falls back to the initial value. Recall the vertical scales on the graphs are different. If you want to see the system, input or output, over several periods, substitute $\text{mod}(x, 1)$ for x while in Func Mode and $\text{mod}(t, 1)$ for t while in DifEqn Mode. By using another initial condition, you can study the system approach to the steady state response.

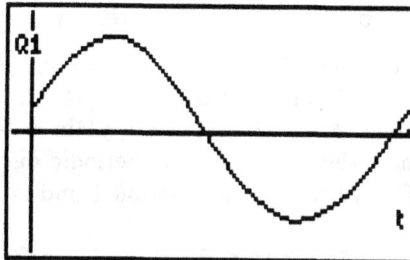

Figure 7.5. Solution of $Q' + r(t)Q = \cos(2\pi t), Q(0) = .04$
$r(t) = (1 + 2t) * (t < .5) + (3 - 2t) * (t \ge .5).$

Example 7.11

Consider again mathematical models for the size of a population with a birth (net) term that is linear in the population size and a term that limits the population representing space and or food restrictions. We want to include in the model a recognition that birth and death rates depend on

natural conditions such as weather. In that regard we study the following
mathematical model:

GRAPH	(DifEqn MODE):
$Q'(t) =$	$Q'1 = (\cos(3.1416 * t)) \wedge 2 * Q1 - .4 * Q1 \wedge 2$, INITC $QI1 =$ see below
AXES	$x = t, y = Q$
RANGE	tMin= 0, tMax= 4, tStep= .05, tPlot= 0, xMin= 0, xMax = 4, xScl= 1, yMin= −.1, yMax= 2, yScl= 1, difTol= .001

Here the first term models the net birth/death growth rate, depends peri-
odically in time and is nonnegative. By graphing the response then varying
the initial condition, we determine that an initial condition of $Q1I = 1.26$
leads to a periodic response that is also approximately the state of other solutions with similar initial conditions. (Such a solution is called stable.) In
this case we could take the input birth/death rate as input and the periodic
response as output.

The population model discussed above contains a Bernoulli differential
equation and can be treated partially by analytical means. This will lead
to the integral in t of a term $\exp(-\sigma t - \mu \cos(2\pi t))$. This of course could
be done numerically, then by using the requirement $Q(1) = Q(0)$, we can
determine the initial condition that leads to the periodic response. However the calculator will allow us to experimentally determine this initial
condition quickly.

Explorations of the Input/Output Graphs
of First Order Problems

1. Use the calculator to graph the input $f(t) = \cos 6.283t + .5 * \cos 3.1415t$
   and the output $Tf(t) = \sin(6.283t + .1578)/6.362 + .5 * \sin(3.1415t + .3082)/3.2968$. Use xMin = 0, xMax= 2, yMin = −1.5, yMax = 1.5.

2. Obtain a graph of the input and the periodic output for the system
   $dQ/dt + Q = f(t)$ where $f(t)$ has period 1 and is given on $0 \le t < 1$
   by

$$f(t) = (8t - 3)(t < .75)/3 + (7 - 8t)(t \ge .75).$$

   Note: The integrals in this exercise and in the Example 7.10 can be
   obtained by using tables; however, in many cases, for example $f(t) = e^{-\sin t}$, the associated integrals are not evaluated by tables.

3. Consider the following population model with periodic birthrate:

$$Q'(t) = (\cos(\pi * t)^2 - .2) - .4 * Q1^2.$$

   Here the net birth/death rate is sometimes negative. Can the population reach a steady periodic state? Try various initial conditions to
   see if you can determine a periodic solution $Q(t)$ with positive values.

7.4. Second Order Problems Using Two First Order Differential Equations

In this section differential equations of the type

$$\frac{d^2w}{dt^2} = g\left(t, w, \frac{dw}{dt}\right)$$

will be treated as a special case of two first order differential equations

$$\frac{dQ_1}{dt} = f(t, Q_1, Q_2), \frac{dQ_2}{dt} = g(t, Q_1, Q_2).$$

The substitution $w = Q_1, dw/dt = Q_2$ in the second order differential equation leads to the system of two differential equations with $f(t, Q_1, Q_2) = Q_2$. Also in this way a second order differential equation with initial conditions $w(t_0) = w_0, dw/dt(t_0) = w_0'$ is treated as an initial value problem for the Q_1, Q_2 system with initial conditions $Q_1(t_0) = w_0, Q_2(t_0) = w_0'$.

Solutions or approximate solutions to an initial value problem that includes two differential equations are studied by constructing graphs of $Q1$ versus t, or $Q2$ versus t and sometimes by constructing a graph of $Q1(t)$ versus $Q2(t)$. We will also be checking to see if the solutions are sensitive to small changes in the initial conditions or to small changes in the problem parameters. It may be important to recognize long term solution patterns and the approach to such a pattern particularly in application problems.

The TI-85 program for obtaining and graphing approximate solutions can be used for several simultaneous differential equations. Other algorithms for obtaining approximate solutions of simultaneous differential equations furnished with initial conditions are straightforward generalizations of the corresponding algorithm for a single differential equation. We will present a calculator program in the last section of this chapter for such an algorithm. Here we will concentrate on the built in program.

Example 7.12

Consider the problem

$$\frac{d^2w}{dt^2} - w = 0, w(0) = \frac{dw}{dt}(0) = 1.$$

It is easy to verify that $w(t) = e^t$ is the solution to this problem. The transformation given above gives the system $dQ1/dt = Q2, dQ2/dt = Q1, Q1(0) = Q2(0) = 1$. Enter the problem into your TI-85 as follows:

GRAPH (DifEqn MODE):
$\boxed{Q'(t) =}$ $Q'1 = Q2, Q'2 = Q1$, $\boxed{\text{INITC}}$ $QI1 = 1, QI2 = 1$, $\boxed{\text{AXES}}$ $x = t, y = Q1$

RANGE   tMin= 0, tMax= 1, tStep= .025, tPlot= 0, xMin= 0, xMax = 1, xScl= 1, yMin= −.1, yMax= 3, yScl= 1, difTol= .001

Press GRAPH, then press EVAL, specify $t = 1$, then press ENTER to get $Q1 = 2.71763\ldots$. This result is an indication of the accuracy of the TI-85 algorithm.

The study of the motion of a mechanical spring and the study of the flow of current in an electrical circuit leads to a linear mathematical model of the form

$$\frac{d^2w}{dt^2} + 2b\frac{dw}{dt} + \sigma^2 w = f(t).$$

Here $b \geq 0$ and will sometimes be called a resistance coefficient: $f(t)$ will be called a *forcing* term. Many of the physical problems modelled by this differential equation contain nonlinear elements that are approximated by terms linear in w or in dw/dt. Our examples and exercises in this section will concern this model and nonlinear extensions of the model.

Example 7.13

Consider the *unforced spring/circuit* model for $b = 1/6, \sigma^2 = 145/144$ and $f(t) = 0$ and graph the solution with $w(0) = 12, w'(0) = 0$ for $0 \leq t \leq 4\pi$. Enter the following settings into your TI-85:

GRAPH   (DifEqn MODE):
$Q'(t) =$ $Q'1 = Q2, Q'2 = -(145 * Q1/144 + Q2/6),$
INITC $QI1 = 12, QI2 = 0,$ AXES $x = t, y = Q1$
RANGE   tMin= 0, tMax= 12.566, tStep= .025, tPlot= 0, xMin= 0, xMax = 12.566, xScl= 1, yMin= −13, yMax= 13, yScl= 1, difTol= .001

Now press GRAPH. By using techniques introduced in the differential equations course you should be able to determine that the analytical solution of this problem is $Q1(t) = 12.04\exp(-t/12)\sin(t + 1.49)$. It is informative to add the graphs of $Q1(t) = \pm12.04 * \exp(\ t/12)$ as an overlay: the solution is seen to oscillate within that envelope. Just as we indicated in the previous sections, you can store the picture, change to Func MODE, draw these envelope curves, and recall the picture to obtain this composite figure.

Example 7.14

Next we consider an *undamped forced spring/circuit problem* with $b = 0, \sigma^2 = 36$ and $f(t) = \cos 4t$. Graph the solution with initial condition $w(0) = w'(0) = 0$ for $0 \leq t \leq 2\pi$. Enter the settings

GRAPH   (DifEqn MODE):
$Q'(t) =$ $Q'1 = Q2, Q'2 = \cos(4 * t) - 36 * Q1,$
INITC $QI1 = 0, QI2 = 0,$ AXES $x = t, y = Q1$
RANGE   tMin= 0, tMax= 6.283, tStep= .025, tPlot= 0, xMin= 0, xMax= 6.283, xScl= 1, yMin= −.15, yMax= .15, yScl= 1, difTol= .001

and press $\boxed{\text{GRAPH}}$. Here the solution is $Q1(t) = \sin(5t) * \sin(t)/10$ and it is informative to add the graphs of $Q1(t) = \pm \sin t$ as an overlay. The solution of the differential equation is seen to oscillate (beat) within these envelope curves.

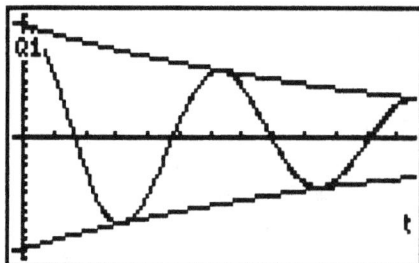

Figure 7.6a. Solutions with Envelop Curves for
$w'' + (1/6)w' + (145/144)w = 0$.

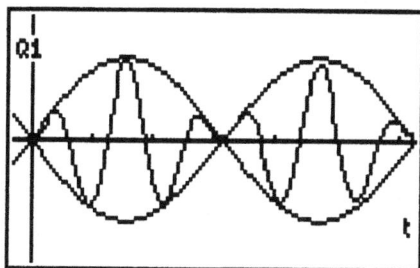

Figure 7.6b. Solutions with Envelop Curves for
(a) $w'' + 36w = \cos 4t$.

Next we will look at a special subproblem of the forced spring/circuit linear model, namely the case, $b > 0$ and $f(t)$ periodic with period P. Your differential equations textbook will show that the solution of an initial value problem for this differential equation is the sum of a term (sometimes called the complementary function) that vanishes as t increases and a particular solution. If the particular solution is periodic, we have isolated the long term pattern of the response.

If $f(t)$ is a simple combination of sine and cosine functions of t, then the method of undetermined coefficients will provide a periodic particular solution. The calculator can be used to study the relationship between such input functions $f(t)$ and the periodic response. For example if $f(t) = a \cos \gamma t$, then $w_p(t) = A(\gamma) \sin(\gamma t + \phi)$, where

$$A(\gamma =) \frac{a}{\sqrt{(\sigma^2 - \gamma^2)^2 + 4b^2\gamma^2}}, \tan \phi = \frac{\sigma^2 - \gamma^2}{2b\gamma}.$$

Moreover, if $f(t)$ is the sum of several such terms, the long term response will be the sum of the corresponding periodic solutions.

Example 7.15

Suppose $b = .5, \sigma^2 = 1.25$. Use the calculator to graph the input $f(t) = \cos t + 3 * (\sin(3t/2)) \wedge 2$ and the output $w_p(t) = \sin(t + .245) + 1.2 - .1805 \sin(3t - 1.201)$. Warning: the reserved name variables in the Func MODE are x and y. In our case $t \to x$ and $f(t)$ and $w_p(t) \to y(x)$.

GRAPH (Func MODE):

$y(x) =$ $y1 = \cos(x) + 3 * (\sin(3 * x/2)) \wedge 2$, $y2 = (\sin(x + .245))/1.031 + 1.5 * (.8 - \sin(3 * x - 1.12)/8.31)$

RANGE xMin= 0, xMax= 6.283, xScl= 1, yMin= −3, yMax= 4, yScl= 1

Does the response $y2$ ever become 0? (Use the trace function of the calculator to make a guess.)

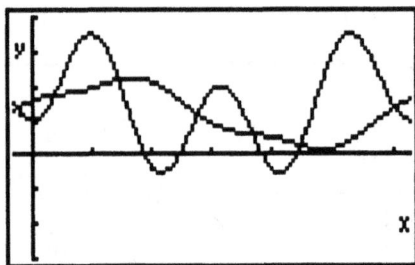

Figure 7.7. Input forcing and periodic output for
$$w'' + w' + 1.25w = \cos t + 3\sin^2 1.5t.$$

Is there a periodic solution to our linear spring/circuit model when the forcing term is more complicated? The $f(t)$ could be made up in pieces by straight lines, or functions with denominators in sine or cosine terms. In these cases the method of variation of parameters can be used to find a periodic solution. The form of the first particular solution that we present will depend on the sign of $b^2 - \sigma^2$. The first case to be considered is $b^2 - \sigma^2 < 0$, and we put $\mu = (\sigma^2 - b^2)^{.5}$. A particular solution is given by

$$w_p(t) = \frac{1}{\mu} \int_0^t e^{-b(t-s)} \sin \mu(t - s) f(s) ds.$$

We want to add to $w_p(t)$ a part of the complementary function (that is, the general solution of the corresponding homogeneous problem) so the resulting new particular solution is periodic. The reader should verify here that the solution $z_1(t)$ of the associated homogeneous problem so that

$z_1(0) = 1, dz_1/t(0) = 0$ is $z_1(t) = e^{-bt}[\cos \mu t + b/\mu \sin \mu t]$: the solution $z_2(t)$ so that $z_2(0) = 0, dz_2/dt(0) = 1$ is $z_2(t) = (1/\mu)e^{-bt} \sin \mu t$. The particular solution that we seek has the form

$$w_q(t) = az_1(t) + bz_2(t) + w_p(t).$$

We choose the numbers a and b so that $w_q(P) = w_q(0)$ and $w'_q(P) = w'_q(0)$. This leads us to the linear system of equations

$$\begin{bmatrix} 1 - z_1(P) & -z_2(P) \\ -z'_1(P) & 1 - z'_2(P) \end{bmatrix} \begin{bmatrix} a \\ b \end{bmatrix} = \begin{bmatrix} w_p(P) \\ w'_p(P) \end{bmatrix}$$

for the numbers a and b. Here we need to evaluate w_p and its derivative at $t = P$ (the period of the system).

We make several notes here: (1) the calculator has an integration key and we can use it to evaluate the integrals numerically, (2) we can use the calculator features to solve the linear system for the numbers a and b and (3) the solution of the spring/circuit system that we seek has value a and derivative value b at $t = 0$. After determining a and b we can graph the solution using the calculator program for solving initial value problems.

Example 7.16

Let us find the periodic solution of the system:

$$\frac{d^2w}{dt^2} + \frac{dw}{dt} + 1.25w = 1 - \sin^4(\pi t).$$

In this problem the period $P = 1$: we use the calculator to evaluate $w_p(P) = .197$. Since

$$w'_p(t) = \frac{1}{\mu} \int_0^t e^{-b(t-s)}[\mu \cos \mu(t - s) - b \sin \mu(t - s)]f(s)ds,$$

we can evaluate this at $t = 1$ using the integration key to get $w'_p(P) = .323$. The matrix given above is

$$\begin{bmatrix} 1 - .5829 & .5104 \\ .6380 & 1 - .0725 \end{bmatrix}$$

and the linear equations for a and b have solution $a = .488, b = .013$. After using the calculator to graph the solution of the system with $QI1 =$

.488, $QI2 = .013$ (with scale $.4 < y < .6$), we can add the graph of the forcing $1 - \sin^4(\pi t)$ (scale $0 < y < 1$).

For the case $b^2 - \sigma^2 > 0$, we can put $\nu = (b^2 - \sigma^2)^{.5}$. Then a particular solution to our spring/circuit problem is given by:

$$w_p(t) = \frac{1}{\nu} \int_0^t e^{-b(t-s)} \sinh \nu(t - s) f(s) ds.$$

To find a periodic solution, just as before, we want to add to $w_p(t)$ a part of the complementary function so the resulting new particular solution is periodic. The reader should also verify the solution $z_1(t)$ of the associated homogeneous problem so that $z_1(0) = 1, dz_1/t(0) = 0$ is $z_1(t) = e^{-bt}[\cosh \nu t + b/\nu \sinh \nu t]$: the solution $z_2(t)$ so that $z_2(0) = 0, dz_2/dt(0) = 1$ is $z_2(t) = (1/\nu)e^{-bt} \sin \nu t$. The reader should work out the remaining calculations for this case and for the case where $b^2 - \sigma^2 = 0$.

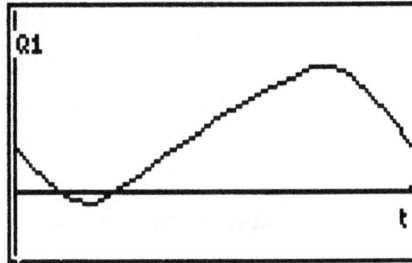

Figure 7.8. Periodic Response $w'' + .7w' + 1.1w = f(t)$ $f(t) = f(t + 1)$, with $f(t) = 100t(t \le .25) + 100(t - 1)(t \ge .75)$ for $0 \le t \le 1$.

The spring/circuit model that we studied can be extended to include nonlinear damping and restoring force terms, that is, to models of the form:

$$\frac{d^2w}{dt^2} + B(\frac{dw}{dt}) + K(w) = f(t)$$

where the functions B and K are increasing functions. Under what conditions will there be a stable periodic solution if the forcing function $f(t)$ is periodic? How would we find it? We will not be able to use the undetermined coefficients or the variation of parameters technique unless the problem is piecewise linear. However, we can hope that a fortunate choice of initial condition will yield a solution that approaches a periodic solution and thus be able to identify approximate initial conditions for the periodic solution.

Example 7.17

Consider $B(v) = v/2, K(w) = w + .5 * w^3$, and $f(t) = \cos t$. (A cubic term $K(w)$ has often been used to model part of the nonlinear spring restoring forces and the $B(v)$ function is a linear function of the type previously considered.)

GRAPH (DifEqn MODE):

$\boxed{Q'(t) =}$ $Q'1 = Q2, Q'2 = COS(t) - .5*Q2 - Q1 - .5*Q1^3,$

$\boxed{\text{INITC}}$ $QI1 = .96, QI2 = .905,$ $\boxed{\text{AXES}}$ $x = t, y = Q1$

$\boxed{\text{RANGE}}$ tMin= 0, tMax= 12.566, tStep= .025, tPlot= 0, xMin= 0, xMax = 12.566, xScl= 1, yMin= −5, yMax= 5, yScl= 1, difTol= .001

Use the $\boxed{\text{EVAL}}$ key at the final value of t to check whether the solution is periodic: if not use the final values as initial conditions for a new solution and repeat. Save the picture as $P1$, then change $K(Q1)$ to $Q1$ and the initial conditions to $QI1 = 0, QI2 = 2$ and overlay the resulting graph on the $P1$. Discuss the changes.

Example 7.18

Next consider $B(v) = (v - .5 + (.5 - 1.5*v)*e^v)*(v < 0) + (v + .5 - (.5 + 1.5*v)/e^v)*(v \geq 0)$, and $K(w) = w + .5*w^3$, and $f(t) = \cos t$. The $B(v)$ function being considered is smooth and has a graph that is approximately a straight line for large $v < 0$ and that is approximately another straight line for large positive v. Also we have $B(0) = B'(0) = 0$.

GRAPH (DifEqn MODE):

$\boxed{Q'(t) =}$ $Q'1 = Q2, Q'2 = COS(t) - (Q2 - .5 + (.5 - 1.5*Q2)*e^{\wedge}Q2)*(Q2 < 0) - (Q2 + .5 - (.5 + 1.5*Q2)/e^{\wedge}Q2)*(Q2 \geq 0) - Q1 - .5*Q1^{\wedge}3$

$\boxed{\text{INITC}}$ $QI1 = 1.1, QI2 = .78,$ $\boxed{\text{AXES}}$ $x = t, y = Q1$

$\boxed{\text{RANGE}}$ tMin= 0, tMax= 12.566, tStep= .025, tPlot= 0, xMin= 0, xMax= 12.566, xScl= 1, yMin= −5, yMax= 5, yScl= 1, difTol= .001

In these exercises, we see a response to periodic forcing that looks periodic and that looks stable just as we guessed from the linear case. The tools needed to prove that there are stable periodic solutions are given in advanced courses. Such results are described in [7].

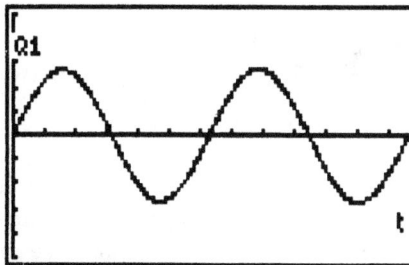

Figure 7.9. Periodic Solution for $K(w) = w$, $B(w) = .25(v - 1)*(v < -1)$ $+ .5v(v \geq -1)*(v < 1) + .25*(v + 1)*(v \geq 1)$.

Explorations on Linear and Nonlinear Spring/Circuit Problems

1. Graph the solution of the damped unforced spring with $b = .25$ and $\sigma^2 = 17/16$ and the initial condition $w(0) = 5, w'(0) = 13/8$ for $0 \leq t \leq 4\pi$. Use the trace function to determine the approximate time when w has a maximum value on this interval.

2. Graph the solution of the undamped forced spring with $\sigma^2 = 49/4$ and $f(t) = 12 \cos 5t$ and the initial condition $w(0) = w'(0) = 0$ for $0 \leq t \leq 4\pi$. This solution has an envelope curve $w = \pm A \sin(\alpha t) * \sin(\beta t)$ Determine A, α, and β and overlay the envelope curve around the solution curve.

3. Determine the periodic solution of the spring/circuit model with $b = .7, \sigma^2 = 1.1$ and forcing with period $P = 1$ with $f(t) = 100t$ for $0 \leq t \leq .25, f(t) = 0$ for $.25 < t < .75$ and $f(t) = 100(t - 1)$ for $.75 \leq t \leq 1$.

4. Consider a nonlinear spring/circuit model in which $B(v) = (.25 * v - .25) * (v < -1) + (.5 * v) * (v \geq -1) * (v \leq 1) + (.25 * v + .25) * (v > 1)$, and $K(w) = w + .5 * w^3$, and $f(t) = \cos t$. Use the calculator to graph the $B(v)$ function for $-3 \leq v \leq 3$. We note that $B(0) = 0$ and B is continuous and has a graph that is composed of three straight lines. Compare with the B function being used in Examples 7.15 and 7.16. Make the usual transformation to a system of two differential equations and graph the solution with $QI1 = .93, QI2 = .89$ for $0 \leq t \leq 4\pi$. Is the solution approximately periodic. How does it compare with the solutions to Examples 7.15 and 7.16?

5. Consider a nonlinear spring/circuit with $B(v) = (-.5 * v^2) * (v < 0) + (.5 * v^2) * (v \geq 0)$ and $K(w) = w + .5 * w^3$, and $f(t) = \cos t$. Notice that the $B(v)$ function being considered is continuous, is piecewise quadratic, and $B(0) = 0$. Make the usual transformation to a pair of differential equations and explore various initial conditions to determine a periodic solution of period 2π.

7.5. Phase Plane Solutions of Differential Equations

Solutions $(Q_1(t), Q_2(t))$ of differential equations

$$Q_1' = f(Q_1, Q_2), Q_2' = g(Q_1, Q_2)$$

may be regarded at parametric equations for a curve in the (Q_1, Q_2) plane. Such curves are said to be in the *phase plane*. We may be able to spot solution characteristics in the phase plane that are not as apparent in Q_1 versus t and Q_2 versus t graphs. Suppose that a solution is given by $(Q_1(t), Q_2(t))$ for $t_0 \leq t \leq t_1$, then because the functions f and g in the differential equations do not depend on t, we have that $(Q_1(t + \sigma)$,

$Q_2(t+\sigma))$ is also a solution for $t_0 \le t+\sigma \le t_1$ for any number σ. Further, if the f and g functions have continuous partial derivatives, solutions to initial value problems are unique; consequently solutions arising from different initial points either coincide or do not intersect.

What are some examples of solution characteristics that may be visible in the phase plane? First we present some background. Solutions (Q_1^e, Q_2^e) of the equations $f(Q_1, Q_2) = 0, g(Q_1, Q_2) = 0$ are constant (in time) solutions of the differential equations and are called **equilibrium solutions**. The phase plane graph of this type of solution is a single point (Q_1^e, Q_2^e). If all solutions that start near such an equilibrium solution approach (Q_1^e, Q_2^e) as time increases, then the equilibrium solution is called (asymptotically) stable and (Q_1^e, Q_2^e) is the steady state solution for such nearby trajectories. Closed trajectories in the phase plane represent **periodic solutions** to the differential equations. If all solutions initiating near such a periodic solution approach the periodic solution as time increases, then the periodic solution is called (asymptotically) stable and is the steady state solution of these nearby solutions.

Example 7.19

Consider the system

$$Q_1' = Q_1 * (2 - Q_1 - Q_2), Q_2' = Q_2 * (3 - Q_1 - 2 * Q_2).$$

Equilibrium points are $(0,0), (0,3), (2,0)$, and $(1,1)$. We want to graph the solutions with initial values at $(Q1, Q2) = (3,2), (Q1, Q2) = (1,3)$, $(Q1, Q2) = (.3, .3)$, and $(Q1, Q2) = (1, .2)$. First enter the settings
[GRAPH] (DifEqn MODE):
[Q'(t) =] see above, [INITC] $QI1 = 3, QI2 = 2$, [AXES] $x = Q1, y = Q2$
[RANGE] tMin= 0, tMax= 3, tStep= .05, tPlot= 0, xMin= −.3, xMax= 4, xScl= 1, yMin= −.3, yMax= 4, yScl= 1, difTol= .001
and press [GRAPH]. Save the graph as $P1$, change the initial conditions to $QI1 = 1, QI2 = 3$, recall $P1$ and store the composite as $P1$ again, repeat for the initial conditions $(.3, .3), (1, .2)$. The composite picture should suggest that $(1, 1)$ is a stable equilibrium point. Since trajectories starting with positive components do not approach any of the other equilibria, such points do not represent stable solutions. Note: these differential equations are discussed in [1] and in [6] as a model for a population problem with two populations that compete for the same food resources.

Example 7.20

Consider the differential equations given below:

$$Q_1' = Q_2, Q_1' = -(Q_1 + \text{sign}\,(Q_2)).$$

Since there is a discontinuity in the differential equation, strange solution curves may occur. In this example (taken from [2]) motion comes to a halt in finite time. Enter the settings [GRAPH] (DifEqn MODE): [Q'(t) =] see above, [INITC] $QI1 = 7.5, QI2 = 0$, [AXES] $x = Q1, y = Q2$ [RANGE] tMin= 0, tMax= 12.566, tStep= .05, tPlot= 0, xMin= -8, xMax= 8, xScl= 1, yMin= -8, yMax= 8, yScl= 1, difTol= .001 Save the graph as $P1$, change the initial conditions to $QI1 = 2.5, QI2 = 0$, change tMax to 3.14 and press [GRAPH]. Recall $P1$ and store the composite as $P1$ again. The composite picture should show $(-.5, 0)$ is at the terminus of both trajectories. See [2] for an analysis.

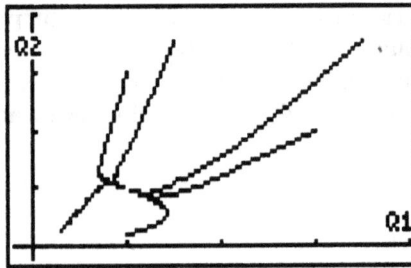

Figure 7.10. Solutions for
$$Q'_1 = Q_1 * (2 - Q_1 - Q_2), Q'_2 = Q_2 * (3 - Q_1 - 2 * Q_2).$$

Next we consider the nonlinear initial value problem

$$\frac{d^2w}{dt^2} + K(w) = 0, w(0) = z, \frac{dw}{dt}(0) = 0,$$

where the number z will be prescribed and the essential feature of the continuous function K is a change of sign from negative to positive as w increases through 0. The system under consideration is thus $dQ_1/dt = Q_2, dQ_2/dt = -K(Q_1)$. By dividing these equations, separating the variables and integrating, we obtain

$$\left(\frac{dQ_1}{dt}\right)^2 = -2\int_0^{Q_1} K(s)ds + 2\int_0^2 K(s)ds.$$

For convenience, we define the function $-F(Q_1)$ as the first integral on the right side and plot $y = F(x)$ versus x. Suppose z is a positive number and we draw a horizontal line at the height $F(z)$. From the equation given above, we see that $Q1(t)$ exists when the vertical difference between the constant $F(z)$ and $F(Q1)$ is positive; dQ_1/dt becomes 0 when the $F(Q_1)$

curve intersects $F(z)$ and then changes sign and reverses direction. In this way we see that a periodic solution exists as shown in Figure 7.11.

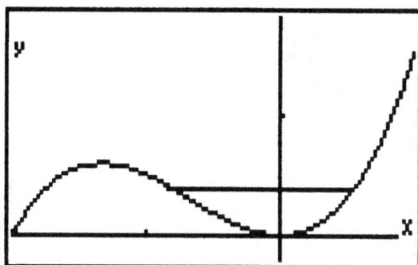

Figure 7.11a. Nonlinear System $w'' + w + .5w^2 = 0$
Construction for Solutions.

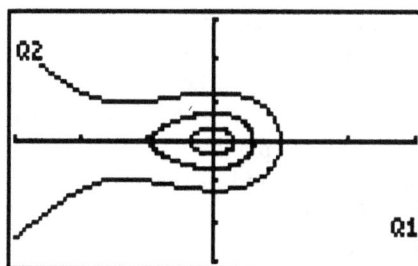

Figure 7.11b. Nonlinear System $w'' + w + .5w^2 = 0$
Typical Solutions.

We can regard the trajectories as solutions of the first order differential equation

$$\frac{dQ_1}{dt} = \pm\sqrt{F(z) - F(Q_1)} : F(x) = 2\int_0^x K(s)ds$$

with $Q(0) = z$. If $P/2$ is the time required to proceed from the starting point to the state $Q_1(P/2) = z_1, Q_2(P/2) = 0$, we have that

$$P = 2\int_0^{P/2} dt = 2\int_{z_1}^z \frac{dw}{\sqrt{F(z) - F(w)}}.$$

An easy example comes from the differential equation $w'' + \sigma^2 w = 0$, where all the solutions have the period $P = 2\pi/\sigma$.

Example 7.21

Consider the piecewise linear system

$$Q_1' = Q_2, Q_2' = -(Q_1 * (Q_1 \le 0) + 1.5 * Q_1 * (Q_1 > 0)).$$

By using the analysis given above for $Q1 \leq 0$ and for $Q1 > 0$, the period $P = P_1 + P_2$ where $P_1 = 2\pi/2 = 3.1416, P_2 = 2\pi/3 = 2.565$. Enter the settings

GRAPH	(DifEqn MODE):
$Q'(t) =$	see above, INITC $QI1 = 3, QI2 = 0$, AXES $x = Q1, y = Q2$
RANGE	tMin= 0, tMax= 5.707, tStep= .05, tPlot= 0, xMin= −4,
	xMax= 4, xScl= 1, yMin= −4, yMax= 4, yScl= 1, difTol= .001

and press GRAPH. The reader will note the curve has a different shape in the regions $Q_1 > 0$ and $Q_1 < 0$. The change in linearity at $Q_1 = 0$ did not destroy two properties of a linear system, viz. each closed trajectory had the same geometric shape and the period around the trajectory was the same for each solution.

Other important examples of nonlinear systems with a family of periodic solutions for the cases: $K(w) = \sin w, K(w) = w + \theta w^3$, etc. are discussed in [7].

Collecting the (Q_1, Q_2) graphs of solutions to a system $Q'_1 = f(Q_1, Q_2)$, $Q'_2 = g(Q_1, Q_2)$ into a composite picture may be difficult, particularly when the f and g functions contain quadratic, cubic or higher order terms in the Q_1, Q_2 variables. This difficulty arises because trajectories that originate in various sectors of the Q_1, Q_2 plane may exhibit entirely different behavior. The following was encountered in the search for a nontrivial example of *orthogonal families of curves*.

Example 7.22

Consider the system: $Q'_1 = Q_2(3Q_2^2 - Q_1^2), Q'_2 = Q_1(Q_1^2 - 2Q_2^2)$ Enter the settings indicated by

GRAPH	(DifEqn MODE):
$Q'(t) =$	see above, INITC $QI1, QI2$ see below, AXES $x = Q1, y = Q2$
RANGE	tMin=0, tMax=see below, tStep= .025, tPlot= 0, xMin= −.2,
	xMax = 5, xScl= 1, yMin— −.5, yMax= 4, yScl= 1, difTol= .001

and by the entries in Table 7.4. Collect the associated solution graphs into a composite picture.

Table 7.4. Settings for Example 7.20, part a.

$Q_1(0)$	$Q_2(0)$	$tMax$
0	4.0	0.15
0	2.5	0.4
0	2.0	0.6
0	1.0	2.5
1	0	4.2
2	0	1.0
3	0	0.3
4	0	0.15

Changes in tMax are necessary because on some solutions the values of Q_1 and/or Q_2 become large and exceed the calculator limits causing an error indication. Experimentation was used to discover appropriate values of tMax. Next we generate a second family of curves orthogonal (at intersections) to those curves in $P1$. These curves are generated as the solution of $Q'_1 = Q_1(Q_1^2 - 2Q_2^2), Q'_2 = Q_2(Q_1^2 - 3Q_2^2)$. These solutions were collected into a composite graph called $P2$ using the same AXES and RANGE settings except as noted in Table 7.5.

Table 7.5. Settings for Example 7.20, part b.

$Q_1(0)$	$Q_2(0)$	$tMax$
1.0	4	3.0
2.0	4	2.0
2.5	4	1.5
3.0	4	3.0
4.0	4	0.015

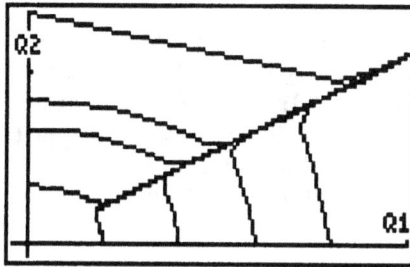

Figure 7.12. Solutions for
$$Q'_1 = Q_2(3Q_2^2 - Q_1^2), Q'_2 = Q_1(Q_1^2 - 2Q_2^2).$$

Explorations with Phase Plane Graphs

1. Graph the phase plane solution $(Q1(t), Q2(t))$ for $0 \le t \le 7$ originating at $(.5, .5)$ of the system

$$Q'_1 = Q_2 + (Q_1 \wedge 3)\cos(Q1 \wedge 2 + Q_2 \wedge 2),$$

$$Q'_2 = (Q_2 \wedge 3)(\cos(Q1 \wedge 2 + Q_2 \wedge 2) - Q1$$

using tStep= .05, xMin= -4, xMax= 4, yMin= -4, yMax= 4. Save the graph as $P1$, change the initial conditions to $QI1 = 0, QI2 = 1.5$, recall $P1$ and store the composite as $P1$ again. The composite picture suggests that there is a stable periodic solution. Notice the origin is not a stable equilibrium point.

2. The differential equations studied by van der Pol have the form

$$\frac{d^2w}{dt^2} + \epsilon(w^2 - 1)\frac{dw}{dt} + w = 0.$$

When the parameter ϵ has a positive value, solutions that initiate at $(w(0), w'(0)) \neq (0, 0)$ spiral clockwise in the (w, w') plane to a periodic solution. For $\epsilon = .2$ and initial condition $(w(0), w'(0)) = (4, 4)$, use the TI-85 program to graph the solution for $0 \leq t \leq 14$. Use xMin = yMin = -5, xMax = yMax = 5, tStep = $.02$. Obtain the graph of the solution for $(w(0), w'(0)) = (.5, .5)$ for $0 \leq t \leq 20$. Try $\epsilon = 1$ and repeat the exercise.

3. Consider the following system that is composed of three linear pieces:

$$Q_1' = Q_2, Q_2' = -((.5 * (Q_1 + 1) - 1) * (Q_1 \leq -1)$$
$$+ Q_1 * (Q_1 > -1) * (Q_1 \leq 1) + (.5 * (Q_1 - 1) + 1) * (Q_1 > 1)).$$

Obtain the $(Q1(t), Q2(t))$ graph of the solution with initial condition $QI1 = 3, QI2 = 0$ for $0 \leq t \leq 7.5$ Save the graph as $P1$, then repeat for $QI1 = 2$, tMax = 7.01; recall $P1$ combine and store again as $P1$. Note: The periods were obtained using the equation given above. In this example $F(w) = w^2$ for $-1 < w < 1$, $F(w) = -1 + (w - 1)^2/2$ for $w < -1$ and $F(w) = -1 + (w + 1)^2/2$ for $w > 1$. In the case starting at $Q1 = 3$, we have

$$P/4 = \int_0^1 \frac{dx}{\sqrt{7 - x^2}} + \sqrt{2} \int_1^3 \frac{dx}{\sqrt{16 - (x_1)^2}} = .3925 + 1.480.$$

The second integral can be evaluated using a trigonometric substitution of the form $x + 1 = 4\sin t$. (The first integral can be handled in a similar way or it can be evaluated numerically. The second integral is improper and the TI-85 has trouble with it.)

7.6. Linear Systems of Differential Equations: Constant Coefficients

In this section we will consider a linear homogeneous system of differential equations with constant coefficients contained in a matrix A with a vector solution $Q(t)$ with, say n components. Such solutions are composed from the eigenvalues and eigenvectors of the matrix A. Eigenvalues and eigenvectors can be quickly determined using the TI-85. The general solution of the system of differential equations can then be used to calculate all solutions of a similar non-homogeneous differential equation especially when the forcing function $f(t)$ is composed of elementary functions. If values of the solution of the nonhomogeneous system are required at only a few points, then the general solution of the homogeneous system may be used in the variation of parameters formula together with the calculator computed value of definite integrals to obtain particular solutions of the nonhomogeneous problem. This includes the case when $f(t)$ is defined piecewise with the pieces made from elementary functions. Of course approximate solutions to vector differential equations can also be obtained using the calculator program for solutions of differential equations initial value problems directly.

Consider the **homogeneous** system of differential equations

$$\frac{dQ}{dt} = AQ$$

where the n by n matrix A is given. The TI-85 calculator has a built in program to compute the eigenvalues and associated eigenvectors of the matrix A. Recall that if λ is an eigenvalue and q is a corresponding eigenvector of A, then $Q(t) = ce^{\lambda t}q$ is a solution of $dQ/dt = AQ$ for any constant c. Thus if A has eigenvalues $\lambda_1, \lambda_2, \ldots, \lambda_n$ and independent corresponding eigenvectors $q_1, q_2, \ldots, q_n$, a general solution of the differential equation is

$$Q(t) = c_1 e^{\lambda_1 t} q_1 + c_2 e^{\lambda_2 t} q_2 + \cdots + c_n e^{\lambda_n t} q_n$$

where $c_1, c_2, \ldots, c_n$ are arbitrary constants. (Also recall that the vectors $q_1, q_2, , \ldots, q_n$ are independent if the eigenvalues are distinct.) This general solution can be put into the form $Q(t) = S(t)c$, where c is the vector with components $c_1, c_2, \ldots, c_n$ and where the matrix $S(t)$, called a **fundamental matrix of solutions**, is given by

$$S(t) = \left[e^{\lambda_1 t} q_1 \; e^{\lambda_2 t} q_2 \, . \, . \, e^{\lambda_n t} q_n \right].$$

If the vector q_0 is given, choose the vector c in the expression given above to satisfy the equation $S(0)c = q_0$. Then $Q(t)$ as so given, will satisfy both the differential equations and the initial condition.

Example 7.23

Press $\boxed{\text{MATRX}}$, then $\boxed{\text{EDIT}}$, type a matrix name such as A, press $\boxed{\text{ENTER}}$, write 2, press $\boxed{\text{ENTER}}$, write 2, press $\boxed{\text{ENTER}}$, continue in this way to enter the elements $1, 1 = -2, 1, 2 = -1, 2, 1 = -1, 2, 2 = -2$, then press $\boxed{\text{EXIT}}$. (If you want to see the matrix, press $\boxed{A}$.) Now press $\boxed{\text{MATRX}}$, press $\boxed{\text{MATH}}$, then $\boxed{\text{eigVl}}$ and $\boxed{A}$ (or whatever you named the matrix), $\boxed{\text{ENTER}}$, and $\boxed{\text{eigVc}}$, $\boxed{A}$, $\boxed{\text{ENTER}}$. You should now have the eigenvalues

$\{-1, -3\}$ and eigenvectors in the display (the eigenvectors appear as columns in a matrix in the same order as the eigenvalues).

In this example, we have a special case with negative eigenvalues λ_1, λ_2 and corresponding eigenvectors q_1, q_2. It is instructive to notice the geometric effect arising from dominant eigenvalues. We suppose the notation is such that $\lambda_1 > \lambda_2$, then since solutions have the form

$$Q(t) = c_1 e^{\lambda_1 t} q_1 + c_2 e^{\lambda_2 t} q_2$$

for those solutions with $c_1 \neq 0$, $Q(t) \approx exp(\lambda_1 t)(c_1 q_1)$ when t increases. If we plot part of the trajectory (for $t > 0$) in the (Q_1, Q_2) plane we will see the solution approaches the origin $(0, 0)$ along the vector q_1 or its negative. Enter the settings

GRAPH	(DifEqn MODE):
$Q'(t) =$	$Q'1 = -(2 * Q1 + Q2), Q'2 = -(Q1 + 2 * Q1),$
INITC	$QI1 = 2, QI2 = 0,$ AXES $x = Q1, y = Q2$
RANGE	tMin= 0, tMax= 3, tStep= .05, tPlot= 0, xMin= -2, xMax
	= 2, xScl= 1, yMin= -2, yMax= 2, yScl= 1, difTol= .001

and press ENTER. Repeat for other initial conditions.

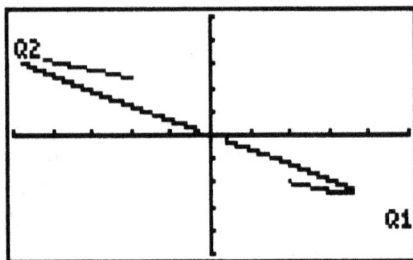

Figure 7.13. Solution of
$Q1' = -(1.35Q1 + 2Q2), Q2' = .6Q1 + .85Q2.$

Example 7.24

The matrix with first row given by $[-.1, 3]$ and second row given by $[-3, -.1]$ has eigenvalues $= -.1 \pm i3$ and associated eigenvectors $q = q_1 \pm iq_2(q_1 = \text{column } [1, 0], q_2 = [0, 1])$. Thus $(q_1 \pm iq_2) \exp((-.1 \pm i3)t)$ are solutions of the system. By expanding this product into its real and imaginary parts, we see that real valued solutions are given by $Q_1(t) = \exp(-.1 * t)[q_1 \cos 3t - q_2 \sin 3t], Q_2(t) = \exp(-.1 * t)[q_1 \sin 3t + q_2 \cos 3t]$. The reader should note that $Q_1(t)$ starts at $t = 0$ on q_1, rotates to q_2 at $t = \pi/6$, proceeds to q_1 at $\pi/3$, etc. That is, the solution cycles in a spiral

toward $(0, 0)$ and the exponential multiplier causes the rotation to contract. Press $\boxed{\text{GRAPH}}$ after entering the settings given below to observe this effect.
$\boxed{\text{GRAPH}}$ (DifEqn MODE):
$\boxed{Q'(t) =}$ $Q'1 = -.1 * Q1 + 3 * Q2, Q'2 = -3 * Q1 - .1 * Q2,$
$\boxed{\text{INITC}}$ $QI1 = 2, QI2 = 0, \boxed{\text{AXES}} \ x = Q1, y = Q2$
$\boxed{\text{RANGE}}$ tMin= 0, tMax= 4, tStep= .05, tPlot= 0, xMin= -2, xMax
       = 2, xScl= 1, yMin= -2, yMax= 2, yScl= 1, difTol= .001
    The reader will recall that a point (Q_1*, Q_2*) for which $f(Q_1*, Q_2*) = g(Q_1*, Q_2*) = 0$ is an equilibrium solution of the possibly nonlinear system

$$dQ_1/dt = f(Q_1, Q_2), dQ_2/dt = g(Q_1, Q_2).$$

If we construct the linear approximation of the system about the point (Q_1*, Q_2*), then the eigenvalues and eigenvectors of the linearized system reveal asymptotic behavior of solution near the equilibrium point.

Example 7.25

    Enter the setting described by
$\boxed{\text{GRAPH}}$ (DifEqn MODE):
$\boxed{Q'(t) =}$ $Q'1 = \sin(2 * Q1) + \cos(Q2) - 2 * Q1, Q'2 = \cos(Q1) - \sin(Q2) -$
       $Q2/10,$
$\boxed{\text{INITC}}$ $QI1 = 1, QI2 = 2, \boxed{\text{AXES}} \ x = Q1, y = Q2$
$\boxed{\text{RANGE}}$ tMin= 0, tMax= 8, tStep= .05, tPlot= 0, xMin= $-.3$, xMax
       = 3, xScl= 1, yMin= $-.5$, yMax= 3, yScl= 1, difTol= .001
and press $\boxed{\text{GRAPH}}$. The trajectory will approach an equilibrium point. Use the $\boxed{\text{EVAL}}$ function key at $t = 8$ to get an approximate location $(Q_1^a, Q_2^a) = (.89, .64)$ of this point. Now define a 2 by 2 matrix A with all zero elements and calculate each of the quantities

$$\frac{\partial f}{\partial Q_1} = 2 * \cos(2 * Q_1^a) - 2, \frac{\partial f}{\partial Q_2} = -\sin(Q_2^a)$$

$$\frac{\partial g}{\partial Q_1} = -\sin(Q_1^a), \frac{\partial g}{\partial Q_2} = -\cos(Q_2^a) - .1$$

and store them as the elements of A by using the STO key as follows. Enter the number $2 * \cos(2 * .89) - 2$,press $\boxed{\text{ENTER}}$, press $\boxed{\text{ANS}}$ $\boxed{\text{STO}}$ A(1,1) $\boxed{\text{ENTER}}$. Modify the other elements of A in the same manner, then press $\boxed{\text{MATRX}}$, $\boxed{\text{MATH}}$, and $\boxed{\text{eigV1}}$ A to get $\{-2.7, -.64\}$ amd press $\boxed{\text{eigVc}}$ A to get column [.9, .4] , column [.3, $-.95$]. The vector resulting from adding [.89, .64] (from the coordinates of the fixed point) to the eigenvector vector [.9, .4] is [1.8, 1]. Change the initial point to $QI1 = 1.8, QI2 = 1$ and press

$\boxed{\text{GRAPH}}$ to see an approximate eigenvector route to the equilibrium point. Repeat for other initial conditions derived from adding $[.89, .64]$ to $\pm$ an eigenvector. We note that the equilibrium point can be located as the limiting point on a trajectory when the equilibrium point is an attractor as above or a repeller use the limit as time decreases.

Suppose a particular real eigenvalue λ is a double root of the eigenvalue equation for the matrix A and there is only one independent vector solution of the system $(A - \lambda I)q = 0$. Then $Q(t) = \exp(\lambda t)q$ is a solution of $dQ/dt = AQ$ and an independent solution can be calculated by assuming the form $Q(t) = t\exp(\lambda t)q + \exp(\lambda t)p$. Then p and q must be nontrivial and satisfy $(A - \lambda I)q = 0$ and $(A - \lambda I)p = q$. By multiplying this last equation on the left by $A - \lambda I$ we get $2(A - \lambda I)^2 p = 0$. We solve for a vector p so that $\{p, q\}$ are independent vectors.

Example 7.26

Enter the matrix A with first row $[-1.75, 1, .25]$, second row $[-.5, 0, -.5]$ and third row $[-.25, 1, -2.25]$ into the calculator and determine that A has eigenvalues with corresponding eigenvectors $\{-1, -1, 2\}$, column $[-.8321, -.5547, -.2773]$, column $[-.8321, -.5547, -.2773]$, and column $[-1, 0, 1]$. For $\lambda = -1$, take $q =$ column $[-.8321, -.5547, -.2773]$. To obtain the p vector proceed as follows: Press $\boxed{\text{ident}}$ $3 + A$, press $\boxed{\text{ENTER}}$. The result is in ANS, then create ANS$\wedge 2$ to get a matrix with row space given by multiples of $[0, .5, -1]$. p can be chosen as any vector in this space so $\{p, q\}$ is an independent set, say $p =$ column $[0, 2, 1]$.

When the vector function $f(t)$ has a special form, a particular solution $Q_p(t)$ of the **nonhomogeneous** linear differential system of differential equations

$$\frac{dQ}{dt} = AQ + f(t)$$

can be found by the method of undetermined coefficients. A general solution of the system is $S(t)c + Q_p(t)$ where $S(t)$ is a fundamental matrix of solutions and c is any n vector.

The variation of parameters solution for a general solution to the non-homogeneous problems with forcing function is

$$Q(t) = S(t)c + \int_0^t S(t - s)S^{-1}(0)f(s)ds$$

where $S(t)$ is a fundamental matrix of solution of the associated homogeneous system and c is any vector. If $f(t)$ is periodic with period T and the eigenvalues of A have nonzero real part, then if c is chosen so that $Q(T) = Q(0)$, i. e.,

$$[S(0) - S(T)]c = \int_0^T S(T - s)S^{-1}(0)f(s)ds$$

then the resulting solution is periodic. It can be shown that when the real parts of the eigenvalues of A have negative real parts, all solutions of the nonhomogeneous system approach this periodic solution. If it is inadvisable to wait for this special solution to emerge at large times, we can choose the initial condition as specified above by calculation and then plot the resulting solution using the built-in algorithm supplied on the TI-85. Of course the calculation of the correct initial condition also may require using the calculator to compute the values of the integrals.

Explorations with Linear Vector Differential Equations

1. What are the eigenvalues and eigenvectors of the matrix with first row given by $[-.5, 1.5]$ and second row $[1.5, -.5]$? Use the settings
GRAPH	(DifEqn MODE):
$Q'(t) =$	$Q'1 = -.1 * Q1 + 1.5 * Q2, Q'2 = 1.5 * Q1 - .1 * Q1,$
INITC	$QI1 = 1, QI2 = -1,$ AXES $x = Q1, y = Q2$
RANGE	tMin= 0, tMax= 4, tStep= .05, tPlot= 0, xMin= -2, xMax

   = 2, xScl= 1, yMin= -2, yMax= 2, yScl= 1, difTol= .001
   to graph the indicated solution. Add the solutions for the other initial condition points in the set

   $$\{(1, -1), (-1, 1), (-1, 1.01), (-1, .99), (1, -.99), (1, -1.01)\}.$$

2. Consider the homogeneous system of differential equations with 3 by 3 matrix A with row 1 given by $[-1.25, -.5, 1.75]$, row 2 given by $[.5, -2, .5]$, and row 3 given by $[.25, .5, -2.75]$. Check by using the calculator to verify that the eigenvalue and eigenvectors are: $\lambda_1 = -1, q_1 =$ column $[3, 2, 1], \lambda_2 = -2, q_2 =$ column $[-1, 2, 1], \lambda_3 = -3, q_3 =$ column $[-1, 0, 1]$. Suppose that the $Q(t)$ is a solution of the system and it is known that $\exp(t) * Q(t) \rightarrow$ column $[3, 2, 1]$*as* $t \rightarrow \infty$. Can initial conditions $Q(0)$ be chosen so that the third component in the resulting solution has two positive roots? Use the differential equation solver to construct the $Q3$ versus t graph of such a solution.

3. Construct a fundamental matrix of solutions to the homogeneous linear system with matrix A given by

$$A = \begin{bmatrix} -.85 & .85 & -2.05 \\ .1 & 0 & -.9 \\ .55 & .15 & -.25 \end{bmatrix}$$

and choose vectors a and b so that $Q_p(t) = a\cos 2\pi t + b\sin 2\pi t$ is a particular solution of $dQ/dt = AQ+$ column $[2\sin 2\pi t, 0, -3\cos 2\pi t]$. Then choose vector c so $S(t)c + Q_p(t)$ solves the nonhomogeneous differential equation and the initial condition $Q(0) =$ column $[1, 2, 4]$.

4. Consider the forcing function $f(t) =$ column $[0, 0, w(t)]$ with $w(t)$ periodic with period 1 and $w(t) = 20t$ for $0 \le t < .25$, $w(t) = 10 * (1 - 2t)$ for $.25 \le t < .75$, and $w(t) = 20 * (t - 1)$ for $.75 \le t < 1$ and the matrix given in the previous problem. Use the calculator to compute the following integral values:

$$EW = \int_0^1 e^{-.5(1-s)}w(s)ds \qquad CEW = \int_0^1 e^{-.3(1-s)}\cos(1 - s)w(s)ds$$

and the quantity SEW with form CEW except the sine function replaces the cosine function. (Answers: $.417, .371, .286$) Use the variation of parameters formula integral to calculate $S(0)c =$ column $[.026, .00174, -.645]$, then plot the solutions $Q1(t), Q2(t), Q3(t)$ using the calculator's differential equation initial value solver. It is interesting to notice here that the third component of the forcing induces very little motion in the first two components even in the long term.

7.7. Concluding Remarks

The material in this chapter has been chosen to illustrate the easy use of a graphics programmable calculator to illustrate concepts present in an elementary differential equations course. The choice of topics and emphasis in such a course is being affected by the availability of technological tools. In this regard see particularly [3] and [5]. Some textbooks (such as [4]) present exercises that require graphical and computing tools: more will do so in the future. This last section presents important topics that did not seem to fall conveniently in the previous material.

In the study of linear problems using LaPlace transform students frequently see a forcing term that depends on a small parameter, say ϵ, that approximates the Dirac delta function as $\epsilon \to 0$. Here the theorem is that the solution of the problem $Q_\epsilon(t) \to$ the solution of a related problem as $\epsilon \to 0$. The following problems give some graphical support for this result.

Example 7.27

In the setting given below use eps $= .2, .1$, and $.05$.

GRAPH (DifEqn MODE):

$Q'(t) =$ $\quad Q'1 = 1(1 - sign(t - eps))/2 * eps - Q1$,

INITC $QI1 = 0$,   AXES $x = t, y = Q1$

RANGE   tMin= 0, tMax= 3, tStep= .005, tPlot= 0, xMin= −.3, xMax = 3, xScl= 1, yMin= −.3, yMax= 1.2, yScl= 1, difTol= .001

Overlay the graphs of the solution of each problem. The graph for the smallest eps should resemble the graph of exp(−t). Notice that the initial condition has been altered in the limiting solution from $Q(0) = 0$ for the approximate problems.

The behavior of various solution trajectories $(Q(t)) : t > 0$ of vector differential equations $dQ/dt = f(Q)$ (called autonomous differential equations), particularly the asymptotic behavior, that is, the *steady state behavior*, is of special interest in scientific models. This includes approach to equilibrium or periodic behavior. If measurements are made on these trajectories only at discrete times $t_1, t_2, \ldots$, then we have a discrete sequence of points $\{(Q(ti)) : i = 1, 2, \ldots\}$. The case of two differential equations is of particular interest because here we have trajectories in the $x - y$ plane that can be studied geometrically.

Discrete time trajectories in the $x - y$ plane can also arise from the solution of a difference equation system of the form $x_{i+1} = f(x_i, y_i)$, $y_{i+1} = g(x_i, y_i)$ with the initial points x_0, y_0 as parameters of the study. Here we can have the points (x_i, y_i) approaching a limit as $i \to \infty$ or the points can have a periodic cycle. We can also have somewhat erratic behavior of the trajectory of points. We construct a simple program for the TI-85 for the much studied system

$$x_{i+1} = x_i \cos \alpha - (y_i - x_i^2) \sin \alpha, \; y_{i+1} = x_i \sin \alpha + (y_i - x_i^2) \cos \alpha$$

with $\cos \alpha = .4$ that was first presented in *Numerical Study of Quadratic Area-Preserving Mappings* in **The Quarterly of Applied Mathematics** in 1969 by M. Henon.

In order for the program to execute we need to store initial values into U and V. To get some idea of the behavior of some trajectories respond with .45 for INIT X (this is called U in the program) and with .45 for INIT y (called V in the program) when executing the program. When the program is complete, the menu for GRAPH appears at the bottom of the screen. Press MORE twice, then press STPIC and name the picture $P1$. Then store another initial condition .5 in U and , 5 in V and execute the program. When the menu appears press MORE, then RCPIC, $P1$ and STPIC $P1$ to get a composite picture of the trajectories. The first trajectory cycles around a closed curve, the second cycles around some islands. Repeat using the other (U, V) initial conditions in the set $\{(.55, .55), (.6, .6), (.7, .7)\}$.

Trajectories that start at $(.75, .75)$ escape to infinity. Trajectories that start
near the origin lie on closed curves. Other composite trajectories arise when
α is given another value. Try $\cos \alpha = .24$. Note: You may want to modify
the program given above by adding inputs for the initial values of x and y.
For example, :Input "INIT X PT= ", U, :Input "INIT Y PT= ", V.

```
PROGRAM: IMAP
:FnOff
:1→xMax
:-1→xMin
:-1→yMin
:1→yMax
:ClDrw
:Input "INIT X",U
:Input "INIT Y",V
:0→S
:Lbl A
:.4*U-(V-U^2)*.916515→W
:.916515*U+(V-U^2)*.4→V
:W→U
:PtOn(U,V)
:S+1→S
:If S<500
:Goto A
```

Program 7.5.

A system $Q_1' = f(Q_1, Q_2), Q_2' = g(Q_1, Q_2)$ can be studied graphically
by using a **direction field**. Line segments with slope $g(Q_1, Q_2)/f(Q_1, Q_2)$
are created with center (Q_1, Q_2) on a grid in the plane. The solution to
the differential equation passing thru any such point is tangent to the line
segment located at that point. Using such a graph the direction of flow, the
presence of equilibrium points may be discovered. A program for creating
a direction field is presented in Program 7.6.

The user should create and enter a program FSD assigning the value
of $f(q1, q2)$ to $W1$, and a program GSD assigning the value of $g(q1, q2)$ to
$W2$ before executing the direction field program. Note $Q1, Q2$ are reserved
variables; thus $q1$, and $q2$ are used for the variable names. **To execute
this program change to Func MODE.**

```
 PROGRAM: DIRF
:FnOff :ClDrw
:Input "xMin=",xMin
:Input "xMax=",xMax
:Input "yMin=",yMin
:Input "yMax=",yMax
:(xMax-xMin)/12→XSTP
:(yMax-yMin)/10→YSTP
:For(I,1,9,1)
:yMin+I*YSTP→q2
:For(J,1,11,1)
:xMin+J*XSTP→q1
:FSD:GSD
:If W1==0
:Then
:If W2==0
:Then
:q1→XB:q2→YB
:q1→XE:q2→YE
:Else
:q1→XB:q2-YSTP/7→YB
:q1→XE:q2+YSTP/7→YE
:End
:Else
:tan⁻¹ (W2/W1)→ANG
:YSTP/5*sin (ANG)→RISE
:XSTP/7*cos (ANG)→RUN
:q1-RUN→XB:q2-RISE→YB
:q1+RUN→XE:q2+RISE→YE
:End
:Line(XB,YB,XE,YE)
:End
:End
```

Program 7.6.

Example 7.28

    Create the five direction fields indicated in the following table:

Table 7.6. Example Direction Fields.

FSD	GSD	xMin	xMax	yMin	yMax
$:1 \to W1$	$:q2*(q2-1) \to W2$	-0.2	4.0	-0.2	1.5
$:1 \to W1$	$:\sin(q1*q2) \to W2$	-0.2	6.0	-0.4	7.0
$:q2 \to W1$	$:-\sin(q1) \to W2$	-3.0	3.0	-3.0	3.0
$:q2-q1 \wedge 2 \to W1$	$:q1-q2+1 \to W2$	-3.0	3.0	-3.0	3.0
$:q1*(2-q1-q2) \to W1$	$:q2*(q1-1) \to W2$	-0.2	2.5	-0.2	2.5

It is instructive to save the pictures, then change to DifEq MODE (use the same xMin, xMax, yMin, yMax settings) and overlay solutions of the same differential equations drawn by the built-in program on the picture.

The Euler algorithm for a pair of differential equations $Q_1' = (t, Q_1, Q_2), Q_2' = g(t, Q_1, Q_2)$ arises from assuming that the slopes Q_1' and Q_2' are approximately constant during a small time step of length H: this gives

$$t_{i+1} = t_i + H, Q_{1,i+1} = Q_{1,i} + H*f(t_i, Q_{i,1}, Q_{2,i}), Q_{2,i+1}$$
$$= Q_{2,i} + H*g(t_i, Q_{1,i}, Q_{2,i}).$$

This algorithm is used in Program 7.7.

As written Program 7.7 does not terminate, that is at the pause statement, the current values of $T, q1$ and $q2$ are on display. When ENTER is pressed, another step is taken, and so forth. To terminate the program, press ON . To execute this program, we need a subprogram producing $f(T, q1, q2) = W1$ that we will store in FSD (abbreviation of f side) and a subprogram producing $g(T, q1, q2) = W2$ that is stored in GSD (abbreviation of g side). An example (for solving $q'' = q$) is

To record Program 7.8 press PRGM , then EDIT , enter the program name, press ENTER , then insert the program as written. To exit press QUIT . To execute the program, press PRGM , press NAME , and press EULER2 .

If we use the FSD and GSD subprograms as given and we enter stepsize $H = .1$, an initial value of 0 for T, initial values of 1 for both $q1$, and $q2$, we get the following results from EULER2:

Table 7.7. Results from EULER2 for $q'' = q, q(0) = q'(0) = 1$.

T	$=$	0.1	0.2	0.3	0.4	0.5	0.6	0.7	0.8	0.9
$q1$	$\approx$	1.10	1.21	1.33	1.46	1.61	1.77	1.95	2.14	2.36

and $q1$ at $T = 1.0$ is 2.59, a crude approximation of the correct solution value $2.71828\ldots$

```
PROGRAM: EULE2
:Input "STEPSIZE=",H
:Input "INIT T=",T
:Input "INIT ⊣1=",⊣1
:Input "INIT ⊣2=",⊣2
:Lbl A
:FSD:GSD
:T+H→T
:⊣1+H*W1→⊣1
:⊣2+H*W2→⊣2
:Disp T
:Disp ⊣1
:Disp ⊣2
:Pause Goto A
```

Program 7.7.

```
PROGRAM: FSD
:⊣2→W1
:
```

```
PROGRAM: GSD
:-⊣1→W2
:
```

Program 7.8.

References

1. Boyce, W. and R. Diprima, *Elementary Differential Equations*, 5th edition, John Wiley & Sons, Inc., New York, 1992.

2. Campbell, S., *An Introduction to Differential Equations and Their Applications*, 2nd edition, John Wiley & Sons, Inc., New York, 1990.

3. Hubbard, J. and B. West, *Differential Equations: A Dynamical Systems Approach*, Springer-Verlag, New York, 1991 and *MacMath: A Dynamical Systems Software Package for the Macintosh*, Springer-Verlag, New York, 1991.

4. Nagle, R. and E. Saff, *Fundamentals of Differential Equations*, 2nd edition, Benjamin Cummings Publishing Company, Inc., Redwood City, CA, 1989.

5. Proctor, T., *Calculator Enhancement for a Course in Differential Equations*, Saunders College Publishing, Philadelpha, 1992.

6. Sanchez, D., R. Allen, and W. Kyner, *Differential Equations*, 2nd edition, Addison-Wesley Publishing Company, Reading, Mass., 1988.

7. Sansone, G. and R. Conti, *Nonlinear Differential Equations*, revised edition, The MacMillan Company, New York, 1964.

8 Explorations in Advanced Engineering Mathematics

D. L. Kreider
Dartmouth College

Previous chapters exploited features of the TI-85 Graphing Calculator to explore problems that arise in elementary mathematics, calculus, probability, linear algebra, and differential equations. The examples chosen show how valuable insights into the mathematical substance of these subjects can be acquired through well-chosen computational activities and show how mathematics and computing march hand-in-hand toward understanding mathematical concepts and deducing useful information from mathematical models. The TI-85's computational and graphing power bring such explorations within reach of students of both elementary and more advanced mathematics.

This chapter considers additional examples that often are part of an "advanced engineering mathematics" course. Application of differential equations to boundary-value problems, determination of critical values, use of infinite series to define and explore non-elementary functions, exploration of periodic phenomena that lead to non-elementary integrals, and problems from vector field theory, are typical topics in such courses. Students often feel helpless when dealing with such problems because so few can be handled adequately with pencil-and-paper computational techniques. The TI-85 provides powerful support in these more difficult problems, helps students to appreciate the enormous economy and power of the mathematical theory involved, and builds students' confidence that such problems, arising from real scientific or engineering needs, are accessible through techniques at the student's hand.

If there was a guiding principle in the selection of examples for this chapter, it was to reveal the potential of the TI-85 calculator in the hands of students. And a significant ingredient in the choice of examples lay in the fun that the author enjoyed as he explored the device and its applications.

Explorations with the Texas Instruments TI-85
John G. Harvey and John W. Kenelly (eds.), pp. 307-338
©1993 by Academic Press, Inc.
All rights of reproduction in any form reserved.
ISBN: 0-12-329070-8

8.1. The Simple Pendulum – Elliptic Integrals

An example is furnished by a well-known periodic system. The motion of a simple pendulum of length L and mass m is governed by the differential equation

$$\frac{1}{2}mL^2 \left(\frac{d\theta}{dt}\right)^2 = mgL(\cos\theta - \cos A) \tag{8.1}$$

where g is the acceleration of gravity, θ is the displacement at time t, and A is the maximum angle of the pendulum's displacement.

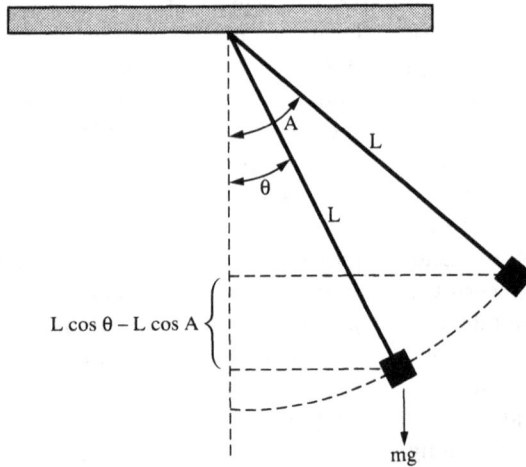

Figure 8.1. The Simple Pendulum.

The student will recognize the left side of Equation 8.1 as the kinetic energy of the pendulum bob at time t. Assuming that energy is conserved this is equal to the right hand side of the equation that expresses the difference in potential energy of the pendulum at the angle θ and its potential energy when at its maximum displacement angle A.

Differentiating Equation 8.1 with respect to t leads to the second-order differential equation

$$\frac{1}{2}mL^2 \cdot 2\dot{\theta}\ddot{\theta} = mgL(-\sin\theta)\dot{\theta}$$

or (when $\dot{\theta} \neq 0$)

$$L\ddot{\theta} + g\sin\theta = 0. \tag{8.2}$$

This second-order differential equation serves equally well to describe the motion of the pendulum. Indeed it is the equation one obtains when applying Newton's second law of motion rather than conservation of energy considerations.

Neither of these equations is easily solved for $\theta(t)$. But Equation 8.2 lends itself well to linearization; that is, to analyzing the pendulum motion under the assumption that the maximum amplitude A (and hence also θ) is small. For then $\sin\theta \approx \theta$, and Equation 8.2 becomes

$$L\ddot\theta + g\theta = 0. \tag{8.3}$$

This is a constant coefficient linear differential equation that we can solve explicitly to obtain the solution

$$\theta(t) = A\cos\sqrt{\frac{g}{L}}\,t.$$

It provides a linear model for the pendulum's motion, and it predicts that the period T of the simple pendulum, at least when the amplitude is small, is

$$T = 2\pi\sqrt{\frac{L}{g}}. \tag{8.4}$$

As an example, when $L = 1$ m, the period $T = 2\pi/\sqrt{g} = 2.0064$ seconds. Indeed this expression can be "typed" directly into the TI-85's home screen and evaluated by pressing $\boxed{\text{ENTER}}$. But since we want to use its value later we shall instead store its value in the variable T by pressing $\boxed{\text{STO}\rightarrow}$ $\boxed{\text{T}}$. The constant $g = 9.80665\ m/sec^2$ is found in the TI-85 menu $\boxed{\text{CONS}}$ $\boxed{\text{BLTIN}}$ $\boxed{\text{MORE}}$.

The linear equation (8.3) also predicts that the period T of the pendulum is independent of the maximum amplitude A. This is not exactly correct, of course. The error, usually called the *circular error of the simple pendulum*, arises from the approximation that led us from the non-linear differential equation (8.2) to the linear differential equation (8.3). We will investigate the circular error using the TI-85.

We return to Equation 8.1, which is the more precise (non-linear) model of the pendulum's motion. Separating the variables in this equation we obtain

$$\sqrt{\frac{L}{2g}}\frac{d\theta}{\sqrt{\cos\theta - \cos A}} = dt.$$

Integrating this equation, we find that the *exact* period of the pendulum can be expressed in the form

$$P = 4\sqrt{\frac{L}{2g}}\int_0^A \frac{d\theta}{\sqrt{\cos\theta - \cos A}}. \tag{8.5}$$

This exact expression for the period P *does* depend on the amplitude. And we can thus study the *circular error* by examining the difference $P - T$ as a function of the maximum amplitude A.

As an example, let us evaluate Equation 8.5 when $L = 1$ and $A = \pi/4$. We attempt this on the TI-85, using the built-in integration function *fnInt* (press $\boxed{\text{CALC}}$ $\boxed{\text{fnInt}}$), setting the values of L and A by executing in the home screen the expressions $1 \rightarrow L$ and $\pi/4 \rightarrow A$, and evaluating the formula (again by typing in the home screen)

$$4\sqrt{(L/2g)}\text{fnInt}(1/\sqrt{(\cos x - \cos A)}, x, 0, A).$$

The TI-85 works valiantly in its effort to evaluate the *improper* integral (the integrand is infinite at the upper limit). It struggles for a considerable period of time but finally terminates with an error. We should never have presented it with this raw problem. Instead, we should first do a bit of mathematics to transform Equation 8.5 into a more suitable form for computation.

As a first step we use the trigonometric identities

$$\cos \theta = 1 - \sin^2 \frac{\theta}{2} \quad \text{and} \quad \cos A = 1 - \sin^2 \frac{A}{2}$$

to rewrite Equation 8.5 in the form

$$P = 2\sqrt{\frac{L}{g}} \int_0^A \frac{d\theta}{\sqrt{K^2 - \sin^2 \frac{\theta}{2}}}, \quad \text{where } K = \sin \frac{A}{2}. \quad (8.6)$$

The integral in Equation 8.6 is still improper, but it invites us to change the variable of integration using the substitution $\sin \theta/2 = K \sin x$. For then

$$d\theta = \frac{2K \cos x\, dx}{\cos \frac{\theta}{2}}$$

$$= \frac{2K \cos x\, dx}{\sqrt{1 - \sin^2 \frac{\theta}{2}}}$$

$$= \frac{2K \cos x\, dx}{\sqrt{1 - K^2 \sin^2 x}}$$

and, substituting into Equation 8.6, we obtain

$$P = 2\sqrt{\frac{L}{g}} \int_0^{\frac{\pi}{2}} \frac{1}{\sqrt{K^2 - K^2 \sin^2 x}} \frac{2K \cos x\, dx}{\sqrt{1 - K^2 \sin^2 x}},$$

or, finally,

$$P = 4\sqrt{\frac{L}{g}} \int_0^{\frac{\pi}{2}} \frac{dx}{\sqrt{1 - K^2 \sin^2 x}}. \tag{8.7}$$

The substitution has worked its magic. The integral in Equation 8.7 is no longer improper, provided that $K^2 < 1$. In our example $K = \sin \frac{A}{2}$ hence, so long as the amplitude of the pendulum is smaller than π, the TI-85 will have no difficulty evaluating the integral. As before we execute in the home screen $1 \to$L, $\pi/4 \to$A, and $K = \sin(A/2)$. (Note that this last equation assigns the *unevaluated* expression $\sin(A/2)$ to the variable K.) We then enter the formula (8.7) into the TI-85 by typing in the home screen

$$P = 4\sqrt{(L/g)}\text{fnInt}(1/\sqrt{(1-K^2 (\sin t)^2)}, t, 0, \pi/2).$$

(Note: We used the variable of integration t instead of x because we subsequently want to draw the graph of P. The variable x is used by the TI-85 as a reserved variable in graphing.) The integral (8.7) can now be evaluated by simply typing P into the home screen and pressing $\boxed{\text{ENTER}}$. With the values of L and A already provided above, the TI-85 quickly gives the value $P = 2.0866$. Thus when $A = \pi/4$ the *circular error* is $P - T = 0.0802$. (Recall that we had earlier typed the formula $T = 2\pi\sqrt{(L/g)}$ into the home screen. This stored in the variable T the period predicted by the linear model.)

Let us examine the circular error $P - T$ as A ranges from 0 to $\pi/2$. The maximum error in this interval occurs when $A = \pi/2$. We store this maximum value in the variable M by typing in the home screen $\pi/2 \to A$ and $P - T \to M$. The value of M, returned by the TI-85, is approximately 0.3618. We will use it to scale the graph of $P - T$ in the examples that follow.

Example 8.1

We graph the circular error $P - T$ on the interval $0 \le A \le \pi/2$ to get a sense of how the error depends on the amplitude A. This can be done on the TI-85 by typing $A = x$ into the home screen (since the TI-85 uses the variable x in its built-in graphing function) and entering the formula $y1 = P - T$ into the TI-85's formula list for graphing. The appropriate range variables for the graph are also entered: xMin $= 0$, xMax $= \pi/2$, yMin $= 0$, yMax $=M$. The graph, drawn by the TI-85, clearly shows the dependency of the circular error (y) on the amplitude (x).

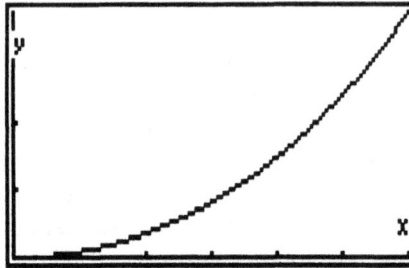

Figure 8.2. Circular Error of the Simple Pendulum.

Example 8.2

We compute the percentage error at each of the 6 points $A = 0$, $\pi/10, 2\pi/10, \ldots, 5\pi/10$. The TI-85 does this easily using the built-in function *seq* for generating a sequence of values. (To access *seq* we type the keystroke sequence $\boxed{\text{LIST}}$ $\boxed{\text{OPS}}$ $\boxed{\text{MORE}}$ $\boxed{\text{seq}}$.) We type into the home-screen

$$\text{seq}(100(\text{P}-\text{T})/\text{P, A, 0, } \pi/2, \pi/10)$$

and the TI-85 responds with the desired list of percentage errors

$$\{0.0 \quad 0.6 \quad 2.5 \quad 5.5 \quad 9.8 \quad 15.3\}.$$

(Note: The TI-85 was set to display one decimal place using the MODE menu.) Thus we see that the circular error is 15% when the maximum amplitude A is $\pi/2$, but it is small when A is small. Indeed we can use the TI-85's SOLVER application to determine the maximum amplitude for which the circular error is less than 1%. Press $\boxed{\text{SOLVER}}$, enter the expression $100(P - T)/P$, and press $\boxed{\text{ENTER}}$. Give *exp* the value 1, place the cursor in the line containing "x=", and press $\boxed{\text{SOLVE}}$. The TI-85 determines that the value of A (i.e. of x) is 0.40016 when the value of the expression is 1. (Recall that we had previously entered the equation $A = x$.) In other words, the circular error is less than 1% when the maximum amplitude is less than 22.9 degrees (or 0.40016 radians).

The computations in the two examples, above, would be extremely difficult, if not impossible, to perform by hand. The TI-85 handles them easily.

The integral in Equation (8.7) is a particular value of the so-called *elliptic integral of the first kind*

$$F(K, A) = \int_0^A \frac{dv}{\sqrt{1 - K^2 \sin^2 v}}. \tag{8.8}$$

For $K^2 < 1$ it is a proper integral that can be evaluated by the TI-85, typing the formula

$$\text{fnInt}(1/\sqrt{(1-K^2(\sin v)^2)},v,0,A)$$

into the home screen and evaluating it.

Explorations with Elliptic Integrals

1. Determine the circular error of the simple pendulum when the maximum displacement angle is 30 degrees.

2. Generate a list (see Example 8.2) of the circular errors corresponding to maximum displacement angles of $0.0, 0.1, 0.2, \ldots, 1.0$ radians.

3. Write a program *EllipticInt* that prompts the user for the values of K and A and then displays the value of $F(K, A)$. (You may want to return to this exercise after studying the examples and programs that follow later in this chapter.)

8.2. Functions Defined by Integrals and Series

The elliptic integral (8.8) can be expressed as an infinite series

$$F(K, A) = A + \frac{1}{2}K^2 \int_0^A \sin^2 v \, dv + \frac{1 \cdot 3}{2 \cdot 4}K^4 \int_0^A \sin^4 v \, dv$$
$$+ \frac{1 \cdot 3 \cdot 5}{2 \cdot 4 \cdot 6}K^6 \int_0^A \sin^6 v \, dv + \cdots$$

and, indeed, the TI-85 can be programmed to sum this series. The series converges reasonably fast when $K^2 < 1$, but it has little advantage over the direct evaluation of the integral using the TI-85's built-in *fnInt* function. Sometimes, however, expansion of functions as infinite series provides the numerical method of choice. In this section we use such techniques to explore *Bessel functions*.

Bessel's equation of order p is

$$x^2 \frac{d^2 y}{dx^2} + x \frac{dy}{dx} + (x^2 - p^2)y = 0. \tag{8.9}$$

Its solutions are often encountered in problems involving damped oscillatory behavior or as fundamental vibrational modes for systems with cylindrical symmetries. Most of the textbook references cited at the end of the chapter develop properties of the most common Bessel functions.

Example 8.3

One solution of Bessel's equation (8.9) can be expressed in the form of a series

$$J_p(x) = \sum_{k=0}^{\infty} \frac{(-1)^k (\frac{x}{2})^{2k+p}}{k!(k+p)!}. \tag{8.10}$$

The function $J_p(x)$ is known as the Bessel function of order p of the first kind. This solution of Equation 8.9, an ordinary Taylor series when p is a non-negative integer, is defined and has a finite limit as $x \to 0$. A second linearly independent solution can also be found, but it does not have a finite limit at $x = 0$.

The Bessel function $J_0(x)$ can be studied using the TI-85 in a number of ways. For example, it can be computed directly as the unique solution of the initial-value problem

$$x\frac{d^2y}{dx^2} + \frac{dy}{dx} + xy = 0; \qquad y(0) = 1, \quad y'(0) = 0. \tag{8.11}$$

In the TI-85's DifEq MODE (set in the MODE menu), by pressing $\boxed{\text{GRAPH}}$ $\boxed{\text{Q}'(\text{t})=}$ we define

$$Q'1 = Q2$$
$$Q'2 = -Q2/t - Q1, \quad \text{and the initial conditions}$$
$$QI1 = 1, \quad QI2 = 0.$$

The differential equation (8.11) is singular at $t = 0$, generating an error if the solution is started with tMin $= 0$. We therefore set the RANGE variables to: tMin $= 0.0001$, tMax $= 10$, tStep $= 0.01$, tPlot $= 0.01$, xMin $= 0$, xMax$= 10$, yMin $= -1.5$, yMax $= 1.5$. The graph of the function $J_0(x)$, drawn by the TI-85 on the interval $0 \le x \le 10$, is shown in Figure 8.3.

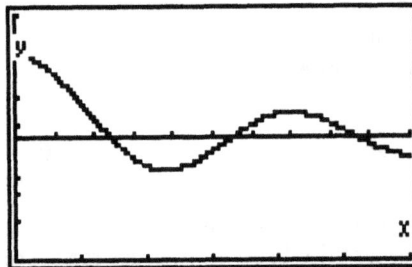

Figure 8.3. Graph of $J_0(x)$.

The graph displays the damped oscillatory behavior that is characteristic of the functions $J_p(x)$. The first three zeros of $J_0(x)$ can be seen to lie approximately at the points $x = 2.4$, $x = 5.5$, and $x = 8.6$. Using the TI-85's TRACE feature we can find them more accurately. For example the first zero occurs between the points $(2.4001, 0.0028442)$ and $(2.4101, -0.0023937)$. Using linear interpolation, a better approximation of this zero can be found. Indeed, if we employ the TI-85's built-in interpolating function *inter* (described in the TI-85 User's Manual) by typing in the home screen

$$\text{inter}(.002844,\ 2.4001,\ -.0023937,\ 2.4101,\ 0),$$

we obtain the value 2.406 as an approximation of the first zero of $J_0(x)$. This value is, in fact, accurate to 1 part in 8000!

Example 8.4

We can also explore $J_0(x)$ through its Taylor series

$$J_0(x) = \sum_{k=0}^{\infty} \frac{(-1)^k (\frac{x}{2})^{2k}}{k!^2}$$

$$= 1 - \frac{x^2}{2^2 1!^2} + \frac{x^4}{2^4 2!^2} - \frac{x^6}{2^6 3!^2} + \cdots. \tag{8.12}$$

Enter the following program into the TI-85: (press $\boxed{\text{PRGM}}$ $\boxed{\text{EDIT}}$, type the name J0, and press $\boxed{\text{ENTER}}$).

```
PROGRAM: J0
:1→J0V
:1→T
:1→I
:Repeat abs T<E-15
: -T*x²/2²/I²→T
:J0V+T→J0V
:I+1→I
:End
:Disp I
:Disp J0V
```

Program 8.1.

In the home screen type

$$1 \to \text{x: J0}$$

and then press $\boxed{\text{ENTER}}$. The value $J_0(x) = 0.765197686558$ is returned, 10 terms of the series being required to obtain this accuracy. Pressing $\boxed{\text{ENTRY}}$ and editing the command line to

$$2.4000 \rightarrow \text{x: J0,}$$

we obtain $J_0(2.4) = 0.002507683297$. A few more trials yields 2.4048 as the first positive zero of $J_0(x)$, accurate to five decimal places.

Finding zeros of functions for which programs have been written is a common activity. It is useful, therefore, to implement a bisection algorithm that serves to automate the sort of trial and error process used above. A TI-85 program to do this follows:

```
PROGRAM: BIS
:Prompt LEFT
:Prompt RIGHT
:LEFT→x
:J0
:sign J0V→SLEFT
:Repeat abs (RIGHT-LEFT)/(1+abs LEFT)<E-12
:(LEFT+RIGHT)/2→ROOT
:ROOT→x
:J0
:If J0V*SLEFT≥0
:Then
:ROOT→LEFT
:Else
:ROOT→RIGHT
:End
:End
:Disp ROOT
```

Program 8.2.

The program BIS applies the bisection algorithm to the function $J_0(x)$. It assumes that we have previously entered the program J0 to compute the value $J_0(x)$ and return its result in the variable J0V. Before executing the program BIS it is useful to edit J0 by eliminating its last two lines that print out the values of the variables J0V and I. Such printing would be distracting, for J0 is called *many* times by the program BIS. Now we execute the bisection algorithm by typing BIS in the home screen and pressing $\boxed{\text{ENTER}}$. The inputs LEFT = 2, RIGHT = 3, are provided when prompted, and the program finds a root of $J_0(x)$ at $x = 2.4048255577$. This root is accurate to the number of decimal places shown!

The program J0 returns the value of $J_0(x)$ in the variable J0V and the number of terms of the series used in the variable I. For example, typing $1 \to x : J0 : I$ into the home screen and pressing $\boxed{\text{ENTER}}$, we see that 10 terms of the series are used to evaluate $J_0(1)$ to 12 digits of accuracy. Pressing $\boxed{\text{ENTRY}}$ and editing the command line to $5 \to x : J0 : I$, we see that 18 terms of the series are needed to compute $J_0(5)$. Again, executing $10 \to x : J0 : I$, we see that 26 terms are needed for $J_0(10)$. To study $J_0(x)$ on the interval $0 \le x \le 10$, therefore, we conclude that 26 terms of the series are sufficient! This enables us to employ yet another feature of the TI-85 to define $J_0(x)$ on the interval $0 \le x \le 10$. Namely we can enter a *defining equation* for $J_0(x)$ into the TI-85's graphing function list. First, making sure that the calculator is in Func MODE, we access the function list by pressing $\boxed{\text{GRAPH}}$ $\boxed{\text{y(x)=}}$, and enter the expression

$$y1 = \text{sum seq}((-1)^\wedge I * (x/2)^\wedge (2I)/I!^2, \ I, \ 0, \ 26, \ 1). \tag{8.13}$$

The TI-85's GRAPH and SOLVER applications can now be used to explore the mathematical properties of the equation. The student should try Exploration problem 1 at the end of this section. For non-integral values of p no such simple formula as (8.13) is available for computing $J_p(x)$. The series (8.10) can still be used, however.

Example 8.5

When $p = 1/4$, for example, we have

$$
\begin{aligned}
J_{\frac{1}{4}}(x) &= \sum_{k=0}^{\infty} \frac{(-1)^k (\frac{x}{2})^{2k+\frac{1}{4}}}{k!(k+\frac{1}{4})!} \\
&= \frac{(\frac{x}{2})^{\frac{1}{4}}}{(\frac{1}{4})!} \left[1 - \frac{(\frac{x}{2})^2}{1 \cdot (1+p)} + \frac{(\frac{x}{2})^4}{1 \cdot 2(1+p)(2+p)} \right. \\
&\qquad \left. - \frac{(\frac{x}{2})^6}{1 \cdot 2 \cdot 3(1+p)(2+p)(3+p)} + \cdots \right]
\end{aligned} \tag{8.14}
$$

Here $\frac{1}{4}!$ must be correctly interpreted as $\Gamma(1 + \frac{1}{4}) \approx 0.9064024740$. The program J0 that computes $J_0(x)$ is easily modified to evaluate $J_{\frac{1}{4}}(x)$ as follows:

```
 PROGRAM: J14
:1→J14V
:1→T
:1→I
:Repeat abs T<E-10
:-T*x²/2²/I/(I+.25)→T
:J14V+T→J14V
:I+1→I
:End
:J14V*(x/2)^.25/.906402474→/J14V
:
```

Program 8.3.

The program J14 returns the value of $J_{\frac{1}{4}}(x)$ in the variable J14V and the number of terms of the series used in the variable I. It is accurate to the same number of digits as the approximation of $\Gamma(\frac{5}{4})$ that it uses, in this case about 10 digits. The student should try Exploration problems 2 and 3 at the end of the section.

Example 8.6

One can also attempt to study $J_{\frac{1}{4}}(x)$ directly as a solution of the initial value problem

$$x^2 \frac{d^2y}{dx^2} + x \frac{dy}{dx} + (x^2 - \frac{1}{4})y = 0, \qquad y(x_0) = y_0, \quad y'(x_0) = y'_0,$$

provided that suitable initial conditions are known. From the series (8.14) we deduce that, for small values of x,

$$J_{\frac{1}{4}}(x) \sim 0.9277296117 x^{\frac{1}{4}}$$

and

$$J'_{\frac{1}{4}}(x) \sim 0.2319324029 x^{-\frac{3}{4}}.$$

Thus we might use as initial conditions $J_{\frac{1}{4}}(0.01) \approx 0.29337$ and $J'_{\frac{1}{4}}(0.01) \approx 7.33427$. With the TI-85 in DifEq MODE, we enter the equations

$$\begin{aligned} Q'1 &= Q2 \\ Q'2 &= -Q2/t - (1 - .25^2/t^2)Q1, \end{aligned} \qquad (8.15)$$

the initial conditions $QI1 = .29337$, $QI2 = 7.33427$, and the range variables

$$tMin = .01$$
$$tMax = 10$$
$$tStep = .1$$
$$tPlot = .01$$
$$xMin = 0$$
$$xMax = 10$$
$$xScl = 1$$
$$yMin = -1.5$$
$$yMax = 1.5$$
$$yScl = .5$$
$$kdifTol = .001$$

Executing $\boxed{\text{GRAPH}}$, the TI-85 produces the following graph of $J_{\frac{1}{4}}(x)$ and its derivative. Note that the first three approximate zeros of $J_{\frac{1}{4}}(x)$ are at the points $x = 2.8$, 6, and 9. (The graph of the derivative can be suppressed, if desired, by *unselecting* the second equation in the function list. Press $\boxed{\text{GRAPH}}$ $\boxed{\text{y(x)=}}$, place the cursor on the second equation, and press $\boxed{\text{SELCT}}$.)

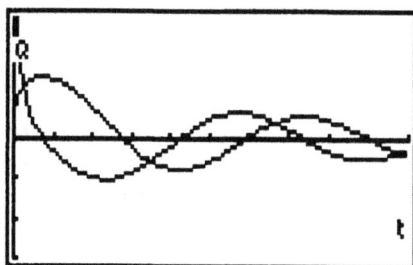

Figure 8.4. Graph of $J_{\frac{1}{4}}(x)$.

Other functions associated with Bessel's equation, studied in most of the textbook references listed at the end of the chapter, can be handled similarly to $J_0(x)$ and $J_{\frac{1}{4}}(x)$. In general, series solutions are developed that converge rapidly over the intervals of interest. Such series are easily programmed on the TI-85, permitting the functions to be investigated, as we have done in the examples above.

Explorations of Series Representations

1. Enter Equation 8.13 into the TI-85's graphing function list, and use the menu items TRACE, MATH, etc., to explore $J_0(x)$. This method is noticeably slower than the previous ones, but it is accurate to 10 decimal places on the *entire* interval $0 \leq x \leq 10$! Find the first and second zeros of $J_0(x)$. Can you find the first positive relative maximum point of $J_0(x)$? Where is its first inflection point?

2. Enter the program given in Example 8.5 into your TI-85 calculator, and experiment with it to compute various values of $J_{\frac{1}{4}}(x)$. How many terms of the series (8.14) are needed to guarantee 10 digits of accuracy for any value of x in the interval $0 \leq x \leq 10$? Having determined this number, repeat the steps we followed for $J_0(x)$ in defining an equation in the HP-85's graph function list (see Equation 8.13). Examine the graph of $J_{\frac{1}{4}}(x)$ on the interval $0 \leq x \leq 10$. What (approximately) is the first positive zero of $J_{\frac{1}{4}}(x)$?

3. Modify the program BIS that implements the bisection algorithm — replacing J0 and J0V by J14 and J14V respectively. Use the modified program to obtain the smallest positive zero μ_1 of $J_{\frac{1}{4}}(x)$. (Answer: $\mu1 \approx 2.780887724$.) We will use this result in a later example.

4. Use the TRACE function to find better approximations of the 2nd and 3rd zeros of $J_{\frac{1}{4}}(x)$. Also use the built-in linear interpolation function *interp* to refine these two zeros. Finally, use the programs J14 and BIS to determine the 2nd and 3rd zeros to a high degree of accuracy.

8.3. Boundary-Value Problems

The Chapter *Exploring Differential Equations with a TI-85* explored in some detail solutions of initial-value problems. Many problems in engineering lead instead to *boundary-value problems* where *initial* conditions give way to conditions imposed at several different points, often the boundary points of an interval of interest. For example

$$\frac{d^2y}{dx^2} + Ly = 0, \qquad y(0) = y(\pi) = 0 \tag{8.16}$$

seeks *non-trivial* solutions of the differential equation that pass through the two distinct points $(0,0)$ and $(\pi, 0)$. In general such solutions exist only for certain discrete values of L, called *eigenvalues* or *characteristic values* or *critical values* of the problem. The corresponding solutions are called *eigenfunctions* or *characteristic modes* of the boundary-value problem.

Example 8.7

Sometimes, as in the case of Equation 8.16, the boundary-value problem can be solved explicitly. For in this very simple case, we can find the general solution of the differential equation and apply the two boundary conditions directly. We obtain:

$$y = C_1 \cos \sqrt{L}x + C_2 \sin \sqrt{L}x$$
$$C_1 \cdot 1 + C_2 \cdot 0 = 0$$
$$C_1 \cos \sqrt{L}\pi + C_2 \sin \sqrt{L}\pi = 0$$

from which we conclude that $C_1 = 0$, and C_2 is arbitrary provided that $sin\sqrt{L}\pi = 0$. Thus, for this simple example, the values

$$L = n^2, \qquad n = 1, 2, 3, \ldots$$

are eigenvalues of the problem, and for the eigenvalue $L_n = n^2$ the corresponding eigenfunction is $\phi_n(x) = \sin nx$. In many physical applications modeled by boundary-value problems the eigenvalues are of central interest and, indeed, it may be only the smallest (positive) one or two eigenvalues that have physical significance. The eigenfunctions are often of lesser interest.

Finding the smallest positive eigenvalue might be done analytically in simple cases such as Equation 8.16 above. When the computations become complicated, however, or when the differential equation itself yields only to numerical methods of solution, one turns to numerical methods for approximating the first several eigenvalues. In such cases we find, once again, that the TI-85 can be helpful.

Example 8.8

Find the smallest eigenvalue of the boundary-value problem (8.16) using numerical methods. Our point of departure is the observation that eigenfunctions of this problem are determined only up to an arbitrary constant (the non-trivial solutions are $C_2 \sin nx$). Thus, to find non-trivial solutions that satisfy the boundary conditions $y(0) = y(\pi) = 0$, it is sufficient to look at solutions of the *initial-value* problem $y(0) = 0$, $y'(0) = 1$. Our aim, then, is to determine values of L for which the solution of this initial-value problem happens also to satisfy the second boundary condition $y(\pi) = 0$.

An analogy comes to mind: that of shooting a bullet at a 45° angle, experimenting with the muzzle velocity required to cause the bullet to land at a predetermined point. This is the same idea that we are proposing to use as a numerical approach to boundary-value problems. Namely, experiment with the value of L that causes the solution of the initial-value problem to pass *exactly* through the predetermined point $(\pi, 0)$. It is not a surprise that such numerical techniques are often called *shooting methods*.

We have seen that the TI-85 can readily solve initial-value problems. And thus we can use it profitably in the present situation. In DifEq MODE we enter the data (press $\boxed{\text{GRAPH}}$ $\boxed{\text{Q}'(\text{t})=}$)

$$Q'1 = Q2$$
$$Q'2 = -LQ1$$
$$QI1 = 0, \quad QI2 = 1$$

and press $\boxed{\text{RANGE}}$ to enter the settings: tMin = 0, tMax = 3.5, tStep = .01, tPlot = 0, xMin = −.5, xMax = 3.5, xScl = π, yMin = −.3, yMax = 1.5. We employ the TI-85 to draw the graph of the solution for various values of L, for example for .5 → L, .9 → L, 1.1 → L, and 1.5 → L. Notice that when $L < 1$ the solution *overshoots* the desired point $(\pi, 0)$, whereas when $L > 1$ it *undershoots* this point. Graphs for these values are generated and superimposed in the pictures below (using STPIC and RCPIC as was explained in the chapter on differential equations).

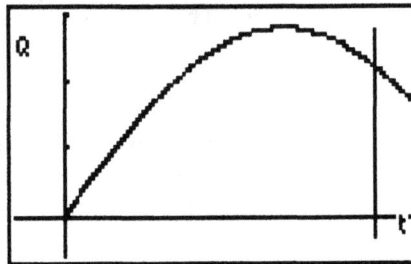

Figure 8.5. Shooting Method: $L = 0.5$.

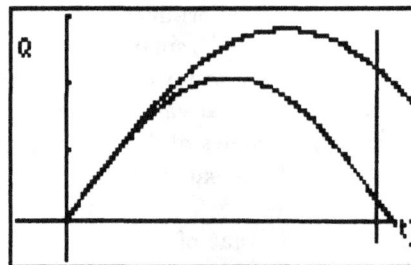

Figure 8.6. Shooting Method: $L = 0.9$.

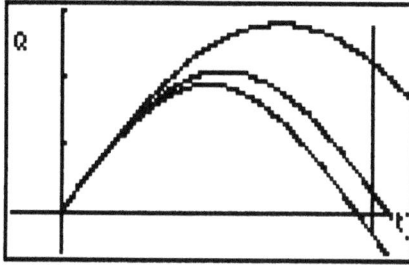

Figure 8.7. Shooting Method: $L = 1.1$.

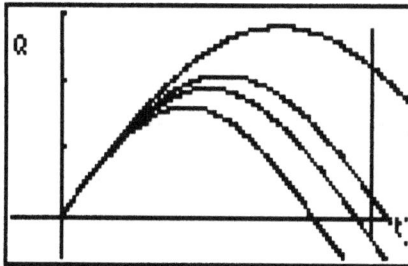

Figure 8.8. Shooting Method: $L = 1.5$.

Further experimentation with the value of L enables us to "zero in" on the eigenvalue $L = 1$. Indeed, an interesting exercise would be to write a program for the TI-85 that graphs the solution on the interval $0 \leq x \leq \pi$, tests the value of $Q1$ at $t = \pi$, and adjusts the value of L up or down according to whether the value of $Q1$ is positive or negative. A variation on the bisection algorithm program BIS from Section 8.3 can be used in this way to determine L to a high degree of accuracy.

Explorations with Boundary-Value Problems

1. Experiment, as in Example 8.8, to find the *second* eigenvalue of the boundary-value problem (8.16). (Answer: $L = 4$) Generate pictures, like those below, that show solutions of the initial-value problem for the values $L = 3.5$, $L = 4.0$, and $L = 4.5$.

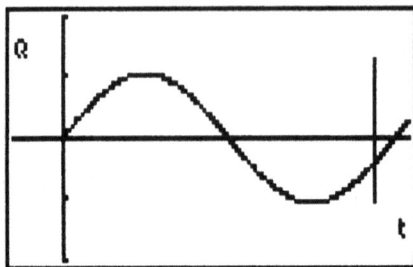

Figure 8.9. Shooting Method: $L = 3.5$.

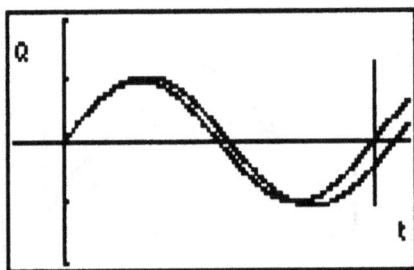

Figure 8.10. Shooting Method: $L = 4.0$.

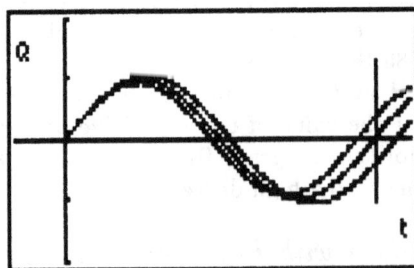

Figure 8.11. Shooting Method: $L = 4.5$.

2. Implement the suggestion, above, that a variation of the program BIS can be used to determine the first (or higher) eigenvalues.

3. Experiment, as in Example 8.8, to find the smallest positive eigenvalue of the boundary-value problem

$$\frac{d^2y}{dx^2} + Lx^2y = 0, \qquad y(0) = y(1) = 0.$$

Here the answer is $L \approx 30$. Can you refine this value?

Note: It can in fact be shown that the positive eigenvalues for this problem are given exactly by

$$L_i = 4\mu_i^2, \qquad i = 1, 2, 3, \cdots$$

where μ_i is the ith positive zero of $J_{\frac{1}{4}}(x)$. We determined in Section 8.2 that $\mu_1 \approx 2.780887724$; hence we have $L_1 \approx 30.933346134$. How well does your result compare with this?

8.4. Exploring Vector Ideas

The TI-85 has the capability of performing several of the mathematical operations on 2 and 3 dimensional vectors usually associated with vector analysis. Combined with the calculus functions of the TI-85, this enables us to perform some of the operations of *vector calculus* as well. We include a few examples in this section.

Example 8.9

Find the angle θ between two vectors $\mathbf{V}_1$, $\mathbf{V}_2$. Using the dot product for vectors we have

$$\cos\theta = \frac{\mathbf{V}_1 \cdot \mathbf{V}_2}{\|\mathbf{V}_1\| \, \|\mathbf{V}_2\|}.$$

The dot product and norm functions are represented on the TI-85 as *dot* and *norm*. They are accessed on the TI-85 by pressing [VECTR] [MATH]. For example, after typing into the home screen $[3, 4, -1] \to V1$ and $[-7, 10, 0] \to V2$, we compute the desired angle θ by evaluating in the home screen

$$\cos^{-1}(\mathrm{dot}(V1, V2)/\mathrm{norm}\, V1/\mathrm{norm}\, V2).$$

This yields the result $\theta = 72.226°$. (Set the TI-85's angle setting to degree mode (accessed through the MODE menu).)

Example 8.10

For the vectors $\mathbf{V}_1$, $\mathbf{V}_2$ of Example 8.9, we can evaluate $\mathbf{V}_1 \times \mathbf{V}_2 = [10, 7, 58]$ by typing *cross (V1,V2)* into the home screen (press [VECTR] [MATH] [cross]). Similarly we can evaluate $\|\mathbf{V}_1 \times \mathbf{V}_2\|$ by executing *norm cross (V1,V2)*, yielding the value 59.27. The latter number is the area of the parallelogram determined by the two vectors $\mathbf{V}_1$ and $\mathbf{V}_2$. The function *norm* is also accessed through the TI-85's [VECTR] [MATH] menu.

Example 8.11

With $V_3 = [2, -5, 17]$ and V_1, V_2 as in Example 8.10, we evaluate $V_1 \times V_2 \cdot V_3$ as $\mathrm{dot}(\mathrm{cross}(V1, V2), V3) = 971$.

Example 8.12

Find the distance between two skew lines in 3-space determined by points P_1, P_2 lying on the lines and vectors V_1, V_2 parallel to the lines.

Since the vector $V_1 \times V_2$ is perpendicular to both lines, and $P_1 - P_2$ is a vector from the point P_1 on the first line to the point P_2 on the second line, we need only project the vector $P_1 - P_2$ onto $V_1 \times V_2$. The length of this projection is dist $= \frac{(P_1 - P_2) \cdot (V_1 \times V_2)}{\|V_1 \times V_2\|}$. Thus we type into the home screen of the TI-85

$$\mathrm{DIST} = \mathrm{dot}(P1 - P2, \ \mathrm{cross}(P1, P2))/\mathrm{norm} \ \mathrm{cross}(P1, P2).$$

Then we need only evaluate the variable DIST after setting the values of P1, P2, V1, and V2. Indeed we can use the TI-85's SOLVER application for this purpose, setting exp=DIST. We also type into the SOLVER's editing window the equations $P1 = [1, 2, 3]$, $P2 = [-2, 3, -5]$, $V1 = [1, 0, 2]$, and $V2 = [-3, 5, 8]$. Placing the cursor on the "exp=" line and pressing ENTER we obtain 1.3395 as the distance between the two lines.

Vector Calculus

The TI-85's calculus operations of differentiation and integration do not apply directly to vector functions. We can apply them, however, to the individual components of vectors and thereby study the geometry of space curves.

Example 8.13

Let a curve $\gamma(t) = [X_1(t), X_2(t), X_3(t)]$ be defined by

$$X_1 = A \cos t$$
$$X_2 = A \sin t$$
$$X_3 = Bt.$$

The curve $\gamma(t)$ is a helix spiraling around the z-axis. We can compute:

$$V = \text{velocity vector} = \frac{d\gamma}{dt}$$

$$v = \text{speed} = \|V\|$$

$$U = \text{unit tangent vector} = \frac{V}{\|V\|}$$

$$\frac{dU}{dt} = v\kappa N, \qquad N = \text{unit normal vector}, \qquad \kappa = \text{curvature}$$

$$BN = U \times N = \text{unit binormal vector}.$$

The vectors **U**, **N** determine the osculating plane of the curve $\gamma(t)$ at a point **P**. The constant $\frac{1}{\kappa}$ is the radius of curvature of the curve at **P**. The vector **BN** is perpendicular to the osculating plane, and its rate of change measures the rate at which the curve tends to twist out of its osculating plane. In fact $\frac{d}{dt}\mathbf{BN} = v\tau\mathbf{N}$, the quantity τ being called the *torsion* of $\gamma(t)$ at **P**.

```
PROGRAM: SPACECUR
:V1=nDer(X1,t)
:V2=nDer(X2,t)
:V3=nDer(X3,t)
:V=[V1,V2,V3]
:U1=V1/norm V
:U2=V2/norm V
:U3=V3/norm V
:norm V→VEL
:U1→U11
:U2→U22
:U3→U33
:[U11,U22,U33]→U
:nDer(U1,t)→N1
:nDer(U2,t)→N2
:nDer(U3,t)→N3
:[N1,N2,N3]→NN
:norm NN/VEL→CURV
:NN/norm NN→N
:cross(U,N)→BN
:Disp "Velocity=",V
:Disp "Tangent=",U
:Disp "Normal=",N
:Disp "Binormal=",BN
:Disp "Speed=",VEL
:Disp "Curvature=",CURV
```

Program 8.4

The TI-85 has rather strict rules limiting the amount of nesting of the calculus operations *der1*, *nDer*, and formation of vectors. The above program skirts some of these restrictions (albeit in a somewhat cumbersome way) and computes, for a given value of t, the vectors **V**, **U**, **N**, **B**, and the speed v and curvature κ.

To use this program, one enters the defining equations for X1, X2, X3 in the homescreen, types SPACECURVE into the home screen, and presses [ENTER]. (For convenience we set the TI-85's display mode to 5 decimal places.)

Example 8.14

As a specific example we take the helix of Example 8.13, with defining equations

$$X_1 = A \cos t$$
$$X_2 = A \sin t$$
$$X_3 = Bt.$$

Then, typing $0 \to t$: SPACECURVE into the home screen (and pressing [ENTER]) we obtain the output

$$\text{Velocity} = [0, 1, 1]$$
$$\text{Tangent} = [0, .70711, .70711]$$
$$\text{Normal} = [-1, 0, 0]$$
$$\text{Binormal} = [0, -.70711, .70711]$$
$$\text{Speed} = 1.41421$$
$$\text{Curvature} = .5 .$$

The student should try the Exploration problems 2 to 5 at the end of the section.

Example 8.15

The line integral $\int_C P\, dx + Q\, dy + R\, dz$, where C is a curve in 3-space and P, Q, R are functions of x, y, z, occurs often in applications of vector calculus. In particular, if

$$\mathbf{F}(x, y, z) = [P(x, y, z), Q(x, y, z), R(x, y, z)]$$

is a vector field that represents a *force* exerted on a particle at the point (x, y, z), and the curve C is represented parametrically by

$$\gamma(t) = [X_1(t), X_2(t), X_3(t)],$$

then the line integral

$$\int_C P\, dx + Q\, dy + R\, dz = \int_C \mathbf{F} \cdot d\gamma$$

expresses the *work done* by the force field on the particle as it moves along C.

As a specific example, let us compute the line integral $\int_C \mathbf{F} \cdot d\gamma$ where C is the helix

$$C: \quad \gamma(t) = [\cos t, \ \sin t, \ t]$$

of Example 8.13 and the force field is given by

$$\mathbf{F}(x, y, z) = [-x, \ y, \ xyz].$$

Then the work done by the force field in one tour around the z-axis is

$$W = \int_C \mathbf{F} \cdot d\gamma = \int_0^{2\pi} \mathbf{F}(\cos t, \sin t, t) \cdot \gamma'(t) \, dt$$

This integral can be evaluated explicitly, with sufficient patience, however we will use the TI-85's vector functions and built-in integration routine to do the job.

Enter into the home screen each of the following commands

$$0 \rightarrow A$$
$$2\pi \rightarrow B$$
$$X1 = \cos t$$
$$X2 = \sin t$$
$$X3 = t$$
$$F = [-X1, \ X2, \ X1 * X2 * X3]$$
$$DX1 = \mathrm{nDer}(X1, t)$$
$$DX2 = \mathrm{nDer}(X2, t)$$
$$DX3 = \mathrm{nDer}(X3, t)$$
$$DR = [DX1, DX2, DX3]$$
$$\mathrm{WORK} = \mathrm{fnInt}(\mathrm{dot}(F, DR), t, A, B).$$

The final command places the unevaluated integral into the variable WORK, thus we can evaluate the work done by typing WORK $\boxed{\text{ENTER}}$ in the home screen. The value 6.28319 is quickly obtained. And we can evaluate other line integrals by changing only the values of A, B, X1, X2, X3, and F. The final five lines (above) do not need to be changed!

Example 8.16

For example, let

$$\gamma(t) = [t + \sin t, \ \cos t^2, \ 2 - t^3], \qquad 0 \leq t \leq 1$$

and
$$\mathbf{F}(x, y, z) = [xye^z,\ 2x + y^2,\ z - \cos xy].$$

We compute the work done by this force field on a particle that moves along the curve from $\gamma(0)$ to $\gamma(1)$. Let he/she who would evaluate by hand the work done in this case please step forward. On the TI-85 we simply enter

$$0 \to A$$
$$1\pi \to B$$
$$X1 = t + \sin t$$
$$X2 = \cos t^2$$
$$X3 = 2 - t\char`\^3$$
$$F = [X1 * X2 * e\char`\^ X3,\ 2X1 + X2^2,\ X3 - \cos(X1 * X2)]$$

and evaluate WORK once again. After a short time we obtain WORK = 5.3283.

Example 8.17

To evaluate a line integral in the plane we use exactly the same equations described above, replacing the 3-dimensional vectors by 2-dimensional vectors. For example the work done by the force field $\mathbf{F} = [-y, x]$ in one tour around the ellipse $\gamma(t) = [2\cos t, 3\sin t]$, $0 \le t \le 2\pi$, is found by entering the following commands into the home screen of the TI-85:

$$0 \to A$$
$$2\pi \to B$$
$$X1 = 2\cos t$$
$$X2 = 3\sin t$$
$$F - [-X2, X1]$$
$$DX1 = \mathrm{nDer}(X1, t)$$
$$DX2 = \mathrm{nDer}(X2, t)$$
$$DR = [DX1, DX2]$$
$$\mathrm{WORK} = \mathrm{fnInt}(\mathrm{dot}(F, DR), t, A, B)$$

This time, typing WORK $\boxed{\text{ENTER}}$ we get WORK = 37.6991.

Example 8.18

Green's Theorem in the plane states that, under suitable conditions of continuity and smoothness, an important relationship holds between a *line integral*

$$\oint_C \mathbf{F} \cdot d\gamma = \oint_C P\,dx + Q\,dy,$$

around a simple closed curve C in the plane, and a *double integral* over the region D of the plane enclosed by the curve:

$$\oint_C \mathbf{F} \cdot d\gamma = \iint_D \left(\frac{\partial Q}{\partial x} - \frac{\partial P}{\partial y} \right) dA. \tag{8.17}$$

This relationship might be viewed as one of several forms of the fundamental theorem of calculus as it extends to multivariable calculus. It has some practical applications to the actual evaluation of line integrals. And it has considerable theoretical significance in characterizing conservative vector fields, i.e. vector fields for which the work done by the field in traversing any simple closed path is zero. (A glance at Equation 8.17 reveals that $\frac{\partial Q}{\partial x} = \frac{\partial P}{\partial y}$ is a sufficient condition for the field to be conservative.)

To apply Green's Theorem to the evaluation of $\oint_C \mathbf{F} \cdot d\gamma$ we need to be able to evaluate double integrals. This can be done by representing the double integral as an *iterated integral*

$$\iint_D \left(\frac{\partial Q}{\partial x} - \frac{\partial P}{\partial y} \right) dA = \int_A^B \int_{C(x)}^{D(x)} \left(\frac{\partial Q}{\partial x} - \frac{\partial P}{\partial y} \right) dy\, dx \tag{8.18}$$

or a sum of several such iterated integrals.

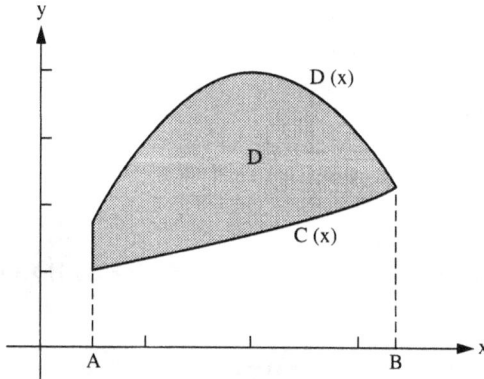

Figure 8.12. Bounded Region for Green's Theorem.

It is an interesting exercise to apply the TI-85 to calculating iterated integrals numerically. One might hope to do this explicitly by entering in the homescreen an expression of the form

$$\text{fnInt}(\text{fnInt}(G, y, C, D), x, A, B).$$

But, again, this does not work because the TI-85 does not support such *nested* use of the integration function. We proceed in a more circumscribed way, therefore. The idea is to use the TI-85's built-in integration function to perform the *inside* integration

$$g(x) = \int_{C(x)}^{D(x)} \left(\frac{\partial Q}{\partial x} - \frac{\partial P}{\partial y} \right) dy,$$

and then to evaluate $\int_A^B g(x)\,dx$ using Simpson's rule.

The following program DBINT accomplishes this task. The program generates values of $g(x)$ at a sequence of points

$$\{A, A + H, A + 2H, \ldots, A + NH\}$$

(where N is an even integer) and then calls a second procedure SIMP to complete the job.

```
 PROGRAM: DBINT
:InpSt ("F(x,y)=",F)
:InpSt ("C(x)=",C)
:InpSt ("D(x)=",D)
:St▸Eq(F,F)
:St▸Eq(C,C)
:St▸Eq(D,D)
:Prompt A
:Prompt B
:(B-A)/50→H
:seq(fnInt(F,y,C,D),x,A,B+H/2,H)→FS
:SIMP
:Fix 5
:Disp "INTEGRAL=",INTEG
```

Program 8.5

The program SIMP expects two inputs from variables FS and H. FS must contain the sequence of values of the function whose integral is to be evaluated by Simpson's rule, and H is the integration step for this numerical integral. (Note: the use of *B+H/2* instead of *B* in generating the sequence in the above program avoids losing the last item of the sequence due to round-off error.)

```
PROGRAM: SIMP
:dimL FS→N
:FS(1)+FS(N)→INTEG
:For(I,2,N-1,2)
:INTEG+4*FS(I)→INTEG
:End
:For(I,3,N-2,2)
:INTEG+2*FS(I)→INTEG
:End
:INTEG*H/3→INTEG
```

Program 8.6.

The student will recognize that this program is evaluating the familiar Simpson formula

$$\text{INTEG} = \frac{h}{3}\left[g_1 + 4g_2 + 2g_3 + 4g_4 + \cdots + g_{n+1}\right]$$

where $g_1, g_2, \ldots, g_{n+1}$ are the values of the integrand at points $a, a+h, a+2h, \ldots, a+nh$. Let us apply the program DBINT to evaluate several instances of $\iint_D f(x, y)\, dy\, dx$. The first few examples are simple enough to calculate the iterated integrals by hand, affording a check on the program.

Example 8.19

With $f(x, y) = 1$ and D the region in the xy-plane bounded by the lines $y = 0$, $x = 1$, and $y = x$, we type DBINT $\boxed{\text{ENTER}}$ in the home screen and respond to the program prompts as follows:

$$F(x, y) = 1$$
$$C(x) = 0$$
$$D(x) = x$$
$$A =? \, 0$$
$$B =? \, 1 \, .$$

With the output mode set to 5 decimal places, the program yields 0.50000. The exact answer $\frac{1}{2}$ in this case is easily calculated by evaluating the iterated integrals.

Example 8.20

Find the area that lies below the parabola $y = 4 - x^2$ and above the hyperbolic cosine curve $y = \cosh x$. (The student should sketch the functions.) In this example it is necessary to find the points of intersection of the two curves. We do this using the HP-85's built-in SOLVER application: press $\boxed{\text{SOLVER}}$, type the expression exp : $4 - x^2 - \cosh x$ and $\boxed{\text{ENTER}}$, set exp $= 0$, set $x = 1$ as an initial guess, and press $\boxed{\text{SOLVE}}$. We obtain the root $x = 1.3761766701915$. The second root is, of course, the negative of this. So we now return to the home screen and type $x \to$ ROOT so that we can use this value in the program DBINT. Finally, we type DBINT $\boxed{\text{ENTER}}$ and provide the inputs

$$F(x, y) = 1$$
$$C(x) = \cosh x$$
$$D(x) = 4 - x^2$$
$$A = ? - \text{ROOT}$$
$$B = ?\text{ROOT}.$$

We obtain the value 5.56470 for the desired area.

Example 8.21

Use Green's Theorem to evaluate the line integral $\int_C \mathbf{F} \cdot d\gamma$ around the curve C defined by pieces of the parabola $y = 4 - x^2$ and the y-axis, as shown in the figure. The curve is traversed in the counterclockwise direction. Let the vector function be defined by

$$\mathbf{F}(x, y) = [2y^3 e^x, \ 3x^2 - 4y^2].$$

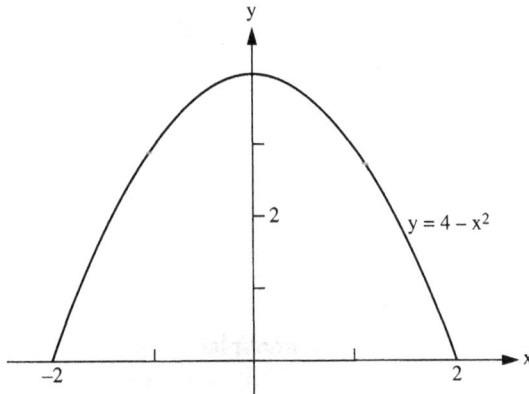

Figure 8.13. Region for $\int_C \mathbf{F} \cdot d\gamma$.

We must evaluate the double integral

$$\iint_D \left(\frac{\partial Q}{\partial x} - \frac{\partial P}{\partial y} \right) dA = \iint_D (6x - 6y^2 e^x) \, dA.$$

Using the program DBINT and providing the inputs

$$F(x, y) = 6x - 2y^2 e^{\wedge} x$$
$$C(x) = 0$$
$$D(x) = 4 - x^2$$
$$A = ? - 2$$
$$B = ?2$$

the value -291.04903 is obtained as the value of the double integral, and this is the desired value of the line integral as well. (See Exploration problem 8 at the end of the section.)

Example 8.22

Situations in which the boundary curves have vertical tangents are numerically troublesome. To explore this in an example where we know the exact outcome, let us find the area in the first quadrant lying under the semicircle $y = \sqrt{4 - x^2}$. Using DBINT directly, with $F(x, y) = 1$, $C(x) = 0$, $D(x) = \sqrt{(4 - x^2)}$, $A = 0$, and $B = 2$, we obtain the value 3.14029, which should be compared with the exact answer π. The trouble arises from the vertical tangent at $x = 2$. We could tackle the difficulty by brute force, using more points in the Simpson's Rule approximation. The program DBINT, as written, employs 50 subintervals. Let us try 800 subintervals instead. (Press PRGM EDIT DBINT ENTER and edit the one line in the program that contains the number "50." Running DBINT again, with the same inputs as above, now yields 3.14157 as the area, a better result but at considerable expense in computing time ($\sim$ 15 minutes). Nonetheless, the problem is solved (assuming that time is plentiful).)

A better approach is to recognize the circular symmetry of the problem and to convert the integral in question to polar coordinates. This gives

$$\iint_D dy\, dx = \int_0^{\frac{\pi}{2}} \int_0^2 \frac{1}{2} r\, dr\, d\theta .$$

Evaluating the new iterated integral using DBINT (reading y for r and x for θ), and giving the inputs

$$F(x, y) = y/2$$
$$C(x) = 0$$
$$D(x) = 1$$
$$A = ?0, \quad \text{and}$$
$$B = ?\pi/2$$

we obtain the area 3.14159 in a few seconds. Changing the mathematical form of problems often pays off handsomely!

Example 8.23

As a final example, we return to Example 8.17. There we evaluated the line integral $\oint_C \mathbf{F} \cdot d\gamma$, where $\mathbf{F}(x, y) = [-y, x]$, and C was the ellipse $\gamma(t) = [2 \cos t, 3 \sin t]$. By Green's Theorem we have, using the symmetry of the problem about the x-axis,

$$\oint_C \mathbf{F} \cdot d\gamma = \iint_D 2 \, dA = 2 \int_{-2}^{2} \int_0^{\frac{3}{2}\sqrt{4-x^2}} 2 \, dy \, dx .$$

Using DBINT to evaluate the iterated integral we get $\oint_C \mathbf{F} \cdot d\gamma = 37.65495$. This agrees with the previous result to only three significant digits, the difficulty being due, once again, to the vertical tangents of the ellipse at $x = \pm 2$. This is corrected either by increasing the number of subintervals in the use of Simpson's Rule (at the expense of a long computation) or, better, making an appropriate change of coordinates. In polar coordinates

$$x = r \cos \theta, \qquad y = r \sin \theta$$

the equation of the ellipse becomes

$$\frac{r^2 \cos^2 \theta}{4} + \frac{r^2 \sin^2 \theta}{9} = 1$$

or, solving for r

$$r = \frac{6}{\sqrt{9 \cos^2 \theta + 4 \sin^2 \theta}}.$$

The double integral thus becomes

$$\iint_D 2 \, dA = \int_0^{\pi} \int_0^{\frac{6}{\sqrt{9 \cos^2 \theta + 4 \sin^2 \theta}}} 2r \, dr \, d\theta,$$

and this is evaluated in a few seconds by DBINT as 37.6991, agreeing with the earlier result in Example 8.17. The inputs to DBINT are

$$F(x, y) = 2y$$
$$C(x) = 0$$
$$D(x) = 6/\sqrt{(9(\cos x)^2 + 4(\sin x)^2)}$$
$$A = ? \, 0$$
$$B = ? \, \pi .$$

Explorations in Vector Calculus

1. Verify that the various forms of the scalar triple product such as $\mathbf{V}_1 \cdot \mathbf{V}_2 \times \mathbf{V}_3$, $\mathbf{V}_2 \times \mathbf{V}_3 \cdot \mathbf{V}_1$, etc., all give the same result. All of them represent the volume of the parallelopiped determined by the three vectors $\mathbf{V}_1$, $\mathbf{V}_2$, $\mathbf{V}_3$.

2. Find the tangent, normal and binormal vectors for the helix (Example 8.14) at the points $t = \pi/4$ and $t = \pi/2$.

3. Explore the geometry of the *twisted cubic curve* — the curve defined by the equations $x(t) = t$, $y(t) = t^2$, $z(t) = z^3$.

4. The student is invited to try simplifying the program given in Example 8.13. It would appear that many of the lines are redundant. But the TI-85's strict nesting rules thwart such efforts.

5. The student is similarly invited to try to include the computation of $\frac{d}{dt}\mathbf{BN}$ in the program of Example 8.13 so that the torsion can then be computed and displayed along with the other quantities. Good luck. The TI-85's nesting rules are not forgiving.

6. Evaluate $\int_C \mathbf{F} \cdot d\gamma$ where C is the ellipse of Example 8.23 and $\mathbf{F}$ is
   (a) $\mathbf{F}(x, y) = [2x^2y, \sin x^2]$
   (b) $\mathbf{F}(x, y) = [xe^y, 3x^2 - y]$ (Can you explain the result?)

7. Referring to Example 8.19, find the area of the region D bounded below by the y-axis and above by the function $y = 4 - x^2$. Sketch the region. Again, the area can be represented by the value of a double integral in which $f(x, y) = 1$. Use the program DBINT to obtain the area. (Answer: 10.66667). Show that the exact area is $10\frac{2}{3}$.

8. Check the result obtained in Example 8.21 by calculating the line integral directly, using the TI-85 and the method of Example 8.17.

9. As suggested in Example 8.23, modify DBINT to implement Simpson's Rule with a greater number of subintervals (50 subintervals are used in the existing program). For example, try 800 subintervals. How long does it take? How accurate is the result?

In each of the following exercises, evaluate the line integral both directly and using Green's Theorem. Use the TI-85 to do these computations, as discussed in this section. When possible, carry out the exact computations mathematically and thereby check your numerical results.

10. Evaluate the line integral

$$\oint_C y \, dx - x \, dy,$$

where C is the boundary of the square $0 \leq x \leq 1$, $0 \leq y \leq 1$, traversed counterclockwise.

11. Evaluate the line integral

$$\oint_C e^x \sin y \, dx + e^x \cos y \, dy$$

around the boundary of the region described in Example 8.21. Explain the result.

References

1. Arfken, George, *Mathematical Methods for Physicists*, Academic Press, New York, 1985.

2. Greenberg, Michael D., *Foundations of Applied Mathematics*, Englewood Cliffs, New Jersey: Prentice-Hall, 1978.

3. Hildebrand, Francis B., *Advanced Calculus for Applications*, Prentice-Hall, New York, 1962.

4. Kreider, D. L., R. G. Kuller, D. R. Ostberg & F. W. Perkins, *An Introduction to Linear Analysis*, Addison-Wesley, Reading, Massachusetts, 1966.

5. Kreyszig, Erwin, *Advanced Engineering Mathematics*, 6th edition, John Wiley and Sons, New York, 1962.

6. O'Neil, Peter V., *Advanced Engineering Mathematics*, Wadsworth, Belmont, California, 1991.

7. Strang, Gilbert, *Introduction to Applied Mathematics*, Wellesley-Cambridge Press, Wellesley, Massachusetts, 1986.

Appendix
Introduction to the
Texas Instruments TI-85

How the TI-85 Is Organized

The *Home screen* is the primary screen of the TI-85, where you enter expressions to be evaluated and see the results, and where you enter instructions to be executed. To return to the Home screen from any other screen, press [2nd] [QUIT].

The TI-85 can display up to eight lines of 21 characters per line. If all text lines of the display are filled, text scrolls off the top of the display. On the Home screen and in the program editor, if a command is longer than one line, it wraps to the beginning of the next line.

An *expression*, for example, $\pi^*\mathbf{radius}^2$, is entered in the same order that it normally is written. It is completed when you press [ENTER]. The entire expression is evaluated according to Equation Operating System™, and the result is displayed on the right side of the next line. If a result is too long to display in its entirety, ellipsis marks (...) are shown at the left or right. Use [▶] and [◀] to scroll the result. If the result is a matrix with more rows than the screen can display, use [▼] and [▲] to scroll the result vertically.

On the TI-85, you can enter *expressions*, which return a value, in most places where a value is required.

An *instruction* is a command that initiates an action. For example, **ClDrw** is an instruction that clears any drawn elements from a graph. Instructions can be used on the Home screen or in a program. They cannot be used in expressions.

Modes

The MODE settings control the way expressions are interpreted and results are displayed. Press [2nd] [MODE] to display the MODE settings. The current settings are highlighted. To change a MODE setting, use [▼] or [▲] to move the cursor to the line of the setting that you want to change, then use [▶] or [◀] to move the cursor to the setting that you want, and press [ENTER].

Table A.1. Editing Keys on the TI-85 Keyboard

▶ and ◀	Moves the cursor within an expression. These are repeating keys.
▼ and ▲	Moves the cursor between lines. These are repeating keys. ▲ on the top line of expression on the Home screen moves the cursor to beginning of the expression. ▼ on bottom line moves the cursor to end.
2nd ▶ or 2nd ◀	Moves the cursor to beginning or end of the expression.
2nd INS	Inserts characters at underline cursor. To end insertion, press 2nd INS or a cursor-movement key.
DEL	Deletes character at the cursor. This is a repeating key.
ENTER	Executes expression or instruction.
CLEAR	• On line with text on the Home screen, clears (blanks) that line. • In an editor, clears (blanks) expression or value where cursor is located; it does not store a zero. • On a blank line on the Home screen, clears everything on the Home screen.
2nd	Enter 2nd operation. Cursor is an ↑. To cancel 2nd, press 2nd.
ALPHA	Enter ALPHA character at **A** cursor. To cancel ALPHA, press ALPHA.
2nd alpha	Enter alpha character at **a** cursor. To cancel alpha, press ALPHA ALPHA ALPHA.
ALPHA ALPHA	Set uppercase ALPHA–lock. To switch from ALPHA–lock to alpha–lock, press 2nd alpha. To cancel ALPHA–lock press, ALPHA.
2nd alpha ALPHA	Set lowercase alpha–lock. To switch from alpha–lock to ALPHA–lock, press ALPHA. To cancel alpha–lock, press 2nd alpha or ALPHA ALPHA.

Normal Sci Eng
Float 0123456789 01
Radian Degree
RectC PolarC
Func Pol Param DifEq
Dec Bin Oct Hex
RectV CylV SphereV
dxDer1 dxNDer

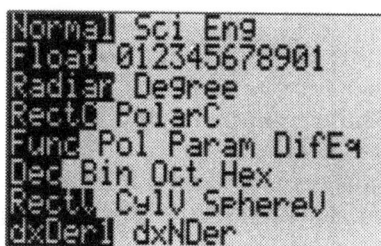

Numeric display format
Number of decimal places
Unit of angle measure
Complex number display format
Type of graphing
Number base
Vector display format
Type of differentiation

Figure A.1. Appendix

Entering and Editing Commands on the TI-85

The TI-85 treats an expression as individual characters, regardless of how it may have been entered. You can type over any character in a name. The TI-85 ignores uppercase and lowercase when it interprets names of functions and instructions. Case is, however, significant for the names of variables and constants. You can enter the names of functions, instructions, variables, and constants in one of several ways:

- Type the characters of the name.
- Press the key or select from a menu to copy the name.
- Copy the name from the CATALOG or VARS screen.

When an expression is evaluated successfully on the Home screen or from a program, the TI-85 stores the result to a special variable, **Ans** (*Last Answer*). You can use the variable **Ans** in most places where its data type is valid. Press [2nd] [ANS] and the variable name **Ans** is copied to the cursor location. When the expression is evaluated, the TI-85 uses the value of **Ans** in the calculation.

When you press [ENTER] on the Home screen to evaluate an expression or execute an instruction, the expression or instruction is stored in a special storage area called *Last Entry*.

To execute Last Entry, press [ENTER] on a blank line on the Home screen; the entry does not display again.

To recall Last Entry, press [2nd] [ENTRY]. Anything on the command line is replaced. You can edit and execute the entry.

When the TI-85 is calculating or graphing, a moving vertical bar, the *busy indicator*, shows in the upper right of the display. (When you pause a graph or a program, the busy indicator is a dotted bar.)

The TI-85 Menus

The TI-85 uses display menus on the bottom two lines (seventh and eighth lines) of the display to access many additional operations. The five redefinable function keys immediately below the display are used to select items from menus. The menu keys are [F1], [F2], [F3], [F4], and [F5]. The

2nd operations of the menu keys are $\boxed{\text{M1}}$, $\boxed{\text{M2}}$, $\boxed{\text{M3}}$, $\boxed{\text{M4}}$, and $\boxed{\text{M5}}$. Menu items are shown on the display above the menu keys. $\boxed{\blacktriangleright}$ at the right of the menu items indicates that there are more items in the menu. Press $\boxed{\text{MORE}}$ to label the menu keys with the next group of menu items. If you are on the final group, $\boxed{\text{MORE}}$ displays the first group.

If you select a menu item that displays another menu, the first menu may move to the seventh line; the new menu displays on the eighth line.

Selecting from Menus

To select a menu item from the eighth line, press the corresponding menu key, $\boxed{\text{F1}}$, ... , $\boxed{\text{F5}}$.

If the item is a character or a name, it is copied to the cursor location. If you select a menu item that accesses another menu, the menu from the eighth-line may move to the seventh line, and the name of the selected menu is highlighted.

If a menu is displayed on the seventh line, to select an item from it:

- Press $\boxed{\text{2nd}}$ and then press the menu key, $\boxed{\text{M1}}$, ... , $\boxed{\text{M5}}$, that corresponds to the item that you want.

- Press $\boxed{\text{EXIT}}$, which causes the menu on the seventh line to move down to the eighth line. Then press the menu key ($\boxed{\text{F1}}$, ... , $\boxed{\text{F5}}$) that corresponds to the item that you want.

Using CATALOG

The CATALOG and VARS pull-down selection screens temporarily replace the current display, but you have not left the application in which you are working. When you press $\boxed{\text{EXIT}}$ or make a selection, the current display and menus are shown again.

The CATALOG contains the names of all the functions and instructions from the keyboard and from menus. When you press $\boxed{\text{2nd}}$ $\boxed{\text{CATALOG}}$, the names of functions and instructions are displayed in alphabetical order. Names that do not begin with an alphabetic character (such as + or ▲Bin) follow Z. To move around the list:

- Press a letter to move quickly to names beginning with that letter. (The keyboard is set in ALPHA–lock.) Uppercase and lowercase names are intermixed.

- Press $\boxed{\blacktriangle}$ on the first name to move quickly to names beginning with special characters at the end of the list.

- Use PAGE↓ and PAGE↑ to move to the next page of names.

- Use $\boxed{\blacktriangledown}$ and $\boxed{\blacktriangle}$ to move down and up the list.

An arrow at the left of the name indicates the selection cursor. To copy a name from the CATALOG to an expression, press $\boxed{\text{ENTER}}$ to select the name to copy. The CATALOG selection screen disappears and the name is copied to the cursor location.

From the CATALOG you can build a CUSTOM menu of 15 functions or instructions that you use frequently.

Using VARS

The VARS selection screen contains the names of all variables and named items. To display the selection screen, select the menu item for the type of variable name. You can move around and select from the VARS selection screen as described above.

Variable types	Entry form
Real numbers	.001 or 1e-3
Complex numbers	$(2,1)$ or $(1\angle 3)$
Real or complex lists	$\{1,2,3,4,5\}$
Real or complex vectors	$[2,5,7,8\}$
Real or complex matrices	$[[1,2,3][4,5,6]]$
Strings	**"HELLO"**
Equations	$\mathbf{AREA}= \pi^*\mathbf{RADIUS}^2$
Constants	constant name

Programs, graph databases, and graph pictures are saved as named items.

The Full-Screen Editors

CONS EDIT	POLY	GRAPH $y(x) =$
LIST EDIT	SOLVER	GRAPH $r(t) =$
MATRX EDIT	SIMULT	GRAPH $E(t) =$
VECTR EDIT	MATH INTER	GRAPH $Q'(t) =$
STAT EDIT	STAT FCST	GRAPH RANGE
PRGM EDIT		GRAPH ZOOM ZFACT

When you select a full-screen editor:

- You leave the Home screen or the application in which you are working, and the appropriate editor displays.
- Any existing menu lines are cleared. The editor menu, if any, displays on the eighth line.

The GRAPH Menu

The TI-85 has extensive built-in functionality for graphing and analyzing functions, polar equations, parametric equations, and differential equations. When the TI-85 is in Function MODE, the GRAPH key accesses the GRAPH menu:

$y(x =)$	Displays the editor where you enter functions to be graphed.
RANGE	Displays the editor where you can change the values of the RANGE variables that define the viewing rectangle.
ZOOM	Operations that allow you to change the viewing rectangle.
TRACE	Allows you to trace along any graphed functions.
GRAPH	Graphs all selected functions.
MATH	Accesses operations such as derivative and integral that allow you to explore the graph mathematically.
DRAW	Accesses operations to draw elements on the graph.
FORMT	Accesses a pull-down selection screen to define how a graph appears.
STGDB RCGDB	Allows you to store a graph database in memory and recall it at a later time.
EVAL	Allows you to evaluate all selected functions at a specified value.
STPIC RCPIC	Allows you to store a picture of the current graph in memory and superimpose it on a graph at a later time.

Index

A

Acceleration 160–161
Amplitude 16
Annuities
 future value 44
 present value 45
ANOVA 126–128
Area
 polar graph 205
 of a region 178–179
Array 230
Asymptotic, asymptotically stable
 289, 297, 301
Attractor 298
Autonomous 301
Average value of a function
 164–165, 170–172

B

Back substitution 236, 238
Bessel's equation
 Bisection method use of 316
 differential equation solution
 314, 318
 of order p 313–314
 of order zero 313–314
 Taylor series expansion 315,
 317–318
 zeros of solutions 315–316,
 319–320, 325
Binomial distribution 77
Bisection method 316, 323–324
Boundary-value problems 320
 eigenvalues 320
 shooting method 321–322

C

Central limit theorem 94, 96,
 102, 103–104
Characteristic polynomial 257
Combination 76–77
Compound interest 11–12
Concave 150
Confidence interval 103–106, 108
Conic sections
 graph 195
 properties 199
Conical sheet 182
Continuity 136–137
Control chart 66
Correlation coefficient 120, 124
Crout reduction 243–244
Critical value 107
Cycloid 208

D

Data
 clearing 54
 deleting 57
 editing 54
 entry 53–55
 naming 54
 one-variable 54, 56, 66
 sorting 56
 two-variable 54
Degrees of freedom 91

Derivative
  first 135, 143–145, 162
  function of one variable 194, 201
  parametric equations 199
  second 135, 143–145
Determinant 232–233, 257
Diagonalization 260
Dirac delta "function" 300
Direction 144–145
Direction fields 267, 302
Discrete system 301
Distortion of graphs 197
Dot product 251
Double integral 331–335
  changing variables 335–336

E

e 215
e^x 12
Eigenvalues 257, 259, 294–299
Eigenvectors 257, 260, 294–299
Electrical circuits 286, 288
Elliptic Integral
  first kind 312
  series 313
Empirical Rule 63, 99
Epicycloid 209
Equilibrium point, equilibrium
  solution 289, 293, 300–302
Error
  type I 107
  type II 107
Error term (infinite series) 225
Euler algorithm 267–269, 304
Expected value 80
Exponential distribution 96
Exponential function 10–14
Extrapolation 123
Extrema 136, 156

F

Falling body 207, 210

Falling body problem 272, 275
Finite mathematics 23
Finite population correction
  factor 102
Flow problems 12–13
Forward substitution 244
Frequency 54, 60
Function
  amplitude 16
  defined by an integral 211
  defined by parametric equations
    199, 207
  exponential 10–14
  graphing 1–18
  inverse 6–7, 12
  linear 2–3
  logarithmic 12–13
  parametric 6–7, 12, 16
  quadratic 3–6
  rational functions 7–8
  symmetric with
    origin 4
    y-axis 4
  translation graph 5–6, 16–18
  trigonometric 15–20
  zero of 3, 190
Function, behavior of
  decreasing 150
  increasing 150
Fundamental matrix of solutions
  295, 298

G

Gamma function 317–318
Gaussian elimination 236
Gram-Schmidt 253
Growth and decay 13–14

H

Histogram 59
Hypocycloid 210
Hypothesis tests 96, 106–113

I

Ill-conditioned 256
Improper integral 184, 217
Improved Euler algorithm 267–269
Indeterminate forms 214
Inflection point 136, 152–153, 158, 274–275
Input function 266, 277–280, 284
Integral
  as an area 136, 171–172, 177
  as an average 136, 171–172
  definite 164–166, 168
Integral approximations 168
  lower average 136
  midpoint estimate 136, 174
  Simpson's Rule 136, 175
  trapezoid estimate 136, 174
  upper-average 136
Interest, compound 11–12
Interval of convergence 221
Inverse 233
  tangent 193, 213
  trig functions 190
Inverse function 6–7, 12
Iterated integral 331

L

ℓn 212
Law of Large Numbers 70
Least squares criterion 116
Least squares solutions 256
Level of significance 107
Limit of a function 135–137
Line graph 80
Linear function 2–3
Linear programming
  graphical solution 34–37
  pivot for Simplex method 39–40
  sensitivity analysis 37–38
Linear systems
  examples 24
  graphical solutions 25–26
  solution using matrix inverse 28
  solution using row reduction 30–31
  solving in SIMULT 24–25
  with infinitely many solutions 30–31
List
  dimension 56–57
  editor 54
  name 52
LIST feature 2–3, 5, 11
Logarithmic function 12–13
LU-factorizations 243–246

M

Markov chains 48–50
Matrix 294–299
  addition 27
  augmented 236
  diagonalizable 260
  ill-conditioned 256
  inverse 27, 233
  lower triangular 243
  multiplication 27
  permutation 243
  powers 232, 235
  reduced row echelon 248–249
  row echelon 247, 249
  skew-symmetric 263
  symmetric 236
  transpose 236
  upper triangular 236, 243, 255
Matrix Editor 231
Maximum 150–152, 155–158, 161–162
Mean 56, 80, 97
Median 57
Minimum 150–152, 155–158
Motion problems 18–20
Multiplier 243

N

Natural logarithm 212
Newton's Method 193

Nonhomogeneous 277, 284, 294, 298
Nonlinear damping, restoring forces 286
Norm 251
Normal distribution 86–91, 96, 102
Normal equations 256
Normal quantile plot 112, 121, 126
Nullspace 261
Numerical integration 168, 278

O

One-tail test 107
Orthogonal
  projection 253–254
  vectors 251
Orthogonal families 292
Orthogonality 251
Orthonormal vectors 251, 254
Outliers 66
Overdetermined systems 256

P

P-value 107
Paired different test 110
Parametric function 6–7, 12, 16
Partial pivoting 242, 248
Pendulum
  circular error 309
  defining equations 308
  linearization 309
  period 309
  simple 308
Percentiles 57
Period of function 16–17
Periodic response, periodic solution 280–281, 289–291
Permutation 77
Permutation matrix 243
Phase plane 288–294
Phase shift 17
Pivot 238
Points of intersection 179

Poisson distribution 83
Polar equations 202
Polynomial approximation 221
Polynomial root finder 258
Population growth 13–14
Population growth equation 274–279
Position 144
Probability 69–70, 81
Probability distribution 79

Q

QR-algorithm 259
QR-factorizations 255
Quadratic function 3–6

R

randM function 233
Random number 70
Random variable 79, 83
Rates 135–136, 143
Rate of rates 135, 143
Rational function 7–8
Reduced row-echelon forms 248–249
Regression
  curvilinear 124
  exponential 124
  linear 116, 124
  logarithmic 124
  power 124
Rejection region 107
Related rates problem 159
Relative frequency 70, 73
Repeller 298
Rescaling 242
Residual 116, 119–122, 126
Root 190–191
Round-off error 245, 249
Row echelon matrix 247, 249
Row operations 29–30

S

Sampling 76, 95

Scatter diagram 63

Second order initial value problem 281

Series, defining functions and integrals 313

Shape of a curve 150

Simpson's Rule 332

Simulation 70

SIMULT application 246

Skewness 61

Solution error 168

Sort 56

Spring motion 286–288

Squeeze Theorem 135, 141

Standard deviation 55

Standard error of the mean 102

Stat editor 54

Steady state 278–280, 289, 301

Switch function 273

Symmetric (function)
  with origin 4
  with y-axis 4

T

T distribution 91, 102

Tail area 88

Tangent line 136, 154, 193, 199, 201
  slope 143

Time series 63

Time value of money 40–48
  future value 40–41
  general loan calculations 46–47
  present value 40–41

Trace 263–264

Translation of graph 5–6, 16–18

Trend 144–145

Trigonometric function 15–20
  cos 16

sin 16
tan 16
unit circle 16

Two-tail test 107

U

Uniform distribution 96, 100

Unit circle 16

V

Variables
  free 247
  pivot 247

Variance 55

Variation of parameters 284, 298

Vector 325
  angle between two vectors 325
  cross product 325
  distance between skew lines 326
  dot product 325
  norm 325

Vector calculus 326
  Green's Theorem 330–331, 334
  line integral 328, 330
  space curves 326
  work 329–330

Velocity 160
  maximum 161

Vertical shift 16

Volume 182
  of revolution 184–185

X

xStat, yStat 56

xyline 64–65

Z

Zero of function 3, 136, 153–156, 163
  complex 156
  real 156

www.ingramcontent.com/pod-product-compliance
Lightning Source LLC
Chambersburg PA
CBHW060803220326
41598CB00022B/2525